ADVANCES IN
GEOPHYSICS

VOLUME 13

Contributors to This Volume

R. T. H. COLLIS
DAVID FINKELSTEIN
PETER HOOD
CONWAY LEOVY
LEONARD B. LOEB
JAMES R. POWELL
S. H. WARD

Advances in
GEOPHYSICS

Edited by

H. E. LANDSBERG

Institute for Fluid Dynamics and Applied Mathematics
University of Maryland, College Park, Maryland

J. VAN MIEGHEM

Royal Belgian Meteorological Institute
Uccle, Belgium

Editorial Advisory Committee

BERNARD HAURWITZ   R. STONELEY
ROGER REVELLE   URHO A. UOTILA

VOLUME 13

1969

Academic Press • New York and London

Copyright © 1969, by Academic Press, Inc.

ALL RIGHTS RESERVED

NO PART OF THIS BOOK MAY BE REPRODUCED IN ANY FORM
BY PHOTOSTAT, MICROFILM, OR ANY OTHER MEANS,
WITHOUT WRITTEN PERMISSION FROM THE PUBLISHERS.

ACADEMIC PRESS, INC.
111 Fifth Avenue
New York, New York 10003

United Kingdom Edition
Published by
ACADEMIC PRESS, INC. (London) Ltd.
Berkeley Square, London, W.1.

Library of Congress Catalog Card Number 52–12266

PRINTED IN THE UNITED STATES OF AMERICA

LIST OF CONTRIBUTORS

R. T. H. COLLIS, *Aerophysics Laboratory, Stanford Research Institute, Menlo Park, California*

DAVID FINKELSTEIN, *Belfer Graduate School of Science, New York and Brookhaven National Laboratory, Upton, New York*

PETER HOOD, *Geological Survey of Canada, Ottawa, Ontario, Canada*

CONWAY LEOVY, *The RAND Corporation, Santa Monica, California*

LEONARD B. LOEB, *Department of Physics, University of California, Berkeley, California*

JAMES R. POWELL, *Brookhaven National Laboratory, Upton, New York*

S. H. WARD, *Department of Mineral Technology, University of California Berkeley, California*

FOREWORD

Good scientific reviews are not easily written and their value has only begun to be appreciated. The editors are continuously surveying the field for contributions in areas of geophysics where progress warrants a substantial review for a broad audience.

In this task we are assisted by our Advisory Committee whose views are of great value to us. We are saddened that one of the members of this Committee, Dr. Walter Lambert, passed away in October 1968. He had been associated with us since the beginning of the series in 1951, and gave valued counsel. This is not the place to eulogize Dr. Lambert, but we want to acknowledge our debt to him again. Dr. Urho A. Uotila has kindly agreed to take his place, and we look forward to a long and pleasant association with him.

We are again privileged to cover a wide variety of topics. Some of them were themes at the 14th General Assembly of the International Union of Geodesy and Geophysics in 1967. Several of the invited speakers agreed to revise their discussions for our series and thus make them available for a much wider audience than can be reached at a meeting.

As this volume leaves our hands we are pleased to inform our readers that we have in preparation a volume that will deal with a single theme: precision radiometry. A total of fourteen reviews put together under the leadership of Dr. A. J. Drummond will treat this important field.

H. E. LANDSBERG
J. VAN MIEGHEM

March, 1969

CONTENTS

LIST OF CONTRIBUTORS v
FOREWORD vii

Airborne Geophysical Methods
PETER HOOD and S. H. WARD

1. Introduction... 2
2. Aeromagnetic Methods... 4
3. Low-Frequency Airborne Electromagnetic Methods.......................... 41
4. Airborne Radiometric Methods... 79
5. Airborne Gravity Methods... 88
6. Other Airborne Remote-Sensing Methods................................. 102

Lidar
R. T. H. COLLIS

1. Introduction... 113
2. The Basic Lidar Technique 114
3. Atmospheric Optical Parameters 117
4. The Significance of Lidar-Measured Optical Parameters.................. 122
5. Application of Lidar Observations to Meteorological Problems and Atmospheric Studies... 125
6. Lidar Contributions to Atmospheric Studies and Meteorological Problems... 133
7. Future Developments... 135
Appendix 136
List of Symbols... 137
References 138

Structure of Ball Lightning
JAMES R. POWELL and DAVID FINKELSTEIN

1. Introduction... 141
2. Earlier Observations and Theories...................................... 144
3. Experimental Evidence 152

4. New Analysis of the Problem 167
5. Summary and Conclusions..................................... 185
List of Symbols... 186
References 188

Energetics of the Middle Atmosphere

Conway Leovy

1. Introduction... 191
2. The Ozone Distribution 193
3. The Water Vapor Distribution 202
4. Infrared Radiative Transfer 204
5. Heat Sources and Sinks above 80 kM 214
List of Symbols... 216
References 217

The Nature and Properties of Gaseous Ions Encountered in Atmospheric Studies

Leonard B. Loeb

Introduction 223
1. Forces Influencing Ionic Behavior 224
2. Mobility Theory and Equations 225
3. Cluster Ions 227
4. Reactions between Ions and Molecules 229
5. Ion Mobility and Ion Identification........................... 233
6. Accuracy of Mobility Determinations as Diagnostic Tools....... 234
7. Summary of Influences Affecting the Presence of Ion Species... 235
8. Normal Ions in the Atmosphere.............................. 236
9. Reaction Rates of Ions...................................... 243
10. Langevin and Intermediate Ions.............................. 244
References 252

Author Index 257
Subject Index.. 265

AIRBORNE GEOPHYSICAL METHODS

Peter Hood*

Geological Survey of Canada, Ottawa, Ontario, Canada

and

S. H. Ward*

Department of Mineral Technology, University of California, Berkeley, California

	Page
1. Introduction	2
1.1. Methods and Applications	3
2. Aeromagnetic Methods	4
2.1. General	4
2.2. The Geomagnetic Field	4
2.3. Airborne Magnetometers	6
2.4. Aeromagnetic Survey Techniques	31
2.5. Aeromagnetic Data Compilation	34
2.6. Gradiometers	37
References	39
3. Low-Frequency Airborne Electromagnetic Methods	41
3.1. Introduction	41
3.2. The Basic Principle of Inductive AEM Systems	42
3.3. Types of AEM Systems	46
3.4. The Design of AEM Systems	49
3.5. Interpretation	68
3.6. Survey Procedures	73
3.7. Possible Future Developments	74
3.8. Conclusion	75
List of Symbols	75
References	76
4. Airborne Radiometric Methods	79
4.1. Introduction	79
4.2. Airborne Scintillation Counters	80
4.3. Airborne Scintillation Spectrometers	81
4.4. Airborne Radiometric Survey Techniques	84
References	86
5. Airborne Gravity Methods	88
5.1. General	88
5.2. Airborne Gravimeters	91
5.3. Airborne Gravity Gradiometers	97
References	100
6. Other Airborne Remote-Sensing Methods	102
6.1. General	102
6.2. Aerial Photography	104

* Peter Hood is the author of Sections, 1, 2, and 4–6. S. H. Ward is the author of Section 3.

    6.3. Ultraviolet Methods 104
    6.4. Infrared Methods 105
    6.5. Microwave Radiometers 107
    6.6. Radar Methods 108
    References 109
    List of Symbols 111

1. Introduction

Developments in airborne geophysics have occurred so rapidly in the past several years that it is a difficult task to follow and record them all. It is probably a fact that as much money and effort is now spent each year on research in the development of airborne geophysical methods as in the ten-year period 1945–1955. For this reason it is very desirable that a critical appraisal be made at this point of the present state of the art.

Geophysics covers a broad spectrum of scientific endeavors and workers in one specialized field are often unaware of important events in another. Unlike any other science, geophysics has become identified not with a particular technology or discipline but with its objective—the study of the earth—and for this reason it embraces many different disciplines and traverses many established lines of communication. Before World War II, geophysics received its main support from commercial concerns engaged in prospecting for buried mineral deposits. The physics of the earth and the upper atmosphere were mostly subjects of academic interest and did not generally involve field instrumentation or surveys. A wartime military application for magnetic methods accelerated developments in this field and subsequently led to their adoption by numerous and diverse groups of workers for a wide variety of applications. Yet it is true to say that while the utility of the airborne magnetometer is universally recognized, applications have been so specialized that the particular methods and practices of one group are not generally appreciated by another.

The period 1945–1955 saw electrical and radiometric instruments successfully adapted by the mineral industries to airborne use. These were pioneering years, with the major achievements made by stubborn and gifted individuals rather than by teams working on organized programs. Research in this period was largely a matter of experimentation, with mineral discoveries providing the yardstick for success.

As the demand for new minerals diminished in the years since 1956, the space age presented a new challenge to geophysicists. While commercial developments continued, largely as a result of the inertia built up during the previous five years, governments took an increasing interest in methods adaptable to space vehicles. Funds became available to research and teaching institutions for work in airborne geophysics. In 1959, airborne gravity was launched successfully through a joint effort by military and commercial geophysicists. Government-supported programs were established to look at the possible applications of microwave, infrared, and other radiation fields in the "remote sensing of environment." The International Geophysical Year and subsequent continuing programs stimulated work in terrestrial and extraterrestrial physics, with consequent improvements in instrumentation for recording small changes in the natural fields. Some of these instruments are now airborne and certainly others will follow in the next few years.

An additional stimulus has been provided by a growing awareness of the importance of geophysics (especially airborne geophysical surveys) in aid programs to the developing nations. Large-scale geological, hydrological, and land-use surveys have been undertaken by the aid agencies in many parts of the world, and geophysics is playing an increasingly prominent role in this work. But there is still a far from perfect understanding of the potential usefulness and the limitations of geophysical methods, not only at the administrative level in the agencies involved, but among the consultants and advisors on whom the agencies rely for their technical orientation.

Space and military scientists have been quicker to record their achievements than their commercial colleagues, with the result that it is easier for the mineral industries to make use of the new technologies than for government workers to benefit from the experiences of the prospecting geophysicists. Commercial security has been partly to blame. Shortage of funds and pressure of other work are probably more significant. Whatever the reason, the few available technical reviews of airborne geophysics are usually not by workers actively concerned with the field applications but by university or government scientists. Though these have often been thorough and detailed in their descriptions of instruments and methods, they have mostly failed to explain the problems and limitations, experience with which is vital to the successful application of a geophysical method.

The following review is an attempt to describe generally and discuss critically some of the airborne geophysical methods that have seen successful application in recent years. Since the author's experience has been mostly confined to the mineral industries, there is a greater emphasis on existing prospecting methods than those utilized for military, geodetic, and other applications. This has been partly intentional, since it is hoped that a greater awareness of the present state of the art will be of benefit to those whose interests lie in the development of new geophysical methods.

1.1. Methods and Applications

In geophysics, "methods" are not nearly so important as "applications." Too often a method is used holus-bolus without any very precise idea of what it is supposed to indicate. One way of subordinating methods is to discuss them as they appear under the headings of applications or objectives. This device is ideally suited to reviews and textbooks on prospecting geophysics. Unfortunately, the applications covered in this review are more numerous than the methods, to avoid complicated cross-referencing, the methods of airborne geophysics have been grouped under the following headings:

(1) aeromagnetic,
(2) low-frequency electromagnetic,
(3) radiometric,
(4) gravity, and
(5) other remote-sensing techniques.

The list could be made much longer, but it has been decided to include the semi-experimental methods such as ultraviolet, infrared, microwave, etc. under "other remote-sensing methods" and to place more emphasis on those which have seen greatest field application.

2. Aeromagnetic Methods

2.1. General

A great deal has been written about the use of aeromagnetic methods in mineral prospecting. A comprehensive review of the state of the art at that time was published in Volume 1 of this series [6][1]. It dealt with the only airborne magnetometer in service then, the fluxgate type, although the earth inductor magnetometer did have a limited use in mineral exploration. Developments over the last decade have been extensive, resulting in a multiplicity of aeromagnetic surveying instruments. This section therefore deals with the present state of the airborne magnetometer art. The next decade will probably see just as many changes in instrumentation as the past one has; two of these can be reliably predicted to be the complete automation of aeromagnetic compilation procedures and the extensive use of aeromagnetic gradiometers.

2.2. The Geomagnetic Field

The basic objective of any aeromagnetic survey is to map the space variation of the geomagnetic field (T) in a given horizontal plane. The geomagnetic field at any particular point in space consists of several time-varying components of widely separated origin, and these have been summarized in Table I.

[1] References cited in each major section of this article appear in a list at the end of the appropriate section.

TABLE I. Components of the geomagnetic field.

Name	Source	Time dependence	Space dependence	Typical amplitude
Dipolar	Deep internal	Slowly decreasing	Approximately dipolar	25,000–70,000 γ
Secular	Earth's core (3000 km)	1–100 years	Random but drifts westward	$\pm$10–100 γ per year
Diurnal	External—associated with sunspots	24 hours (Period 27 days, 12 months, 11 years)	With magnetic latitude and sunspot activity	10–100 γ
Micropulsations	External	0.002–0.1 Hz	Same as diurnal plus magnetic storms	Usually 1–10 γ but up to 500 γ
AFMAG	External	1–1000 Hz	Same as diurnal plus thunderstorm activity	0.01 γ per sec
Telluric current effects	Shallow internal	0.002–1000 Hz	Geologic	0–0.01 γ per sec
Induced magnetization of rocks	Shallow, internal down to Curie point geotherm at 13 miles	Secular	Geologic—varies with % magnetite in rocks	0–50,000 $\times$ 10^{-6} emu/cc
Remanent magnetization of rocks		Some decay with geologic time	Geologic	0–200,000 $\times$ 10^{-6} emu/cc

In prospecting surveys, the only anomalies of interest are those due to the induced and remanent magnetization of rocks, whereas in geomagnetic surveys measurement of the variation of the earth's main magnetic field is of primary interest. For this reason prospecting surveys are flown at low elevations, usually 1000 ft or less, and geomagnetic surveys are flown at high elevations, usually in excess of 10,000 ft. In both cases, the short-term time variations of the earth's field produce unwanted interference which has to be removed. These are often the limiting factor affecting the final accuracy of a given aeromagnetic survey.

In surveys of the magnetic field in space, the magnetometer used must be capable of measuring very low fields and must be insensitive to orientation.

2.3. Airborne Magnetometers

The magnetometers used in airborne surveys are either absolute instruments or variometers. The former measure the entire ambient field whereas the latter record only the variations relative to an arbitrary datum. The measurements may be vector or scalar, and in the analog recording mode, continuous or intermittent, depending on the particular instrument used.

In the following description of airborne magnetometers, most space is devoted to those that have had the greatest utilization in airborne surveys, although some new developments are mentioned that will undoubtedly be successfully applied in the near future. The categories of magnetometer discussed are as follows:

(1) Earth inductor,
(2) Fluxgate,
(3) Proton precession—free and spin,
(4) Optical absorption,
(5) Electron beam, and
(6) Hall effect.

Descriptions of the various types have been summarized in Table II.

2.3.1. Earth Inductor Magnetometers. The earth inductor has the distinction of being the first magnetometer flown in an airborne survey [30]. The basic principle depends upon the well-known fact that when a conductor moves through a magnetic field an electric current will be induced in the conductor, a phenomenon discovered by Faraday in the middle of the nineteenth century. The sensitivity of the original equipment flown in the U.S.S.R. in 1936 was about 100 γ, and the apparatus measured the vertical intensity of the earth's magnetic field because the axis of rotation of the coil was maintained in a horizontal plane using a gimbal suspension. Subsequently, Lundberg [31] built a helicopter-borne earth-inductor magnetometer (Fig. 1) in the years following World War II in which the rotating coil or rotor was driven by compressed air at 1800 Hz. The rotor was not connected to an outside circuit, i.e., use of slip rings was avoided, but with the appropriate capacitor it formed a closed, tuned circuit. A secondary pickup coil had an alternating voltage induced in it whose amplitude was dependent on the ambient vertical field because the axis of the rotor was horizontal. The main part of the vertical field was annulled by a Helmholtz coil system. Sensitivity of the equipment appeared to be about 25 γ, which is sufficiently sensitive for mining exploration purposes.

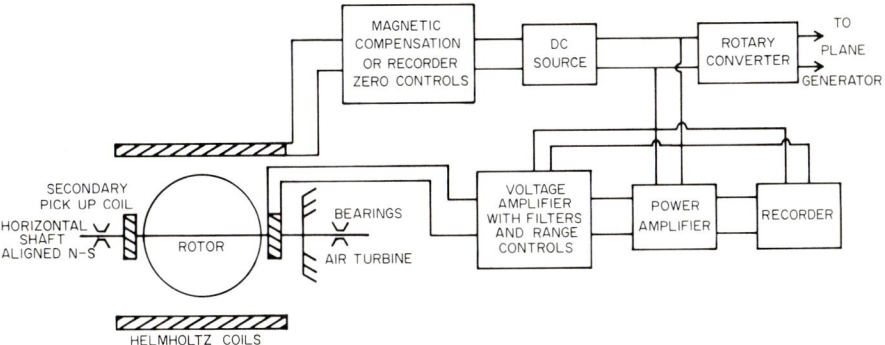

FIG. 1. Earth inductor magnetometer

2.3.2. Fluxgate Magnetometers. The early history of the fluxgate magnetometer and its use in antisubmarine warfare during World War II has been outlined by Fromm [13]. Balsley [6] and Jensen [23] have described some of the first aeromagnetic surveys carried out using the fluxgate instrument.

The sensitive element of a saturable-core or fluxgate magnetometer consists of a short length of high-permeability ferromagnetic material having a narrow hysteresis loop which acts as a core for one or more windings connected to ac exciting and indicating circuits. There are two basic fluxgate circuits, the peak voltage and second-harmonic types [52], which are both vector variometer instruments.

2.3.2.1. Peak-voltage fluxgate magnetometer. The basic circuit of the first type, as used in the highly successful Gulf Mark III instrument, is illustrated in Fig. 2. It consists in actual practice of a rod of Permalloy around which are wound two coils—the dc bias current coil and the main magnetometer coil. Half of the windings on the main coil are wound in one direction and the other half are wound in the opposite direction, to produce

TABLE II.

Magnetometer	Principle	Type	Manufacturer
Earth inductor (vector variometer)	Rotating coil		Lundberg (Can)
Fluxgate (vector variometer)	Saturable core	Peak voltage	Gulf Oil Co. (USA) Mark III
		Second harmonic	AN/ASQ-3, 8, 10 (USA) AM-13 (USSR) NOL VAM-1, -2 (USA) D.O. (Can)
Proton precession (scalar absolute)	Free precession of hydrogen proton	Reciprocal	Varian Assoc. (USA) V-4910, V-4912 V-4914
			Elliot Bros. (UK) EMD 14 and EMD 21 Elsec-Wisconsin (UK-USA)
		Direct-reading	Barringer Research AM-101 (Can) Geological Survey of Canada Telmag Bureau of Mineral Resources (Australia) Prackla (W. Germany) PM-22 and PM-24 Varian Assoc. (USA) V-4937A
			AYaAM-6 (USSR)

Types of airborne magnetometer.

Component measured	Readout A-Analog D-Digital	Sensitivity	Status (1967)	Mount
Z	A Continuous All fluxgates continuous	25 γ	Outdated	Boom on helicopter
T	A	< 1 γ	Current use	Inboard
ΔT filtered	A	< 1 γ	Current use	Inboard
T	A	2 γ	Current use	Bird
X, Y, Z	D and A		Current use	Inboard
X, Y, Z	D and A		Current use	Inboard
T	A Every 0.5 sec	1 γ	Superseded	Bird
T	D and A Every 0.6 sec	< 1 γ	Current use	Bird or inboard
T	D and A Every 0.5 sec	1–3 γ	Current use	Bird or inboard
T	D and A From 1 per sec to 1 per 7 sec	2–0.5 γ	Current use	Bird
T	D, A and visual Every 1 sec	1 or 5 γ	Current use	Bird or inboard
T	D and A Every 2 sec	0.1 γ	Current use	Bird or inboard
T	A Every 0.5 sec	1 γ	Current use	Bird or inboard
T	D and A Every 1 sec	2 γ	Current use	Bird
T	D and A Every 1 sec or 0.5 sec	1 or 2 γ	Current use	Bird
T	A Every 1 sec	1 γ	Current use	Bird

TABLE II.

Magnetometer	Principle	Type	Manufacturer
Proton precession (scalar absolute)	Forced precession of hydrogen proton	Overhauser	Sud Aviation (France) MP121 and MP122
Optical absorption (scalar absolute)	Optical pumping	Cesium vapor	CSF/CGG (France)
		Rubidium vapor	Varian Assoc. (USA) V-4916
		Metastable helium	Texas Instr. (USA)
Electron-beam or aspect tube (vector variometer)	Deflection of moving electron		Elliott Bros. (UK) EMD 13

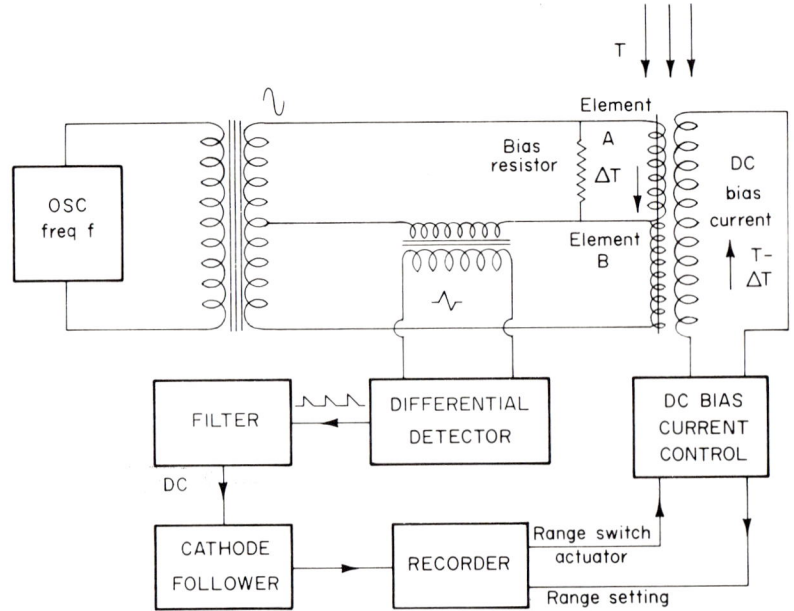

FIG. 2. Basic circuit of a peak-voltage fluxgate magnetometer.

Types of airborne magnetometer (*continued*).

Component measured	Readout A-Analog D-Digital	Sensitivity	Status (1967)	Mount
T	D and A Every 1 sec	$0.1\ \gamma$	Current use	Bird
T	D and A Continuous	$0.01\ \gamma$	Current use	Bird
T	D and A Continuous	$0.05\ \gamma$	Current use	Bird
T	D and A Continuous	$0.05\ \gamma$	Under development	Bird
T	A Continuous	$10\ \gamma$	Current use	Inboard

two elements. The main coil is excited with an audio frequency current from an oscillator, and the resultant output of the system is in the form of positive and negative spikes if an ambient magnetic field is acting. A dc current is passed through the bias coil to annul most of the earth's field (T). Figure 3 illustrates the mode of operation. The dc bias coil annuls all but a small part (ΔT) of the earth's field. The field applied to the cores of coils A and B will at any given instant be ($\Delta T + H_e \sin 2\pi$ ft) and ($\Delta T - H_e \sin 2\pi$ ft) where H_e is the maximum value of the applied field due to the exciting current which oscillates at frequency f. The most common explanation of the circuit is that because the residual field ΔT biases one of the cores with respect to the other, one of the cores will reach the saturation knee of the magnetization curve before the other in one half cycle, while the other core reaches saturation first in the second half of the cycle. This causes the resultant induction curves to be asymmetrical, and their resultant induction ($B_A + B_B$) consists of a wave form which has an amplitude proportional to ΔT. The emf induced in a coil is proportional to the rate of change of flux; voltage spikes are therefore produced in the main coil which coincide with the steep side of the resultant induction ($B_A + B_B$).

However typical excitation fields applied to the fluxgate cores are often about 50 Oe amplitude at 1000 Hz. This means that the excitation field is changing at a rate of about $3 \times 10^{10}\ \gamma$/sec. An ambient field of $1\ \gamma$ would thus produce a difference in the time of saturation of the two cores of the order of 10^{-10} sec, which could hardly produce an observable pulse, or enough energy to be detected by any normal electrical circuit. Furthermore, if the fluxgate is to produce a null for an ambient field of less than 1 γ, the two cores would have to be matched to something like 1 part in 10^6 in their magnetic properties, which would make a workable fluxgate instrument almost impossible to construct [44a].

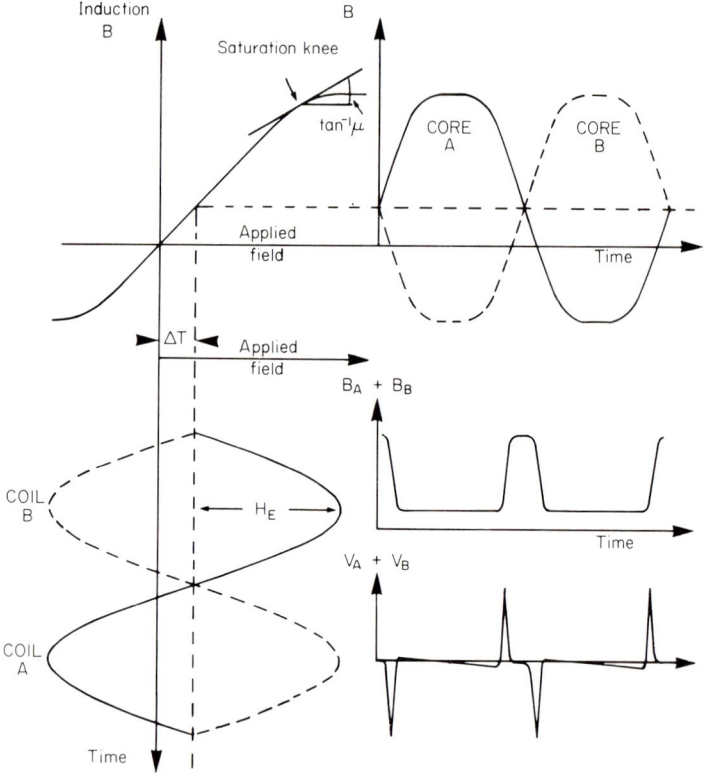

FIG. 3. Mode of production of spikes from a two-element fluxgate magnetometer.

A better explanation is that when the cores are unsaturated, their high permeability μ increases the flux due to ΔT through the secondary winding. When the cores become saturated (as they do $2f$ times a second), their permeability μ (the slope of the BH curve in Fig. 3) decreases, and the flux through the secondary due to ΔT decreases to the value it would have if the Permalloy cores were not there [43], i.e., the flux through the secondary fluctuates at a frequency of $2f$. Thus voltage spikes are induced in the secondary winding whose amplitude is proportional to the flux and thus to ΔT, the residual external field. The spikes will be alternately positive and negative in accordance with the sign of the change of $(B_A + B_B)$. It will be noted that in essence the action of the fluxgate is that of a magnetic amplifier. A bias resistor is usually inserted across one of the elements which unbalances the circuit and produces alternate positive and negative pulses of equal magnitude in zero field. When an ambient field ΔT is acting, a difference in the relative amplitudes of the positive and negative spikes is produced and is measured by a differential detector. This arrangement avoids the noise problem in zero ambient field.

It is essential in the airborne instrument that the fluxgate element be kept accurately aligned parallel to the earth's field. This is achieved by the use of two additional orienting inductors mutually perpendicular and in a plane at right angles to the measuring element. The error signal registered by these inductors, when the measuring element becomes misaligned with the magnetic field, is applied to one or both of two servomotors which drive the gimbal-mounted inductor assembly back into alignment. The orientation accuracy achieved by this mechanism is approximately $\pm 0.1°$, resulting in a possible measuring accuracy of about $\pm 0.08\ \gamma$ in a 50,000 γ field. In normal field use, the actual error for relative measurements is about $\pm 0.2\ \gamma$. Calibration of the compensating circuits can establish the absolute value of the magnetic datum to about $\pm 0.5\%$ or approximately $\pm 250\ \gamma$.

The Gulf Mk. III 410 Hz fluxgate instrument, which has seen widespread use for prospecting and geologic mapping, has a range of 250,000 γ, weighs 219 lb (for inboard installation), and operates from a 28 V, 360 W, dc supply. The 10-in. chart recorder has selectable full scale deflection subranges of 300, 600, 1200, 2400, or 4800 γ with stepping switches 5/6 of the subrange. Chart speed can be varied from 1 to 12 in/min or can be driven by Doppler. Instrumental drifts do not exceed 6 γ/hr [2a].

2.3.2.2. Second-harmonic fluxgate magnetometers. In the more recently developed magnetometers, only the second-harmonic component of the unbalanced voltage is utilized. The system used in the Naval Ordnance Laboratory Vector Airborne Magnetometer type 2A (VAM-2A) which has been described by Schonstedt and Irons [41], is shown in Fig. 4. Most of the second-harmonic content in the 1000 Hz oscillation from the driver is removed by a bandpass filter as this could produce a false zero-ambient field. The second-harmonic signal generated in the fluxgate element is separated from the excitation voltage by a 2000 Hz bandpass filter. The second-harmonic signal is then amplified, phase detected, and the resultant output is fed back to cancel the field acting along the inductor. This current is recorded on a zero center continuous chart. Most of the earth's field however is canceled by a much larger current from the bias control circuit. The Vector Airborne Magnetometer was designed to carry out high-level (20,000 ft) geomagnetic surveys and is the instrument used in Project Magnet, which is a worldwide survey of the earth's magnetic field undertaken by the U.S. Naval Oceanographic Office. The magnetometer system is pendulously suspended, enabling the inclination and declination, in addition to the total field, to be measured. The heading of the aircraft is determined by astronomical means, enabling the declination to be ascertained.

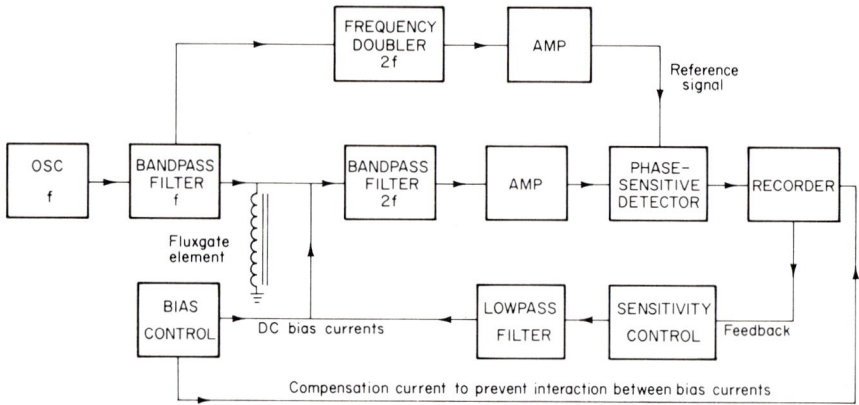

Fig. 4. Second-harmonic fluxgate magnetometer.

Actually the Vector Airborne Magnetometer uses much of the circuitry contained in the military magnetic airborne detector (MAD) AN/ASQ-3A, which has been described by Rumbaugh and Alldredge [40]. Earlier MAD equipment used for submarine detection during World War II has also been described by Fromm [13]. Later unclassified military magnetometers working on the second-harmonic principle are the AN/ASQ-8, and its transistorized replacement, the AN/ASQ-10, which is much lighter, smaller, and consumes only one-sixth the power [4]. The military magnetometers cannot be used without modifications because a high-pass filter is incorporated to discriminate against low frequency anomalies such as those produced by the underlying geology. Both the Geological Survey of Canada and the Scandinavian governments have used modified AN/ASQ-3 magnetometers for their surveys. The U.S.G.S. [12] and the Australian Bureau of Mineral Resources are presently employing modified AN/ASQ-10 instruments. The latter organization also uses a shaft encoding disk to record the magnetometer data digitally from an analog recorder.

Figure 5 shows a fluxgate magnetometer system developed at the Canadian National Aeronautical Establishment for survey use [17]. The basic instrument is an AN/ASQ-8 magnetometer which has been modified to a null-type instrument, using a servomotor as a phase-sensitive detector. The 800 Hz second-harmonic signal from the fluxgate element is amplified and fed to one winding of the two-phase servomotor. The other winding gets its second-harmonic signal via an amplifier and frequency doubler directly from the 400 Hz driver oscillator. The servomotor is thus made to drive a slide-wire potentiometer which controls the magnitude of the dc current fed back to the detector winding, and cancels any unbalanced field at the fluxgate element. The angular position of the feedback potentiometer, which is proportional to the feedback current and hence to the external field, is measured

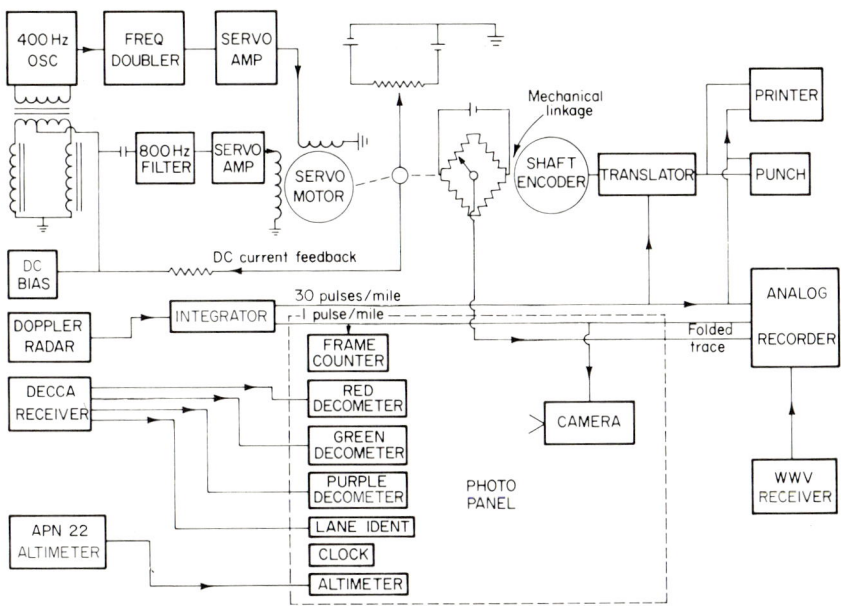

FIG. 5. National Aeronautical Establishment (Canada) digital fluxgate magnetometer.

by means of a shaft encoder. The output of the encoder is converted by a translator unit to the binary decimal codes necessary to operate a paper punch and printer. The resultant punched paper output is in 5-hole binary-coded decimal format, and the printer output is a 5-digit number, the least significant figure being 0.1 γ. The digital data are read out at 1/30 mi intervals on command from a Doppler radar set. A folded analog profile is also obtained, together with the time signals from a WWV receiver and Doppler fiducials. The time signals are useful for subsequent comparison of the trace with the diurnal record obtained from a ground station. When the instrument was used for a survey of the Nova Scotian continental shelf, the Decca navigational system was used for positioning. The mile pulse from the Doppler radar set triggered a camera which photographed the deccometer dials, frame counter, clock, and altimeter.

A second-harmonic airborne fluxgate magnetometer, the AM-13, has been developed in the U.S.S.R. Three-orthogonal elements excited at 500 Hz comprise the fluxgate assembly which is gimbal-mounted in a bird but has only two degrees of freedom to maintain the orientation of the fluxgate head. The operating range of the instrument is 11,000 γ, and the sensitivities of the analog recorder chart are 2 γ/mm on the 400 γ subrange, and 10 γ/mm on the 2000 γ subrange. The drift of the system is less than 5 γ/hr.

A geomagnetic airborne magnetometer has also been built at the Dominion Observatory, Ottawa [42]. It was found possible to avoid the use of filters by tuning the fluxgate element with a suitable capacitor to the second-harmonic of the excitation frequency. This produces a great increase in sensitivity and a suppression of the other harmonics. The magnetometer head consists of three orthogonal fluxgate elements which are mounted on a gyro-stabilized horizontal platform and aligned vertically, along and at right angles to the axis of the aircraft. The azimuth reference for the instrument is provided by a directional gyroscope mounted on the platform whose drift is corrected by astronomical measurements. The outputs from the inductors and the heading of the aircraft are fed into an analog computer which calculates the declination, and the horizontal and vertical components of the earth's magnetic field. The average value of these components over 5 min intervals is also computed. The accuracy of measurement is estimated to be 0.1° in declination and 20 γ for the components.

2.3.3. Proton Free- and Spin-Precession Magnetometers.

2.3.3.1. Proton free-precession magnetometers—reciprocal type. For approximately a decade after World War II, the only total-intensity magnetometer used in airborne geophysical surveys was the fluxgate type. In 1955, Varian Associates of California introduced a proton free-precession magnetometer, the V-4910. The principle of operation of this scalar magnetometer and Varian's subsequent models, the most recent reciprocal-type being the V-4914 [26] is illustrated in Fig. 6. It was found by Packard and Varian [33] that after a polarizing field is quickly removed from a sample of water, an audio-frequency signal which persists for a second or so may be detected in the same coils used for the polarizing current. This is caused by the hydrogen proton (but not oxygen) which has a magnetic moment owing to its spin. When an external field is applied, the magnetic moment vectors will align themselves parallel to the field. When the external field is cut off, the protons are under the influence of the earth's field and they will precess rather like gyroscopes. Because the frequency of precession (f) is directly proportional to the ambient field (T), it is only necessary to measure the former in order to ascertain the latter.

The relevant formula for T in gammas and f in Hertz is:-

(2.1) $$T = Kf$$

where the constant $K = 23.4874$, and actually the constant $K = 2\pi/$gyromagnetic ratio of the proton, so that proton precession magnetometers are absolute instruments.

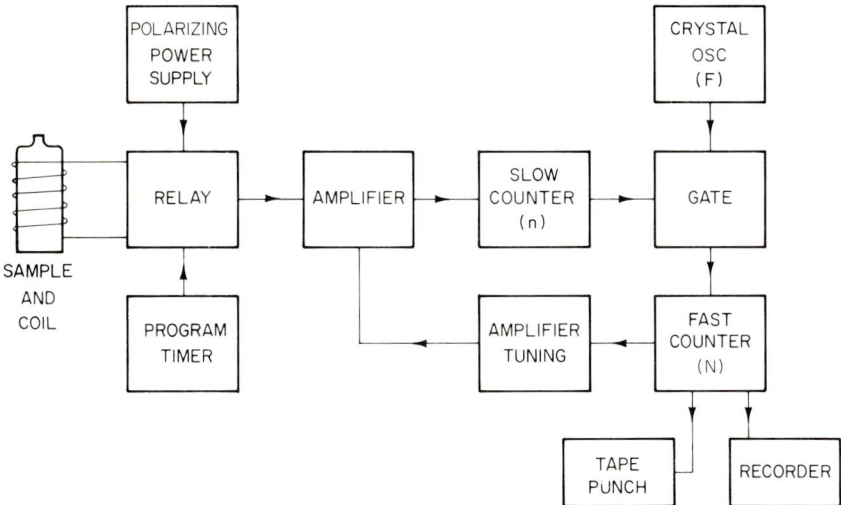

Fig. 6. Proton free-precession magnetometer—reciprocal type.

An accurate determination of the gyromagnetic ratio of the proton (γ_p) by Driscoll and Bender [11] yielded a value of $\gamma_p = 2.67513 \pm 0.00002 \times 10^4$ Oe^{-1} sec^{-1}. The result was subsequently confirmed by Vigoureux [48, 49].

The precession signal will last for several seconds, its amplitude decaying exponentially with time. The rate of decrease of amplitude is dependent on

(1) the gradient of the ambient field which causes loss of phase coherence because protons in different parts of the sample precess at slightly different frequencies;

(2) radiation damping owing to the energy loss caused by the inducing of currents in the coil system;

(3) true spin-spin relaxation of the sample [36].

Waters and Phillips [51] found that with their particular equipment in which the precession frequency was counted for 1 sec, the measurement became unreliable when the gradient exceeded 60 γ/ft, i.e., about 2γ/cm.

If the axis of the polarizing/detecting coil is at angle θ to the total field T, the amplitude of the precession signal is proportional to $\sin^2 \theta$. Thus the amplitude for $\theta = 45°$ is only half that for $\theta = 90°$, and it will fall to very low values as θ approaches zero, so that there is a dead zone along the axis of the coil.

To avoid this difficulty, two mutually perpendicular coils are commonly used. With this configuration, if the total field T makes an angle θ with the normal to the plane defined by the axes of the coils, then the signal amplitude is proportional to $(1 + \cos^2 \theta)/2$ [52]. Thus between $\theta = 0°$ and $\theta = 90°$, the signal amplitude is only reduced by a factor of 2. Alternatively, a toroidal coil may be used to avoid the possibility of low precession signal amplitudes. It follows that for a proton free-precession magnetometer used in low magnetic latitudes, the head should be designed so that the normal to the plane defined by the axes of the coils is horizontal, because the flight lines should be flown in a north-south direction in low magnetic latitudes (see Section 2.4.2).

If the proton head rotates during the count period, the precession frequency recorded will be either increased or decreased depending upon the sense of the rotation with respect to the coil system. It is readily apparent from Eq. (2.1) relating the total field to the precession frequency that the precession shift is 23.4874 γ/rotation/sec, i.e., 0.0652 γ/degree/sec.

In the reciprocal-type of proton-precession airborne magnetometer (Fig. 6), a field of approximately 100 Oe is applied to the sample, e.g., water, alcohol, or kerosene, in the bottle for about 0.2 sec. The program timer then switches off the polarizing current, the hydrogen protons in the sample begin to precess, and the electronic unit begins to count. The small signal induced in the coil is amplified and limited so that it is essentially a square wave when presented to the binary counters in the Slow Counter. The Slow Counter opens the electronic Gate upon receipt of the first square wave and closes it after the nth. The Gate, when open, allows the output of a crystal oscillator of frequency F to reach the binaries of a Fast Counter. When the Slow Counter has counted n cycles of the proton-precession signal, the Gate is closed, and the Fast Counter will register N counts. The complete polarize-count cycle usually takes about 0.7 sec. The time (t) that the Gate is open is given by

$$(2.2) \qquad t = \frac{n}{f} = \frac{N}{F}$$

Now $T = Kf = KnF/N$ so that for a given instrument the count N of the Fast Counter is inversely proportional to the earth's magnetic field T. The sensitivity (S) of the instrument is ± 1 count of the Fast Counter, so that the sensitivity $S = T/N = (T^2/KnF) \, \gamma$. Thus the sensitivity of reciprocal-type proton free-precession magnetometers is better when the instrument is operated in areas where the earth's field has a lower value. This is readily seen on Fig. 7, which is a graph showing the relationship between T, n, and N for $F = 100$ kHz. The count time t is also given on the figure (N.B.: $t = N/F$).

The Elsec–Wisconsin proton-precession magnetometer consists of an Elsec ground instrument, manufactured by the Littlemore Scientific Engineering Co. Ltd. in the U.K., which obtains the basic magnetic field values, and a digital-recording system designed and constructed at the University of Wisconsin [57]. The reciprocal magnetic field values and time are recorded in binary coded format on punched paper tape for direct input into a computer.

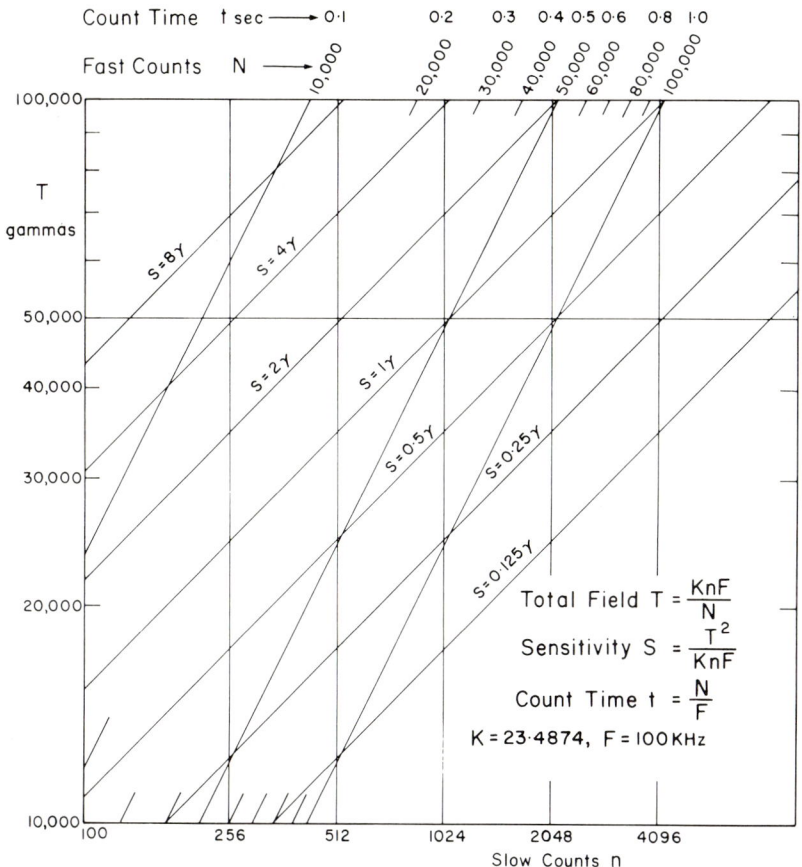

FIG. 7. Graph showing the relationship between the total field (T), slow (n), and fast (N) counts, count time (t), and sensitivity (S) for reciprocal-type proton free precession magnetometers using a 100 kHz fast oscillator.

An analog output is also obtained on a suitable recorder. Elliott Brothers, Ltd., in the U.K. manufacture a transistorized reciprocal-type of proton-precession magnetometer designated the EMD-14. A solution of N/7000 $FeCl_3$ is used in the detector head.

2.3.3.2. Direct-reading proton free-precession magnetometers. Figure 8 is a schematic showing the basic circuitry used in direct-reading instruments. After the polarizing field has been switched off, the precession signal

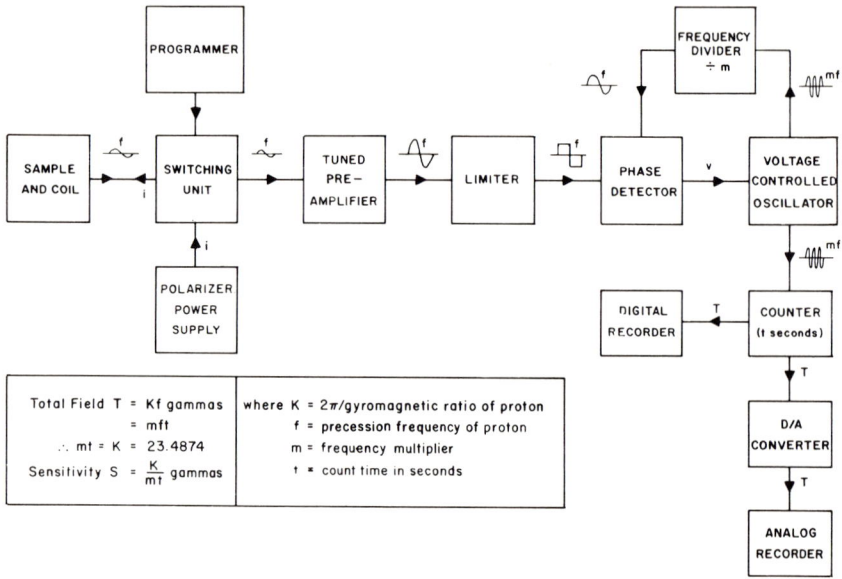

FIG. 8. Direct-reading proton free-precession magnetometer.

from the coil is first amplified, then squared and fed to a phase detector. The output of the phase-detector governs a voltage-controlled oscillator (VCO), which generates a frequency approximately m times the precession frequency f. This frequency is fed back to the phase detector through a $1/m$ frequency divider so that it is reduced almost to f. If there is a frequency difference between the original input and the fed-back frequency to the phase detector, then the resultant output voltage of the phase detector will change until the VCO frequency is exactly mf. The multiplied frequency mf from the VCO is then counted for an exact time period t so that the resultant count is the same number as the total field reading in gammas. Thus $T = mf \cdot t = Kf$ so that $mt = K = 23.4874$. For a sensitivity S gammas, $S = K/mt$, and the relationship between m, t, and S for proton precession magnetometers is shown in Fig. 14. The count is then recorded in digital and/or analog form.

The Geological Survey of Canada (GSC) [50] has developed a 0.1 γ direct-reading instrument which is illustrated in Fig. 9. Approximately 100 Oe is passed through a toroidal coil which polarizes a sample of distilled water in a stinger mounted on the tail of the aircraft. One advantage to using a toroidal coil is that external interference from the electrical system of the aircraft is minimized. After suitable amplification, the audio signal is telemetered to a ground station by a frequency-modulated transmitter. Audio frequency fiducial tones are also transmitted to the ground, where they are fed to the recording outputs to aid in subsequent track recovery as the same tone actuates a positioning camera in the aircraft. In order to synchronize the polarize-count cycle of the aircraft and ground system, the telemetered proton precession signal is fed to the programmer of the ground magnetometer and causes the unit to switch from the polarize to the count state.

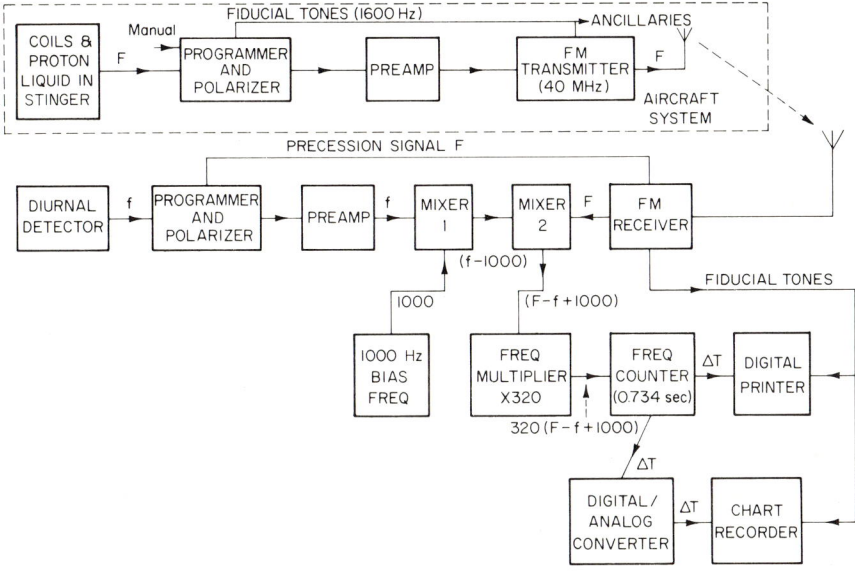

Fig. 9. GSC Telmag—a direct-reading proton free-precession magnetometer with automatic diurnal correction.

The audio-frequency outputs from a 1kHz oscillator and the ground-diurnal signal are mixed and the difference is fed to a second mixer together with the telemetered signal from the survey aircraft. The difference frequency is multiplied by 320, using a circuit developed by Serson [44], and the resultant frequency is counted for 0.734 sec to give a diurnally-corrected total-intensity magnetic field value directly in gammas. The 1 kHz oscillator thus increases the resultant output of the system by 23,487.4 γ ensuring that no polarity changes occur in the recorded output because negative anomalies of this magnitude are not usually recorded during airborne surveys except in rare instances.

The operational difficulties involved in telemetering the proton signal to the ground station may be avoided by simultaneously recording the airborne and base station values, using synchronized crystal-controlled clocks. If the magnetometer outputs are recorded on digital tape, then the two sets of data may be subsequently subtracted, using a computer to produce diurnally-corrected total intensity data.

The Australian Bureau of Mineral Resources has also constructed a direct-reading transistorized proton magnetometer for use in light aircraft. The coil system is mounted in a bird and about 250 Oe are used to polarize the kerosene proton source every half second. Rapid removal of the last few oersteds of the polarizing field is accomplished by critically damping the coil with the appropriate shunt resistor. The bandpass of the tuned detector coil is increased from approximately 25 to about 500 Hz by mismatching the detector coil to the preamplifier input. Manual tuning of the precession amplifier is then adequate to cope with normal operating conditions. The amplifier signal is clipped and actuates a zero-crossing Schmidt trigger. Pulses from the Schmidt trigger are then fed to the input of a phase-locked oscillator, the function of which is to multiply the input frequency by a factor of 200 so that the effective noise bandwidth is reduced, enabling the required resolution to be obtained for relatively low signal-to-noise ratios. The phase-locked loop contains a low-pass phase-lag filter, the time constants of which are chosen for optimum operation, taking into account the following characteristics of the BMR magnetometer: (a) a pull-in range, i.e., the range of frequencies over which the system will lock when the input signal is suddenly applied, of 2300 ± 820 Hz. This corresponds to a magnetic field range of 35,000 to 73,000 γ; (b) a flicker time, i.e., the time taken to achieve pull-in, less than 100 msec; (c) a natural frequency, i.e., the largest time rate-of-change of input frequency which can be tracked by the loop, of 35 Hz. This enables the system to track a field changing at the rate of 820 γ/sec; (d) a noise bandwidth of roughly 1 Hz; and (e) a maximum count error due to imperfect filtering of ± 1. The total magnetic field is then obtained by counting the frequency of the phase-locked oscillator for 0.1174 sec. This is accomplished by counting the multiplied proton frequency during the time required for 10,000 cycles of a 85.152 kHz crystal oscillator to be counted by a second counter (N.B.: 85.152 kHz = 10,000 $m/23.4874$ and $m = 200$).

Transistorized direct-reading proton-precession airborne magnetometers have also been produced by Barringer Research Ltd. of Toronto, the AM-101 ($m = 64$ and 128) (Fig. 10) by Varian Associates of Palo Alto, the V-4937 ($m = 32$ and 64), and by Prakla of Hannover, the PM-22 and PM-24 ($m = 128$).

FIG. 10. Barringer AM101 direct-reading proton free-precession magnetometer installation in light twin-engine aircraft.

A direct-reading proton-precession airborne magnetometer, the AYaAM-6, has been built in the U.S.S.R. The Larmor precession frequency is multiplied 24 times and counted for 0.97867 sec to obtain a 1-γ accuracy [58]. The relevant characteristics of all the direct-reading instruments are listed in Table II.

2.3.3.3. The Overhauser or spin-precession magnetometer. This magnetometer [1] is somewhat similar to the proton free-precession type in that the sensitive element consists of a coil system enclosing a liquid sample. The liquid sample is a solvent such as water containing protons in which is dissolved a parametric substance having a hyperfine spectrum of the stationary type which includes a narrow electronic resonance line whose frequency is not zero even in a null field, and which is saturable by a high-frequency alternating field. Examples of parametric substances are the metallic salts of the transition group or free radicals such as potassium or sodium nitrosodisulfonate.

The coil system consists of a high-frequency energizing coil whose plane is perpendicular to the low-frequency pickup coil. Abragam *et al.* [2] have used two high-frequency (55 MHz for nitrosodisulfonate) coils at right angles to one another and to the pickup coil in order to produce a circularly polarized energizing field. The high-frequency alternating field saturates the electronic resonance line of the parametric substance and it is found that the spin energy of electrons is transferred by coupling to the nuclear (proton) spin of the solvent. This mechanism is called the Overhauser [32] effect. Consequently the polarized protons will precess continuously at the Larmor frequency which is governed by the ambient magnetic field and which is detected by the pickup coils. After amplification, the Larmor frequency may be determined by suitable electronic circuitry such as that shown in Figs. 6 and 8. Sud Aviation of France has developed an airborne Overhauser magnetometer designated the MP 121 [46]. It operates over the range 27,000–75,000 γ. The sampling rate is 1 reading/sec, and a sensitivity of 0.1 γ is possible with digital recording on punched paper tape. The analog subranges are 100, 200, 500, and 1000 γ. Total weight of the airborne installation with towed bird is 280 lb. An analog record is also made with a maximum sensitivity of about 0.5 γ.

One of the difficulties of Overhauser magnetometers hitherto has been that the precession characteristics of the liquid used in the detector head deteriorated over a relatively short period. However, several permanent hydrogenated liquids, such as triacetoneamine oxide and its derivatives, and ditertiarybutyl nitroxide, have now been perfected by workers at the Grenoble Laboratory of the French Atomic Energy Commission [29].

2.3.4. Optical Absorption Magnetometers. The high sensitivity atomic-resonance magnetometers are the latest type to be used in airborne geophysical surveys and are currently (1968) under intensive development. Table II shows that three varieties have been produced, namely the metastable helium, rubidium- and cesium-vapor magnetometers. All these magnetometers make use of the optical-pumping technique originally devised by Kastler [25]. The basic physics of these magnetometers is complicated, but some idea of the principles involved may be gained from the following explanation.

The alkali metals all possess a single, unpaired electron in their outermost electron shells. This optical or valence electron spins about its own axis and rotates about the nucleus of the atom. The electron possesses a magnetic moment because of its spin. According to the laws of quantum mechanics, when the atom is placed in an external magnetic field, two energy sublevels of the atom are created, depending on whether the magnetic moment is parallel or antiparallel to the magnetic field. The difference in energies of these sublevels is proportional to the applied magnetic field (Zeeman effect). Photons of the appropriate radio frequency ν can provide the energy for the electron spin to change to the higher sublevel. If ΔE is the difference in energies between the sublevels, then the required frequency is given by the equation

(2.3) $$\Delta E = h\nu$$

where h is Planck's constant. For Rb_{85} the separation is approximately 4.667 Hz/γ [7]. Absorption of photons of light will also provide the energy necessary for the electron to go to a higher orbit. From this orbit it falls spontaneously to either of the two possible fundamental energy states, emitting a photon of light at a characteristic frequency.

Figure 11 shows the self-oscillating alkali-vapor magnetometer which was based on the work of Dehmelt [9]. The light from the electrodeless alkali-vapor lamp is first collimated, filtered to obtain a given optical line in its

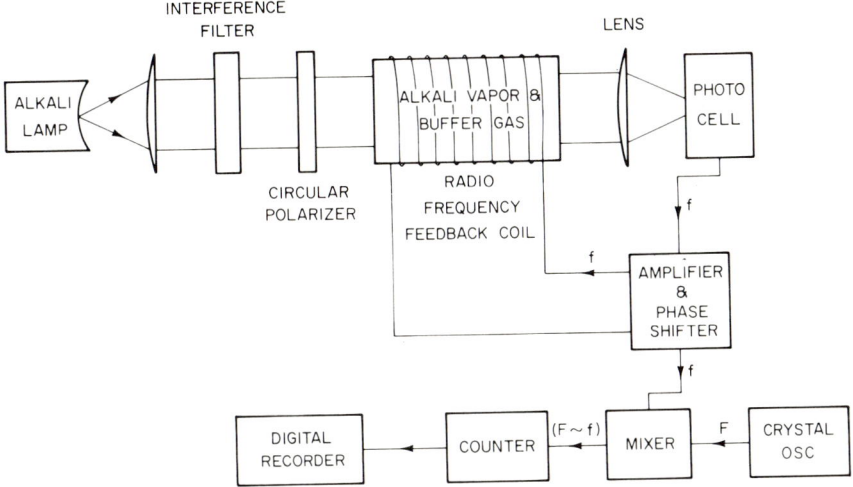

FIG. 11. Self-oscillating alkali-vapor magnetometer.

spectrum, and circularly polarized using standard optical techniques. The optical line selected by the interference filter corresponds to that emitted when the electron falls to one of the two energy states in the fundamental orbit (S) from the first excited orbit (P). Thus absorption of photons of this frequency by the valence electrons causes them to jump from the $^2S_{1/2}$ ground state to the $^2P_{1/2}$ state (Fig. 12). During the jump in orbits and owing to the polarization of the alkali-vapor line, the quantum number (m) associated with each excited electron increases by one. An excited electron soon falls back to any of the ground state sublevels, emitting its potential energy in the form of light at the same frequency as that initially absorbed. This process continues repeatedly but because the $^2P_{1/2}$ state contains no m level above

$m = +3$, transitions from the $^2S_{1/2}$ state when $m = +3$ are not possible (see Fig. 12), so that the atoms are fixed in this state, and cannot absorb any more light. When the majority of the atoms reach this condition, the gas cell becomes transparent, the transmission of the light through the gas cell reaches a maximum, and it has therefore been "pumped" into this energy level by the incoming light. If an alternating magnetic field at the Larmor frequency is applied at right angles to the ambient field, the "pumped" atoms will be

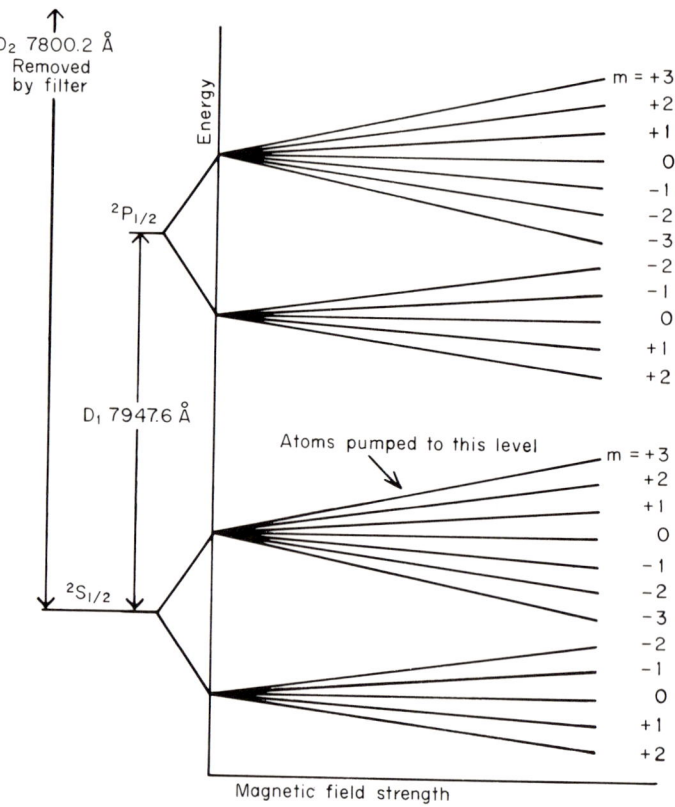

FIG. 12. Energy states of rubidium 85 vapor.

redistributed to all the m sublevels. The cell then becomes opaque to the incident light. The change in amplitude of the component of circularly polarized light perpendicular to the incident light is intensity modulated at the Larmor frequency. This signal may be amplified, shifted in phase by 90°, and fed back to the solenoid wound around the gas cell, and the system will oscillate at the Larmor frequency, which is proportional to the ambient magnetic field value, and will change as the field varies.

There is a second basic method of monitoring the Larmor frequency by sweeping either the magnetic field or the RF oscillator frequency through resonance, using a servoloop to correct the RF oscillator frequency. Parsons and Wiatr [34] have described such a self-tracking magnetometer.

There is a difference in energy between pairs of Zeeman sublevels proportional to T^2 because of the Back–Goudsmit effect. As the angle between the optical axis and the magnetic field is changed, the relative population of the various sublevels changes, producing an appreciable asymmetry in the local pattern when the average line is used for detection. For a typical value of the earth's field of 50,000 γ, this may amount to as much as 20 γ under the worst orientation conditions for the rubidium vapor magnetometer. Another problem with the single cell magnetometer is that the correct sign for the phase shift also depends upon the angle between the optical axis and the direction of the total field. Both these problems may be overcome to a great extent by using a dual-cell magnetometer having a common rubidium lamp. Thus for one-half of the dual-cell, the optical-pumping process is made to populate the $m = +2$ sublevel, whereas in the other half, the $m = -2$ sublevel of the atoms is used.

The further disadvantage of the single-cell instrument, which has dead zones, i.e., will not operate, when the field is either parallel with or at right angles to the optical axis, may be also overcome in the dual-cell system by having the optical axes of each half noncoaxial.

Varian Associates of Palo Alto have successfully developed a self-oscillating airborne rubidium-vapor magnetometer [20, 24] which has been designated the Model V-4916 (Figs. 13(a) and (b)). This transistorized instrument has a sensitivity of 0.05 γ and consists of four units: the sensor assembly, controller, recorder, and data processor. The sensor assembly contains the coaxial dual-cell rubidium-vapor detector and gimbal assembly for orientation of the thermostated magnetometer head, which provides 360° rotation about the vertical axis, and 180° rotation about the horizontal axis. The controller, assembly converts the magnetometer frequency to a dc voltage to drive a Varian chart recorder. This is accomplished using a crystal-controlled oscillator, mixer, and discriminator. The controller unit also contains the orientation control system for actuating the gimbal system in the magnetometer head. The rubidium frequency is fed to a phase-lock multiplier which increases the frequency by a factor of 8 and this higher frequency is then counted for about 0.5 sec. Navigational data are provided by a Bendix Doppler and Rosemount pressure altimeter. The magnetometer and navigational data together with time are recorded on a single digital magnetic tape by programing through a shift register.

Strome [45] has described the earlier developments of a rubidium-vapor airborne magnetometer at the Canadian National Aeronautical Establishment using basic Varian X-4935 dual-cell instruments mounted in bird and

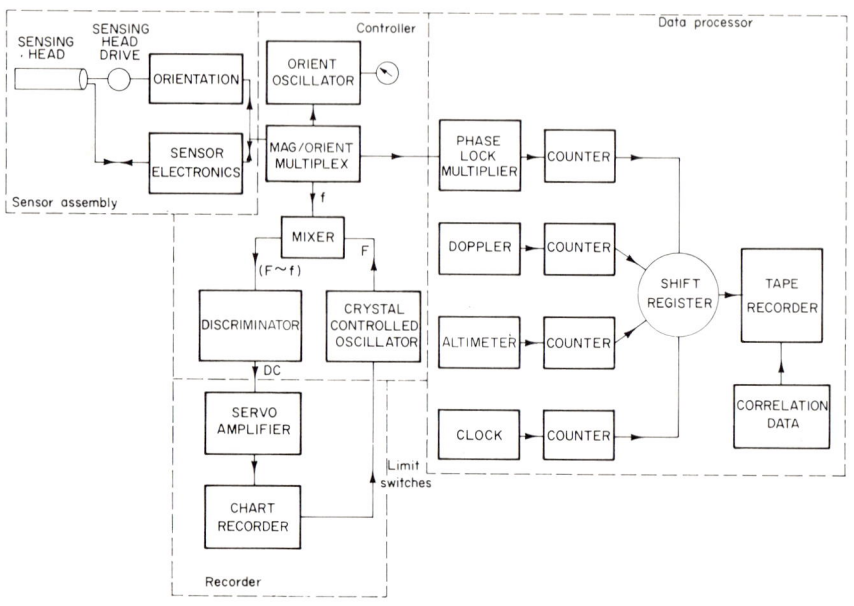

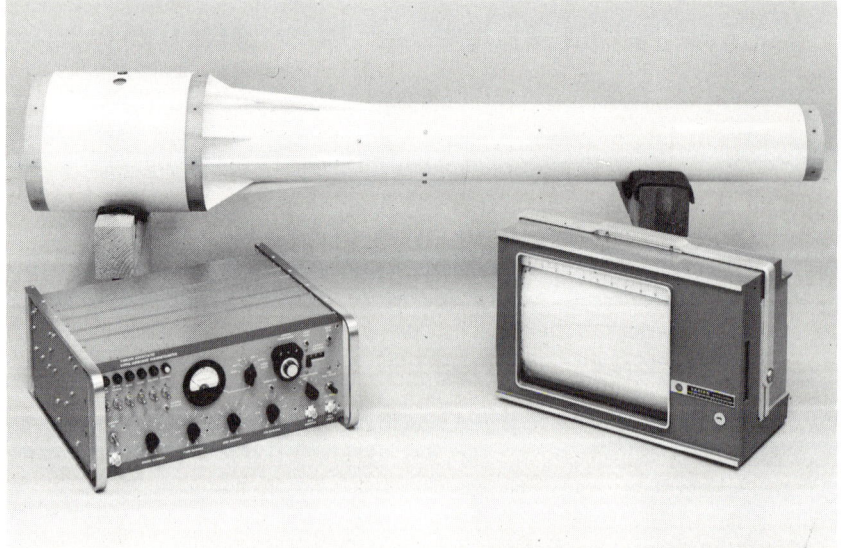

Fig. 13. Varian V-4916 rubidium-vapor airborne magnetometer (diagram and photo).

tail boom installations. This high-sensitivity magnetometer has a digital recording system which utilizes both magnetic tape and printer tape, and also has a seven-channel analog monitor system. More recent modifications have enabled the readings from both magnetometers, Decca navigation coordinates, and time from a digital clock to be recorded on a single digital magnetic tape by the use of a shift register.

Compagnie Générale de Telegraphie sans Fils (CSF) and Compagnie Générale de Géophysique (CGG) have jointly developed a cesium-vapor airborne magnetometer which also utilizes the optical pumping technique [14, 15]. The sensor and associated electronics are contained in a towed bird assembly. The resultant cesium-vapor frequency, which is 175 kHz for a 50,000-γ field (see Table III), is first multiplied by 30, and then counted over a 0.953-sec period by a frequency meter. The magnetic field values are recorded on a digital magnetic tape recorder with ± 0.01 γ sensitivity and are also monitored on an analog recorder.

Texas Instruments has also built an airborne magnetometer using a metastable helium sensor but to date no details have been released.

TABLE III. Comparison of optical absorption magnetometers.

	He^4	Rb^{85}	Cs^{133}
Approximate Hz/γ	28.024	4.667	3.498
°C Temperature for optimum vapor pressure of 10^{-6}mm.		34	23
Back–Goudsmit effect in gammas at 50,000 γ for single cell		7.7	1.9

Figure 14 is a graph of the frequency multiplication factor (m) plotted against the count time (t) for various sensitivities (S) for proton precession and optical absorption magnetometers. It is readily apparent from Fig. 14 that optical-absorption magnetometers have a sensitivity approximately two orders of magnitude greater than the proton-precession type. In the case of optical-absorption magnetometers, the graph is still valid even if the Larmor frequency is first reduced by heterodyning with a stable crystal oscillator (see Fig. 17), before multiplication and counting.

2.3.5 *Electron-Beam or Aspect-Tube Magnetometer*. In one type of electron-beam magnetometer [8], the deflection of an electron-beam by a magnetic field is automatically compensated by a voltage applied to the electrostatic deflection plates of a miniature cathode-ray tube. The output of a photomultiplier which senses the position of the light spot produced by the electron-beam is fed back, after suitable amplification, to the electrostatic deflection plates to keep the position of the light spot unchanged on the face of the cathode-ray tube.

Elliott Brothers Ltd. in the United Kingdom have developed a slightly different total-field magnetometer utilizing this principle, which was installed in an Otter aircraft belonging to Rio Tinto (Canada) Ltd. together with the Mullard coplanar vertical-coil EM system. In the Elliott EMD13 magnetometer, the electron-beam current is equally shared between two collector electrodes in zero ambient magnetic-field conditions. To measure the strength of an ambient field, the beam deflection induced by it is sensed by connecting

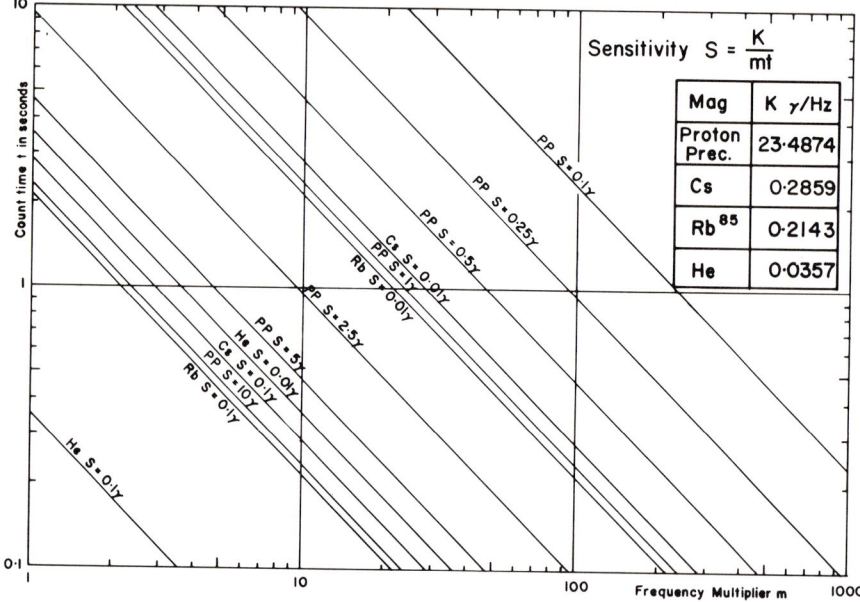

FIG. 14. Graph comparing the relationship between the frequency multiplication factor (m), count time (t), and sensitivity (S) for proton-precession and optical absorption magnetometers.

the collectors across the input of a differential amplifier. The resultant current output of this amplifier is used to restore the beam to the equilibrium condition by feedback through a coil on the detector tube. The steady state value of this current is then a direct measure of the ambient field component in the direction parallel to the gap between the collectors.

The outputs from two orienting detector tubes is used to orient a third measuring tube into parallel alignment with the axis of the total magnetic vector. This is accomplished by gimbal-mounting the three tubes orthogonally and driving the gimbal by a servoloop which maintains the orientor tubes along axes of zero field component. Additional coils are used on the detector head to backoff the stray fields from the aircraft and demagnetize the installation.

The sensitivity of the Elliott aspect electron-tube magnetometer is around 10 γ in actual field use, and its main function is to ascertain whether there is an associated magnetic anomaly with the EM anomalies detected. The system was acquired by Canadian Mineral Surveys Ltd. in 1960.

2.3.6. Hall Effect Magnetometers. In 1879, Hall [19] discovered that when a magnetic field is applied at right angles to a conductor in which an electric current is flowing, a transverse potential is produced across the conductor at right angles to both the current and the magnetic field which is directly proportional to the orthogonal magnetic field. Indium antimonide appears to be the best material to use at the present time, but the ultimate sensitivity for a 1-sec response time is limited by the thermal noise output of indium antimonide crystals to 10 γ [52]. Although a Hall-effect compass has been developed in the United Kingdom [39], no reports of a survey airborne magnetometer based on the phenomenon have appeared in the literature.

2.4. Aeromagnetic Survey Techniques

2.4.1. Aircraft Installation. Since about 1960 there has been a trend to the use of light twin-engined airplanes such as the Aero Commander, Beechcraft Queenair, and Piper Aztec [37] in aeromagnetic surveys because of their cheapness to operate.

Nowadays the magnetometer head is usually mounted in an inboard installation (stinger) although with the alkali-vapor instruments some companies have reverted to the towed bird because of magnetic compensation difficulties with high sensitivity magnetometers.

Inboard installations have been made possible by the improvement in the magnetic compensation of aircraft. There are three different sources of interference owing to the aircraft itself. The first is the permanent magnetism of the various components made of steel, such as the engines, whose direction remains fixed with respect to the aircraft. The second source is the induced field owing to the magnetic susceptibility of these same components and the earth's field. Its polarity and magnitude depend upon the orientation of the aircraft with respect to the earth's magnetic field. The third source of interference is that caused by the magnetic effect of eddy currents generated in the skin and other conducting parts of the aircraft by their motion in the earth's magnetic field.

In passive compensation systems, the permanent magnetism is eliminated by the use of a set of three-orthogonal compensating coils mounted near the magnetometer head of the roll axis of the aircraft [4] through which the appropriate dc currents may be passed. The induced components are eliminated by the use of permalloy strips, and the eddy currents are compensated for by the use of coils of wire mounted in close proximity to the sensitive element. It is necessary to carry out a set of pitches and rolls of the aircraft in low gradient areas in order to separate the effects of the various components [28]. The excellence of compensation of a given aircraft is measured by its "figure of merit." This index is obtained by summing, without regard to sign, the amplitudes of the 12 magnetic anomalies recorded when the aircraft carries out 20° rolls, 10° pitches, and 10° yaws peak-to-peak on north, east, south, and west headings.

In recent years, active 9-term and 16-term aircraft magnetic compensation systems have been developed by Canadian Aviation Electronics (CAE) Industries Ltd., Montreal. The use of these active compensation systems improves the "figure of merit" of a given aircraft considerably, and drastically reduces the time required for aircraft compensation.

2.4.2. Surveying Procedure. Airborne magnetometer surveys in mining exploration are usually carried out at a mean terrain clearance of 500 ft and a line spacing of 1/4 mi or less. In Canada, the federal government uses a line spacing of 1/2 mi and a survey altitude of 1000 ft. The choice of survey altitude and line spacing are interrelated because, in contouring an aeromagnetic map, features have to be followed across the flight lines. Lower flight elevations will produce more detailed profiles [35], i.e., greater resolution of the anomalies, and it is therefore necessary to reduce the distance between the flight lines in order to maintain the accuracy of the map. That there is some inaccuracy on all aeromagnetic maps is obvious from the fact that the magnetic highs and lows on aeromagnetic maps always occur on flight lines. The probability of missing an anomaly of given shape and dimensions has been discussed by Agocs [3].

In petroleum exploration, the magnetic effects of the basement rocks buried beneath several thousands of feet of nonmagnetic sediments are the primary interest. It is therefore advisable to fly high enough so that surface effects (often man made) are negligible. Thus, in oil exploration, the survey aircraft is flown at a constant barometric altitude which is usually 1000 ft. The common rule of thumb for line spacing is to space the flight lines no further apart than the depth to the basement [38].

Aeromagnetic survey traverses are usually flown using an azimuth within 30° of the direction at right angles to the main magnetic anomaly trend, which usually corresponds to the strike of the basement rocks. At low magnetic latitudes however, magnetic anomalies tend to be elongated in an east-west direction, so that it is advisable to use a flight-line direction within 45° of north, irrespective of the main geological strike.

In order to tie the survey lines to a common datum, a series of control lines about 6 mi apart are usually flown at right angles to the main flight-line direction through areas of low magnetic gradient. The factors which limit the accuracy of an aeromagnetic survey may be summarized as follows:

(a) Internal causes

(1) Magnetic miscompensation of survey aircraft.
(2) Magnetometer noise due to electronic components and misorientations of detector if a vector magnetometer is used. The error is approximately $T\theta^2/2$ where T is the total field value and θ is the misorientation angle in radians. For 1 γ accuracy in a field of 50,000 γ this amounts to 22 min.
(3) Drift of magnetometer readings because of aging of components, e.g., ambient field cancellation batteries in fluxgate type, and temperature drift of components, e.g., crystal frequencies.
(4) Imperfect calibration of magnetometer.
(5) Too slow response time or sampling rate of the instrument.

(b) External causes

(1) Diurnal variation of the earth's magnetic field.
(2) Variations in terrain clearance especially in mountainous country.
(3) Positioning errors owing to mistakes in plotting the track, or imperfections in the electronic positioning system [10].
(4) Incorrect choice of flight-line spacing and direction.

2.4.2.1. Positioning. The cheapest method of recovering the aircraft track is still by the use of 35 mm cameras, of either continuous strip or the more usual single frame (every 2 sec) variety. In areas where the base maps are poor or nonexistent, Doppler navigation [56] has been used. Over water, electronic positioning systems, such as Decca, Shoran and its derivatives Hiran and Shiran, and Raydist [27] may be used. For reconnaissance surveys, the World War II Loran A [5] system has also been used, and the more accurate Loran C system is now available for navigational use in many areas.

Jensen [24] has considered the stringent flight-path restrictions necessary in the conduct of high-sensitivity aeromagnetic surveys. He concludes that all high-resolution magnetic surveying should be under the control of Doppler navigation, precision pressure altimeters, and modern autopilots.

2.4.2.2. Ground monitors for diurnal variation of the earth's magnetic field. Magnetic diurnal stations [55] are usually set up to monitor the earth's field during the survey operations, and the resultant records are used to decide on any reflights necessary because of magnetic disturbances. The ground monitor is often located at the airstrip because of logistical convenience.

Sometimes the data obtained from the ground monitor are used to correct the airborne data by direct subtraction [22]. There are various objections to this procedure which include the following:

(a) The amplitude and phase of the diurnal variation changes with the geographic location. The size of error (B) in gammas to be expected is given in the following relationship suggested by Whitham and Niblett [53]:

(2.4) $$B = Kad$$

where $K =$ constant for a given area, e.g., for the Edmonton area of Alberta, Whitham and Niblett [53] estimated an approximate value of $K = 0.01/\text{mi}$

$d =$ separation of survey aircraft and ground station in miles (the total separation being less than 100 mi)

$a =$ magnetic activity in gammas

The value of the magnetic activity a during a survey flight may be obtained from the formula

(2.5) $$a = \frac{1}{N}\left[\sum_{n=0}^{N}(T_n - \bar{T})^2\right]^{1/2}$$

where N is the number of readings obtained from the ground monitor record during the flight. If t is the survey flight time and t' is the sampling time of the ground monitor, then $N = t/t'$; T_n is the instantaneous value of the total field measured by the ground monitor at a given instant, and $\bar{T}$ is the average (undisturbed) value of the total field measured by the ground monitor.

(b) Amplification of the diurnal variation at the ground station is caused by anomalous electromagnetic induction in the ground. There will be an additional effect if the ground station is located near a geologic body having significant magnetic susceptibility [18].

2.5. Aeromagnetic Data Compilation

There are four raw end products at the completion of aeromagnetic survey operations. These are the magnetometer, positioning (including altimeter) and diurnal variation data, and the magnetometer operator's log. Up to the present time, most survey companies have recorded the variations in the total intensity of the earth's magnetic field as an analog trace on a suitable paper chart. The radio altimeter profile is often recorded on the same chart. Over land areas, a 35 mm film of the strip of ground below the aircraft is used to position the aircraft track, whereas over water an electronic positioning system is necessary. Numbered fiducials are used to relate the film to the magnetometer record. After editing the data for consistency, the flight lines are plotted on base maps using recognizable points spaced about every 2 mi. Air photographs are often an essential aid in this step.

The next step is the control of the magnetics which eliminates the differences in magnetic level between the flight lines and corrects for the diurnal variation of the earth's magnetic field and the drift of the magnetometer. First a number of control lines are chosen from the main flight lines which pass through areas of low gradient (as do the tie lines which cross the main flight lines at right angles) and were flown on quiet diurnal days. Often a set of four control lines is chosen at the boundaries of the surveyed area. On a suitable diagrammatic chart of the flight lines, the magnetometer values occurring at the intersection of the control and tie (or base) lines are written. For a given loop on the resultant chart formed by pairs of control and tie lines, if A, B, C, and D (Fig. 15) are the intersection values on the control

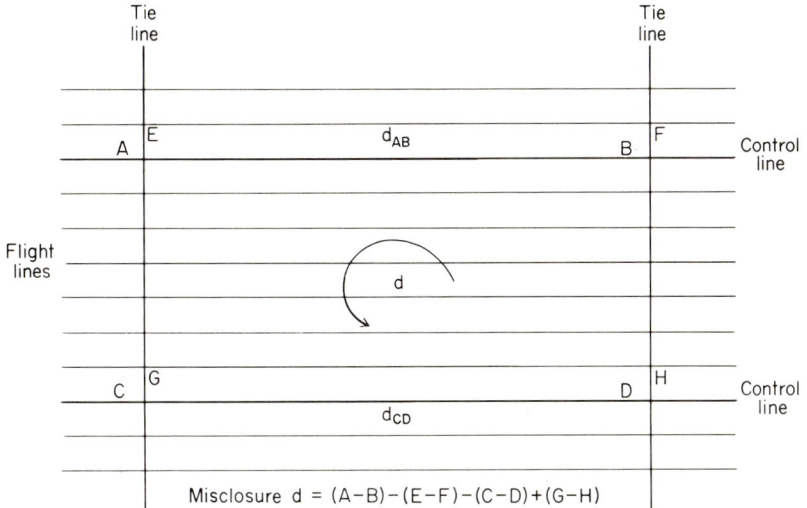

FIG. 15. Aeromagnetic compilation procedure.

lines, and E, F, G, and H are the corresponding intersection values on the tie lines, then the misclosure or drift around the loop is given by

$$(2.6) \quad \text{Misclosure } (d) = (A-B) - (E-F) - (C-D) + (G-H)$$
$$= d_{AB} + d_{CD}$$

These misclosures are then distributed along the control lines in a manner which appears to give the most reasonable values (usually the least). It is usually assumed no drift occurs along the tie lines between the control lines. Inspection of the appropriate ground diurnal tapes is made at this stage to help in assigning the drifts along the control lines.

If a nonabsolute magnetometer is used, it is necessary to establish the value of the total field at a point in the survey area, preferably on a tie or control line. An arbitrary value around 2000 γ is then assigned to this point so that none of the values on the map will be negative, and it is only necessary to add an even amount (56,000 γ in Fig. 16) to get total field values on the

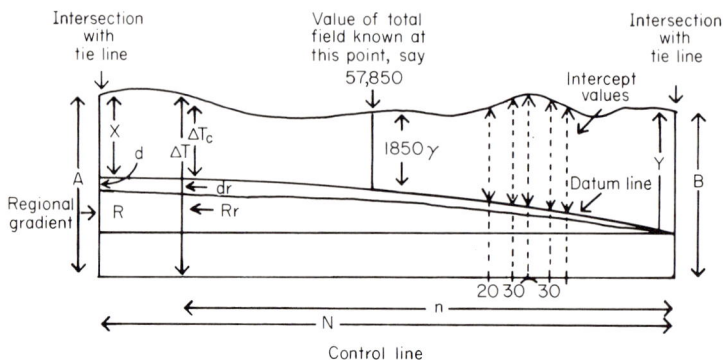

FIG. 16. Intersection of aeromagnetic charts.

We have drift d + Regional $R = (A - X) - (B - Y)$. If the drift and regional gradient are assumed to be linear between the tie lines, by similar triangles $(dr + Rr)/(d + R) = n/N$; also $\Delta T_c + d_r + R_r + (B - Y) = \Delta T$, the value at any point. Therefore, the intercept value $\Delta T_c = \Delta T - (B - Y) - (d_r + R_r) = \Delta T - (B - Y) - (n/N)[(A - X) - (B - Y)]$.

map. From this known point, using the drift and the magnetic values (see Fig. 15), a datum line may be drawn between the intersections of the control and tie lines. Subsequently, using the appropriate drift values, the datum may be carried through the control system. Sometimes the regional gradient of the earth's magnetic field is also removed at this stage. There is usually little point in subtracting anything other than a plane regional gradient. The published maps and tables showing the regional variation of the earth's magnetic field are consulted and the gradient along the survey lines established.

The next step is to intercept the tapes at the map contour interval, usually 5 or 10 γ (see Fig. 16), transfer the intercepted values to the appropriate flight line drawn on the base map, and contour the magnetic values. As the scale of the magnetometer tape is generally different from (being usually greater than) that of the map compilation scale, it is necessary to change the scales by some technique. Some survey companies use a mechanical pantograph-type instrument to transcribe the intercept values directly on to the base map, but there are other, simpler geometric conversion methods which are equally satisfactory.

When the magnetometer data have been recorded digitally, it is possible to automate the procedure considerably [12]. Because it is much cheaper to use aerial photography rather than an electronic positioning method to position the flight path, this system, with or without the aid of a Doppler navigation system, will probably remain in general use for many years to come. This would mean that the control analysis would be a manual procedure, and in any case, it is a somewhat subjective procedure and it would be advisable to check this critical step before proceeding further. A print-out of the digital magnetic values is necessary at this point to obtain the crossover values A, B, C, etc. It is probably advisable to have a printer-tape output facility for the airborne magnetometer to provide a print-out of the magnetometer values, as an alternative record of the magnetometer data in case of a malfunction. The primary recording system will either be punched paper or digital magnetic tape. The preferred system will probably be the latter because it is less mechanical and the physical bulk of the magnetic tape is less. Once the intersection values on the tie lines (X and Y in Fig. 16) have been established, the corrected magnetic field values may be obtained from the computer, using the formula given in Fig. 16. Computer programs are also available for automatic contouring of the resultant values.

2.6. Gradiometers

There appear to be few attempts to measure the gradient in airborne magnetometer surveys. During World War II, the British tested a MAD gradiometer [13] using two coaxially-mounted coils of wire separated by about 8 ft. The work was continued in the U.S., but the successful development of the saturable-core magnetometer terminated further work.

The first proposals for an aeromagnetic gradiometer seem to be by Fromm [13] who envisaged that a gradiometer might be constructed either by stabilizing each of two magnetometers independently, so that their axes remained parallel, or by aligning their axes parallel and connecting them by a rigid framework. Actually Vacquier [47] applied for a gradiometer patent (No. 2,407,202) in 1941 along with his original U.S. patent on the fluxgate element.

Both Wickerham [54] and Glicken [16] have described the measurement and use of the first-horizontal derivative in aeromagnetic surveys. This is readily obtainable from the records and does not require any elaborate modifications to the airborne magnetometer equipment, although Wickerham did describe the modifications necessary to convert the Gulf airborne fluxgate magnetometer to a so-called time-derivative gradiometer. The feedback current to the fluxgate element was controlled by a servomotor geared to a rate generator, whose rectified output was then proportional to the horizontal gradient of the total field. Witham [52] has noted that the accuracy of such a gradiometer far exceeds that feasible by using two fluxgate elements separated in space.

The recent development of high-sensitivity magnetometers such as the optical-pumping varieties has made feasible the measurement of the first-vertical derivative of the total field in aeromagnetic surveys. This is accomplished by using two sensitive magnetometer heads separated by a constant vertical distance, and by recording the difference in outputs. Figure 17 is a block diagram of a vertical gradiometer system which uses optical absorption

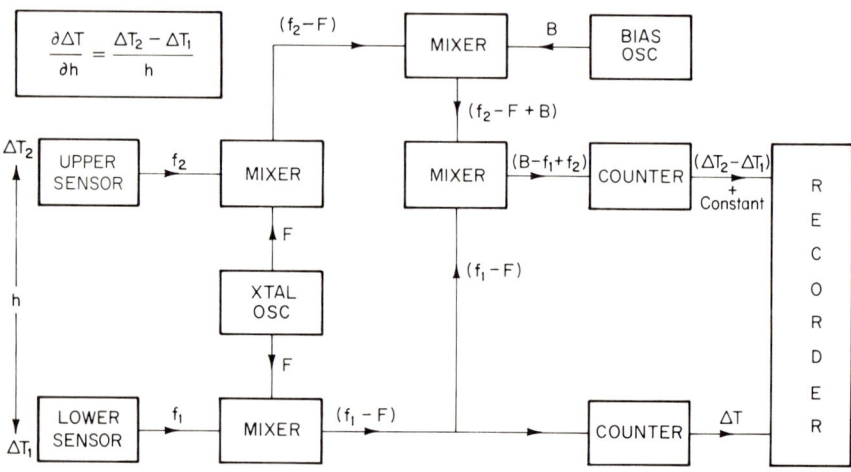

FIG. 17. Block diagram of vertical gradiometer system using optical absorption magnetometers.

magnetometers. The two outputs (f_1 and f_2) from the sensors are first mixed with the crystal oscillator frequency (F). The difference frequency ($f_2 - F$) of the upper sensor is then increased by a bias frequency B before being mixed with the difference frequency ($f_1 - F$) of the lower sensor. The resultant frequency ($B - f_1 + f_2$) is thus a measure of the difference ($\Delta T_2 - \Delta T_1$) in the total intensity values of the sensors plus a constant. The bias frequency B is added to avoid loss of sign (positive or negative) of the vertical gradient measurement because f_1 and f_2 do not differ greatly. Both the frequencies ($B - f_1 + f_2$) and ($f_1 - F$) are counted for a set time period, as described earlier in the article, in order to yield the gradient and ΔT values, respectively. The effect of diurnal variation is thus eliminated in the resultant differential output, and this is an especially desirable feature in northern Canada where the diurnal variation is usually much greater than that found in more southerly magnetic latitudes. Moreover steeply dipping geological contacts in high magnetic latitudes are outlined by the resultant zero-gradient contour

[21]. The data reduction necessary to produce a vertical-gradient map is much simpler that in the total-field case because no datum leveling is necessary. As the aircraft track will be available from the main compilation, it is only necessary to plot the resultant vertical-gradient values on the track map and contour. Thus two maps will be obtained for little more than the cost for one but with a greatly increased gain in geophysical information concerning the geometry of the causative bodies. Actually a first-derivative map is difficult (and therefore costly) to produce by any other means.

REFERENCES

1. Abragam, A., Combrisson, J., and Solomon, I. (1962). Method and device for measurement of magnetic fields by magnetic resonance. U.S. Patent No. 3,049,662.
2. Abragam, A., Combrisson, J., and Solomon, I. (1963). Magnetic resonance magnetometer. Can. Patent No. 661,807. 22 pp.
2a. Affleck, J. (1967). Private communication.
3. Agocs, W. B. (1955). Line spacing effect and determination of optimum spacing illustrated by Marmora, Ontario magnetic anomaly. *Geophysics* **20**, 871–885.
4. Anonymous (1959). Aviation Electronics Technician 3 & 2. *Bur. Naval Personnel*, NAVPERS 10317-A, 464 pp.
5. Anonymous (1959). Air Navigation, U.S. Air Force, Manual 51-40, **1**, 420 pp.
6. Balsley, J. R. (1952). Aeromagnetic surveying. *Advan. Geophys.* **1**, 313–349.
7. Bloom, A. L. (1962). Principles of operation of the rubidium vapor magnetometer, *Appl. Optics* **1**, 61–68.
8. Cragg, B. G. (1955). An electronic magnetometer. *J. Sci. Instr.* **32**, 385–386.
9. Dehmelt, H. G. (1957). Modulation of a light beam by precessing absorbing atoms. *Phys. Rev.* **105**, 1924–1925.
10. Domzalski, W. (1957). Some problems of the aeromagnetic surveys. *Geophys. Prospecting* **5**, 469–479.
11. Driscoll, R. L., and Bender, P. L. (1958). Proton gyromagnetic ratio. *Phys. Rev. Letters* **1**, 413–414.
12. Evenden, G. I., Frischknecht, F. C., and Meuschke, J. L. (1967). Digital recording and processing of airborne geophysical data. *U.S. Geol. Surv.* Prof. Paper 575-D, D79-D84.
13. Fromm, W. E. (1952). The magnetic airborne detector. *Adv. Electron.* **4**, 257–299.
14. Giret, R., and Malnar, L. (1965). Un nouveau magnétomètre aérien; le magnétomètre à vapeur de caesium, *Geophys. Prospecting* **13**, 225–239.
15. Giret, R. I. (1965). Some results of aeromagnetic surveying with a digital cesium-vapor magnetometer, *Geophysics* **30**, 883–890.
16. Glicken, M. (1955). Uses and limitations of the airborne magnetic gradiometer. *Mining Eng.* **7**, 1054–1056.
17. Godby, E. A. (1963). A survey magnetometer with digital output. *NCR (Can.), DME/NAE Quart. Bull.* **1963** (3), 37–46.
18. Goldstein, N. E. and Ward, S. H. (1966). The separation of remanent from induced magnetism in situ. *Geophysics* **31**, 779–796.
19. Hall, E. H. (1880). On a new action of the magnet on electric currents *Phil. Mag.* **9**, 225–230.

20. Herbert, R., and Langan, L. (1965). The airborne rubidium magnetometer. Varian Assoc., *Geophys. Tech. Mem.* **19**.
21. Hood, P. (1965). Gradient measurements in aeromagnetic surveying. *Geophysics* **30** 891–902.
22. Hoylman, H. W. (1961). How to determine and remove diurnal effects precisely. *World Oil* **153**, 107–112.
23. Jensen, H. (1961). The airborne magnetometer, *Sci. Am.* **204**, 151–162.
24. Jensen, H. (1965). Instrument details and applications of a new airborne magnetometer. *Geophysics* **30**, 875–882.
25. Kastler, A. (1950). Quelques suggestions concernant la production optique et la detection optique d'une inegalité de population des niveaux de quantification spatiale des atomes. *J. Phys. Radium* **11**, 255–263.
26. Langan, L. (1961). The airborne Varian magnetometer, Varian Assoc., *Geophys. Tech. Mem.* **9**.
27. Laurila, S. (1960). "Electronic Surveying and Mapping," 294 pp. Ohio State Univ. Press, Columbus, Ohio.
28. Leliak, P. (1961). Identification and evaluation of magnetic field sources of magnetic airborne detector equipped aircraft. *IRE Trans Aerospace and Navigational Electron* **8**, 95–105.
29. Lemaire, H., Rassat, A., Brière, R., Dupeyre, R. M., and Salvi, A. (1967). Magnetometers for measuring the earth's magnetic field and its variations. 40 pp. Can. Patent 764,597.
30. Logachev, A. A. (1946). The development and application of airborne magnetometers in the USSR. *Geophysics* **11**, 135–147.
31. Lundberg, H. (1947). Results obtained by the helicopter-borne magnetometer. *Trans. Can. Inst. Mining Met.* **50**, 392–400.
32. Overhauser, A. W. (1953). Paramagnetic relaxation in metals. *Phys. Rev.* **89**, 689–700.
33. Packard, M., and Varian R., (1954). Free nuclear induction in the earth's magnetic field. *Phys. Rev.* **93**, 941.
34. Parsons, L. W., and Wiatr, Z. M. (1962). Rubidium vapor magnetometer. *J. Sci. Instr.* **39**, 292–300.
35. Paterson, N. R. (1962). Geological mapping by magnetometer surveys. *Proc. Benedum Earth Mag. Symp., Pittsburgh, Pennsylvania*, pp. 139–157.
36. Powles, J. G., and Cutler, D., (1958). Audio frequency nuclear resonance echoes. *Arch. Sci. (Geneva)* **11**, 209–214.
37. Reford, M. S. (1964, 2nd ed. 39 pp.) Airborne magnetometer surveys for petroleum exploration. Aero Service Corp., Philadelphia, Pennsylvania.
38. Reford, M. S., and Sumner, J. S. (1964). Aeromagnetics. *Geophysics* **29**, 482–516.
39. Ross, E. M., Saker, E. W., and Thompson, N. A. C. (1957). The Hall-effect compass. *J. Sci. Inst.*, **34**, 479–484.
40. Rumbaugh, L. H., and Alldredge, L. R. (1949). Airborne equipment for geomagnetic measurements. *Trans. Amer. Geophys. Un.*, **30**, 836–848.
41. Schonstedt, E. O., and Irons, H. R. (1955). NOL Vector Airborne Magnetometer Type 2A. *Trans. Amer. Geophys. Un.*, **36**, 25–41.
42. Serson, P. H., Mack, S. Z., and Whitham, K. (1957). A three-component airborne magnetometer. *Publ. Dom. Obs., Canada*, **19**, 97 pp.
43. Serson, P. H. (1957). An electrical recording magnetometer. *Can. J. Physics*, **35**, 1387–1394.
44. Serson, P. H. (1961). Proton-precession magnetometer. Canada Patent 618,762. U.S. Patent 3,070,745 (1962).

44a. Serson, P. H. (1966). Private communication.
45. Strome, W. M. (1964). Magnetometer installation in a North Star aircraft. *NRC (Can.) DME/NAE Quart. Bull.* **1963**, pp. 23–30.
46. Thomas, J. (1965). Le magnétomètre aéroporte MP 121 Sud-Aviation. *Geophys. Prospecting* **13**, 22–36.
47. Vacquier, V. V. (1946). Apparatus for responding to magnetic fields. U.S. patent 2,407,202.
48. Vigoureux, P. (1962). A determination of the gyromagnetic ratio of the proton. *Proc. Roy. Soc.*, **A270**, 72–89.
49. Vigoureux, P. (1963). Gyromagnetic ratio of the proton. *Nature*, **198**, 1188.
50. Washkurak, S., and Sawatzky, P. (1966). The Serson direct-reading proton free-precession magnetometer: Part II—Airborne use with telemetering and automatic diurnal correction. *Geol. Surv. Canada.* Paper 65–31, 33–75.
51. Waters, G. S., and Phillips, G. (1956). A new method of measuring the earth's magnetic field. *Geophys. Prospecting*, **4**, 1–9.
52. Whitham, K. (1960). Measurement of the geomagnetic elements, in "Methods and Techniques in Geophysics," pp. 104–167. Wiley (Interscience), New York.
53. Whitham, K., and Niblett, E. R. (1961). The diurnal problem in aeromagnetic surveying in Canada. *Geophysics*, **26**, 211–228.
54. Wickerham, W. E. (1954). The Gulf airborne magnetic gradiometer. *Geophysics*, **19**, 116–123.
55. Wickerham, W. E. (1955). Magnetic storm monitor. *Mining Eng.*, **7**, 966–968.
56. Wilson, R. A. R., and English, J. E. (1959). New instruments in the airborne search for oil; the Doppler navigator system. *Petrol. Times*, **63**, 427–428.
57. Wold, R. J. (1964). The Elsec-Wisconsin digital recording proton-precession magnetometer system. Geophys. and Polar Res. Cent., Univ. Wisconsin, Res. Rept. Ser. 64-4, 83 pp.
58. Yanovskii, B. M. (1963). Earth's magnetism. 461 pp. Leningrad Univ.

3. Low-Frequency Airborne Electromagnetic Methods[1]

3.1. Introduction

Airborne electromagnetic (AEM) methods have been used in mineral exploration since 1950. Many different systems have been placed in operation and a detailed discussion of them all would not serve the purpose of the present article. Instead we shall set down the basis of all inductive AEM systems and illustrate our points by reference to specific systems. For previous surveys of AEM systems, the reader might wish to refer to discussions by Grant and West [16], Keller and Frischknecht [21], Parasnis [25], Pemberton [32], and Ward [50, 51]. Results with specific systems have been presented by Barringer [2], Boniwell [4], Boyd and Roberts [6], Brant *et al.* [7], Cheriton [8], Collett [9], Fleming and Brooks [12], Gaur [14], Geleynse and Barringer [15], Hedstrom and Parasnis [17, 18], Joklik [20], Khomenyuk [22], MacKay

[1] The list of symbols used in this section will be found at the end of the section.

and Paterson [23], Mizyuk [24], Paterson [26–29] Pemberton [30, 31], Podolsky [33], Rattew [34], Schaub [35–41], Slichter [42], Tikhonov and Dmitriev [43, 44], Tornquist [45], Ward [46–49], Ward and Barker [52], Ward et al., [53, 54], White [55], and Wieduwilt [56].

Up to the present, AEM systems have been used primarily in the search for massive sulfide-ore deposits, but recent limited success has been obtained in applying the method to groundwater problems [4, 9–11] and in geologic mapping [54]. This review is intended only to discuss the principles of AEM methods applied to ore prospection. Other applications probably will require an expansion in the current concept of an AEM system prior to general acceptance.

3.2. The Basic Principle of Inductive AEM Systems

The basic principle of all active inductive AEM systems is measurement of the change in mutual impedance of a pair of induction coils as this pair is moved relative to a subsurface conductor of electricity. If the subsurface conductor may be simulated by a small closed circuit, consisting of an inductance L and resistance R, then its effect on the mutual impedance of the coil pair is readily studied. Thus in Fig. 18, energy is transferred from the transmitting coil to the receiving coil via the direct path M_{TR} and via the indirect path $M_{TC} - M_{CR}$. The voltage induced in the receiving coil via the direct path is

$$(3.1) \qquad e = -\mu_0 N_R \mathbf{A}_R \cdot d\mathbf{H}_T/dt$$

where μ_0 is the permeability of free space, A_R and N_R are the area and turns, respectively, of the receiving coil, while H_T is the field of the transmitting coil at the position of the receiving coil. The influence of the host medium surrounding the subsurface conductor is ignored in the first approximation. If the field of a transmitting coil, located at the origin, is dipolar, we may describe it by

$$(3.2) \qquad \mathbf{H}_T = \frac{N_T A_T I_T}{4\pi} \left[\frac{(2x^2 - y^2 - z^2)\mathbf{i} + 3xy\mathbf{j} + 3xz\mathbf{k}}{\rho^5} \right] e^{i\omega t}$$

where $\rho^2 = x^2 + y^2 + z^2$; $\mathbf{i}, \mathbf{j}, \mathbf{k}$ are unit vectors in the x, y, z directions; $N_T A_T I_T$ is the turns-area-current product of the transmitting coil; $\omega = 2\pi f$ is the angular frequency. The axis of the transmitting dipole is in the x direction. The receiving coil is located at the point x, y, z. Then the expression (3.1) may be written

$$(3.3) \qquad e = -\frac{i\mu\omega}{4\pi} N_T A_T I_T N_R A_R f(x, y, z) e^{i\omega t}$$

where

$$f = \alpha \cdot \left[\frac{(2x^2 - y^2 - z^2)\mathbf{i} + 3xy\mathbf{j} + 3xz\mathbf{k}}{\rho^5} \right]$$

and α is a unit vector normal to the area A_R.

A similar voltage induced in the subsurface conductor will be

(3.4) $$e_C = -\frac{i\mu\omega}{4\pi} N_T A_T I_T N_C A_C f_C(x_C, y_C, z_C) e^{i\omega t}$$

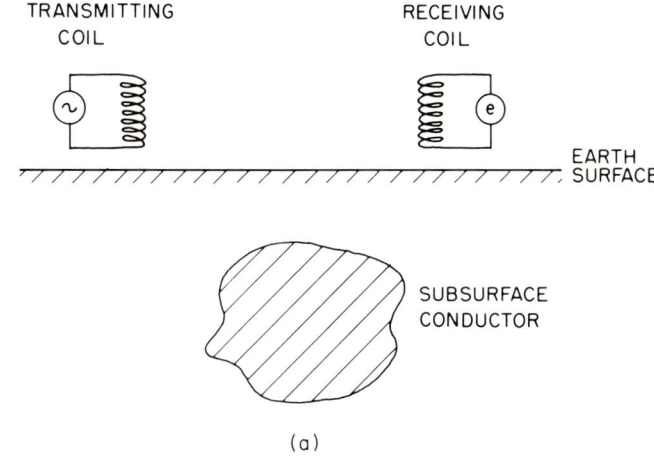

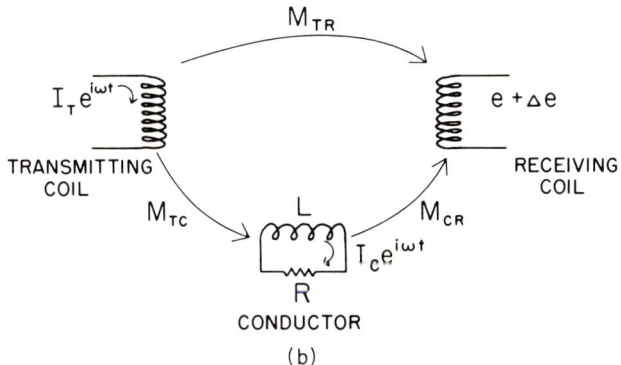

FIG. 18. Schematic representation of energy transfer paths and elements of an AEM system in the presence of a conductor.

where

$$f_C = \boldsymbol{\beta} \cdot \left[\frac{(2x_C^2 - y_C^2 - z_C^2)\mathbf{i} + 3x_C y_C \mathbf{j} + 3x_C z_C \mathbf{k}}{\rho_C^5} \right]$$

and $\boldsymbol{\beta}$ is a unit vector normal to the area A_C of the coil which simulates the subsurface conductor whose center is located at (x_C, y_C, z_C). The number of turns in the simulating coil is N_C. The voltage described by (3.4) will cause a current to flow in the subsurface conductor, the amplitude and phase of which are described by

(3.5) $$I_C = -\frac{i\mu\omega}{4\pi} N_T A_T I_T N_C A_C \frac{f_C}{R + i\omega L}$$

This current causes an incremental voltage to be developed in the receiving coil

(3.6) $$\Delta e = \frac{-i\mu\omega}{4\pi} N_C A_C I_C N_R A_R f_{CR}\, e^{i\omega}$$

where

(3.7) $$f_{CR} = \boldsymbol{\alpha} \cdot \left[\frac{\{2(x-x_C)^2 - (y-y_C)^2 - (z-z_C)^2\}\mathbf{i} + 3(x-x_C)(y-y_C)\mathbf{j} + 3(x-x_C)(z-z_C)\mathbf{k}}{(\rho - \rho_C)^5} \right]$$

and where now the axis of the coil simulating the conductor has been placed in the x direction for simplicity, i.e. $\boldsymbol{\beta} = \mathbf{i}$. Then the incremental voltage of (3.6), expressed as a fraction of the primary voltage e is

(3.8) $$\frac{\Delta e}{e} = \frac{1}{4\pi} \left(\frac{-i\mu\omega}{R + i\omega L} \right) N_C^2 A_C^2 \frac{f_C f_{CR}}{f}$$

The phase of this ratio then is controlled by the ratio of the equivalent R and L in the subsurface; its amplitude is controlled by the angular frequency ω, by the magnitude of R and L, by the square of the equivalent NA product of the conductor, and by the transmitter-conductor-receiver configuration. The more resistive the subsurface body, the greater will be the imaginary component, while the larger the conductor, the larger the amplitude of the response. From this discussion, we may visualize that measurement of the real and imaginary parts of the normalized incremental voltage, $\Delta e/e$, can provide information on the location, depth, size, and resistivity of the buried conductor.

The mutual impedance between the transmitting and receiving coils in the absence of a subsurface conductor is

(3.9) $$Z = e/I_T$$

while the change in mutual impedance brought about by the conductor is

(3.10) $$\Delta Z = \Delta e / I_T$$

and the normalized change in mutual impedance is then

(3.11) $$\Delta Z / Z = \Delta e / e$$

The quantities given by (3.11) are complex; the real part is referred to as the in-phase anomaly and the imaginary part is referred to as the quadrature anomaly. Anomalies are usually measured in parts per million (ppm) since (3.11) is dimensionless.

The above elementary development has been presented in the frequency domain so that the complex ratio

(3.12) $$f(\omega) = \Delta e(\omega) / e(\omega)$$

is a function of angular frequency. Measurement of the real or imaginary part of $f(\omega)$, or both, is made continuously at one or more frequencies.

If measurements are to be made in the time domain, a repetitive impulsive source of current is applied to the transmitting coil and measurements of a function $g'(t) = \Delta e(t)$ are made during intervals when the transmitter current is off. This function may be normalized by the voltage $e(t)$ induced in the receiving coil when the transmitter current is on, i.e., $g(t) = \Delta e(t)/e(t)$, and then a quantity comparable to $f(\omega)$ is obtained with which to study the system.

An example of the function $f(\omega)$ recorded by a helicopter-borne AEM system is contained in Fig. 19. This data was obtained with the Varian Associates–Texas Gulf Sulphur Company system, used at an altitude of 150 ft over the Whistle Mine of the International Nickel Company. The full scale deflection was 2000 ppm, for each of the in-phase and quadrature anomalies for this recording. An example of the function $\Delta e(t)$ obtained with the INPUT

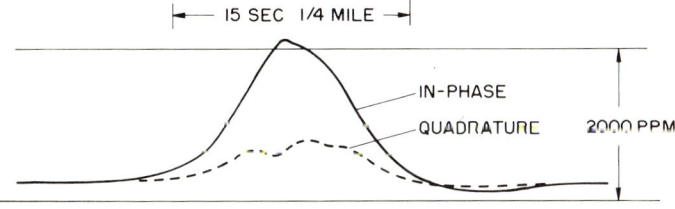

Fig. 19. In-phase and quadrature anomalies recorded by Varian Associates–Texas Gulf Sulphur Company helicopter electromagnetic system over the Whistle Mine, Sudbury, Ontario, Canada.

AEM system, developed by Barringer Research Limited, is illustrated in Fig. 20.

There is one passive inductive AEM system which relies upon magnetic fields from atmospheric electrical discharges as energy sources. This passive system, AFMAG, is not amenable to the above analysis. Its principle was described by Ward [47].

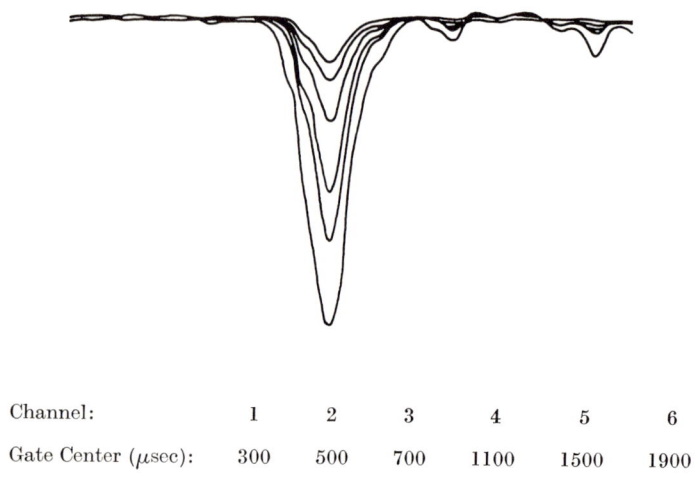

Channel:	1	2	3	4	5	6
Gate Center (μsec):	300	500	700	1100	1500	1900

FIG. 20. Anomalies recorded in each of the six channels of the INPUT system over a sulfide deposit. The anomaly of largest amplitude corresponds with channel 1, which has a gate center occurring 300 μsec after cessation of the current pulse. The amplitude of the anomaly decreases with increasing channel number since the higher numbered channels represent samples farther down the transient decay curve.

3.3. Types of AEM Systems

Active AEM systems are of two types: (1) those in which both the transmitting and receiving coils are rigidly mounted on the airframe, on a structure attached to the airframe, or on a structure towed from the aircraft, and (2) those in which the transmitting coil is fixed to the airframe while the receiving coil is mounted in a "bird" towed beneath and behind the aircraft. It is convenient to label these *rigid boom* and *towed bird* systems, respectively. Each of these divisions may be subdivided according to the configurations of transmitting and receiving coils employed. Figures 21 and 22, respectively, illustrate subdivisions of rigid boom and towed bird systems. The letters T and R refer to transmitting and receiving coils, respectively, while the subscripts x, y, and z refer to the orientations of the coil axes. The x direction is

AIRBORNE GEOPHYSICAL METHODS 47

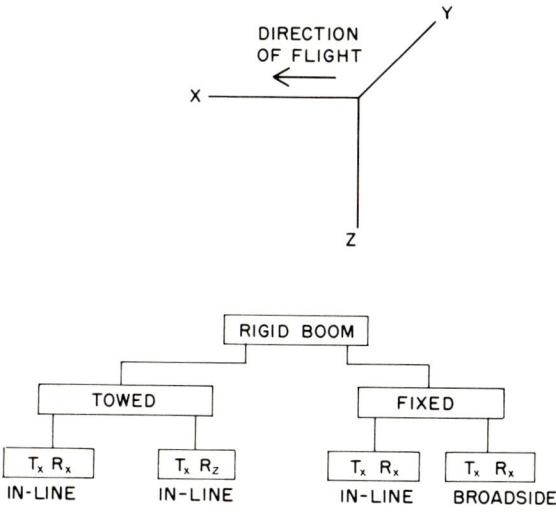

FIG. 21. The classification and characteristics of rigid boom AEM systems.

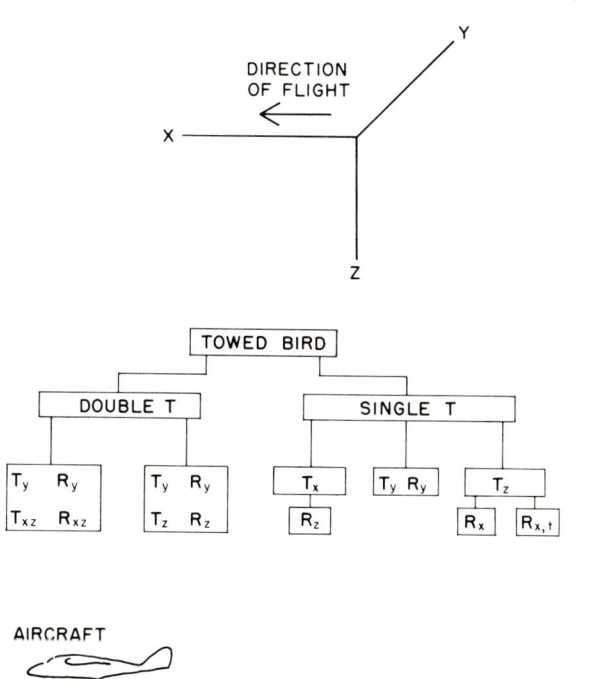

FIG. 22. The classification and characteristics of towed bird AEM systems.

the flight direction while the z direction is positive downward. Then the y direction forms the third member of a right-hand set. Some rigid boom systems are operated with one coil behind the other in the direction of flight (*in-line*) while other rigid boom systems involve a wing-tip coil location so that the system is flown *broadside*. All *towed bird* systems are in-line, since the receiving coil trails the transmitting coil as Fig. 22 illustrates. At least 11 rigid boom systems and 7 towed bird systems have been used in surveys, although not all of these systems are in use at this time. Figures 23 and 24 are pictorial representations of the Aero Canso rigid boom system and the INPUT system, respectively.

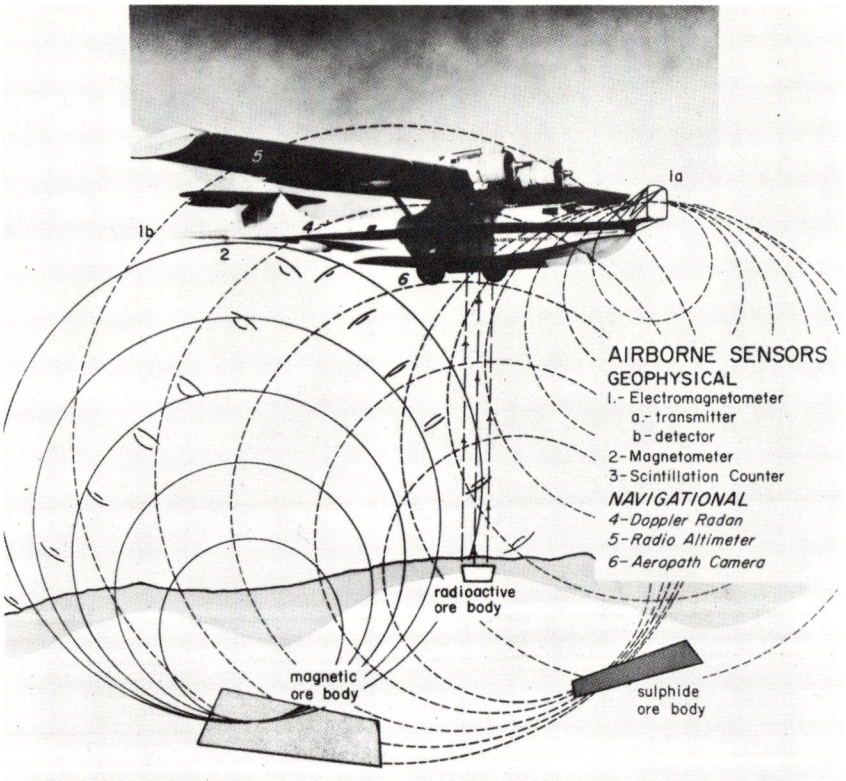

FIG. 23. The Aero Canso rigid boom AEM system.

Operating frequencies for frequency-domain systems range from 140 to 4000 Hz. Coil separations range from 20 to 83 ft for the rigid boom class and from 200 to 500 ft with the towed bird class. Noise levels may be as low as 10 ppm per 50 ft of coil separation with the rigid boom devices to 500 ppm per 500 ft of coil separation with the towed bird devices.

The only time domain system, INPUT, is of the towed bird type, and the received signal is sampled at six intervals after cessation of the current pulse in the transmitting coil. The pulse repetition rate for this system is 286 pulses/sec.

The rotary field system, which may be classed with the towed bird systems, sometimes locates the transmitting coil in one airplane and the receiving coil in a second airplane. The AFMAG system may also be classed with towed bird systems.

FIG. 24. The Barringer INPUT towed bird AEM system.

Because of the great variety of coil configurations, operating frequencies, electronic systems, etc., it is not practical in an article of this type to describe every system in detail. Rather than attempt this, we shall list the variables available to the designer and indicate the manner in which the design problem is approached. Reference is made to Ward [50, 51] for earlier discussions of particular systems and to Table IV which contains a summary of all AEM systems known to the author.

3.4. The Design of AEM Systems

3.4.1. Factors in Design. The factors which should be considered in the design of an AEM system include (1) enhancement of signal-to-noise ratio, (2) the operating frequency or frequencies in the frequency domain, or the times,

TABLE IV. Summary of AEM Systems.[a]

SYSTEM NAME	AIRCRAFT	OPERATING FREQUENCY Hz	IN-PHASE I / QUADRATURE Q / AMPLITUDE A / PHASE θ	COIL SEPARATION ρ-FT	NOISE PPM	NOISE REDUCED TO ρ=50 FT	RESPONSE TO UNIFORM EARTH	ELEVATIONAL VARIATION OF RESPONSE TO UNIFORM EARTH	NORMAL TERRAIN CLEARANCE	COIL CONFIGURATION	REMARKS
BARRINGER #1	HELICOPTER	400	I, Q	30	3	14	2[b]	1[b]	100 (COILS)	Tx Rx IN-LINE	TOWED
LOCKWOOD	HELICOPTER	4000	I, Q	30	10	46	2	1	100 (COILS)	Tx Rx IN-LINE	TOWED
NUCOM	HELICOPTER	1000	I, Q	20	10	156	2		100 (COILS)	Tx Rx IN-LINE	TOWED
SANDER	HELICOPTER	1050	I, Q	23	10	97	2	1	100 (COILS)	Tx Rx IN-LINE	TOWED
SCINTREX HEM 701	HELICOPTER	1600	I, Q	30	3	14	2	1	100 (COILS)	Tx Rx IN-LINE	TOWED
BARRINGER #2	HELICOPTER	400	I, Q	30	3	14	1	4	100 (COILS)	Tx Rz IN-LINE	TOWED
NEWMONT/ AERO	HELICOPTER	390	I, Q	60	40	23	2	1	150	Tx Rx IN-LINE	MOUNTED
VARIAN/ TGS	HELICOPTER	400	I, Q	50	10	10	2	1	150	Tx Rx IN-LINE	MOUNTED
AERO/ CANSO	CANSO PBY	390	I, Q	83	20?	5?	2	1	200	Tx Rx IN-LINE	MOUNTED
AERO/ OTTER	OTTER	320	I, Q	62	20	10	3	3	175	Tx Rx BROADSIDE	MOUNTED
BOLIDEN	-	3200	I, Q	52	20	18	3	3	175?	Tx Rx BROADSIDE	MOUNTED
SCINTREX/ AIRESOURCE	OTTER	320	I, Q	62	20	10	3	3	175	Tx Rx BROADSIDE	MOUNTED

RIGID BOOM (applies to the MOUNTED systems from NEWMONT/AERO downward)

System	Aircraft		Measurement							Coil Config	Coil Separation
INCO	ANSON/TWIN OTTER	1800/2000	A, Φ	500	1500	1500	2-4	1-3	450	Ty Ry plus Txz Rxz	x ≈ 400; z ≈ 225
ABEM (2 PLANE) (NOW CRAELIUS TERRATEST AB)		880	A, Φ	800	3000	1500	3-4	2-3	300	Ty Ry plus Tz Rz	x ≈ 800
MC PHAR F-400	BEAVER	340/1070	Q (2)	400	1000	2000	1	4	400	Tx Rz	x ≈ 200; z ≈ 300
AERO-PHYSICS	ANSON	140	Q	500	4500	4500	3	3	450	Ty Ry	x ≈ 400; z ≈ 225
HUNTING (NOW LOCKWOOD)	CANSO (NOW DC3)	400/2300	Q (2)	500	850	850	1	4	450	Tz Rx	x ≈ 420; z ≈ 220
INPUT	CANSO	TIME DOMAIN	6 SAMPLES	350	-	800?	1	4	400	Tz Rx (Ry + Rz) (TO BE ADDED)	x ≈ 250; z ≈ 250
AFMAG	ANY	140/470	VOLTAGE RATIO	NO TRANSMITTER	2%	NA	0	0	100-1000	TWO ⊥ COILS, AXES 45° TO x-DIR'N	VARIABLE

← TOWED BIRD →

[a] Data compiled from Ward [51] and Hood [19].
[b] 0 = least; 4 = most.

numbers, and intervals of samples in the time domain, (3) coil configuration, (4) the number of transmitting and receiving coils, (5) measurement of in-phase or quadrature components in frequency domain systems, or both, (6) the efficiency of search and depth of exploration of the system, (7) the response of the system to the various shapes and orientations of the conductors to be mapped by the system, and (8) the geological and topographical environment in which the system is to be used. No one system can be optimal in all respects, and hence any practical system represents a compromise of the optimums for each of the above constraints and others.

3.4.2. Sources of Noise. The depth of exploration obtainable with any electromagnetic method, is governed by the signal-to-noise ratio (S/N) at the receiving coil. Hence, when designing an AEM system, we must attempt to identify all of the sources of noise, to evaluate their characteristics, and then to attempt to minimize them within the passband and risetime of the receiving system. Four basic types of noise sources are recognized: *instrument noise, disturbance field noise, geologic noise,* and *cultural noise*. A discussion of each of these four types follows.

3.4.2.1. Instrument noise. This type of noise arises in the receiving circuits, the transmitting circuits, or in the airframe and machinery associated with the AEM system. A subdivision of this noise source type is:

1. Johnson noise in the detection coils.
2. Johnson noise in the input circuit of the first amplifier.
3. Noise mechanically induced in the detecting coil. This may include magnetostriction in iron-cored coils, microphonic vibrations, and oscillations of the coils in the earth's main magnetic field, relative motion between the receiving coil and the airframe, boom, wings, etc., in which eddy currents have been induced by the transmitter, and changes in the effective area of the transmitting or receiving coils.
4. Aircraft ignition, aircraft auxiliary electrical system, or aircraft machinery noise.
5. Thermal drift of the parameters of the transmitting or receiving coils or associated electronics.
6. Mutual impedance changes due to relative rotation between transmitting and receiving coils.
7. Mutual impedance changes due to change in separation between transmitting and receiving coils.

3.4.2.2. Disturbance field noise. There is a general background of ambient electromagnetic fields upon which the transmitted AEM field is superimposed. There are two main sources of this noise type as follows:

1. Alternating magnetic fields resulting from artificial sources such as transmission lines, telephone lines, radio transmitters, etc.

2. Alternating magnetic fields resulting from natural sources such as thunderstorms and other atmospheric discharges.

The combination of instrument noise and disturbance field noise is often referred to as *system* noise, since it is the noise which remains when the system is flown at a sufficiently high altitude to be out of the influence of the geologic noise and cultural noise to be described below.

3.4.2.3. Geologic noise. Geologic noise is the background of minor anomalies of no significance upon which the significant anomalies are superimposed.

Geologic noise is a relative term, for what may be signal in one search problem may become noise in another. This noise may be of longer or shorter space wavelength than the signal, or there may be partial or complete spectral overlap between signal and geologic noise. Enhancement of the ratio of signal to geologic noise may be effected through judicious selection of the coil separation ρ, the operating frequency f, and the coil configuration. Generally, the fixed boom systems produce much less geologic noise, when searching for massive sulfide-ore deposits, than do the towed bird systems. While most past systems have been designed with the intent to suppress geologic noise, there is considerable justification for recording all the geologic information and separating the signal from the geologic noise in the interpretation process.

3.4.2.4. Cultural noise. Pipelines, transmission lines, telephone lines, railroad tracks, fences, buildings, and other cultural features produce anomalies when an AEM system is flown over them because these features form one part of an electrical circuit which is closed via an earth return. Thus the anomalies obtained over cultural objects may be governed in amplitude and phase by the electrical conductivity of the ground and the effectiveness of the coupling between the object and ground. Complete spectral overlap between these anomalies and the ones sought can occur, so that AEM surveys are frequently of limited value near cultural developments.

3.4.3. Sources of Signal.

3.4.3.1. The response of a homogeneous half-space. If the earth were homogeneous electrically, and we were concerned with an AEM survey at low elevation over a small fraction of its surface, we would be prepared to approximate the earth by a homogeneous half-space. This half-space would modify the mutual impedance between a pair of coils in a characteristic manner. Figure 25 computed from tabular data presented by Frischknecht [13] is an Argand diagram for a pair of coaxial coils whose planes are vertical, based on calculations of the real and imaginary parts of the mutual impedance ratio $\Delta Z/Z$. The convex-up curves are lines of constant h/ρ, where h is the common height of the coil pair, and ρ is their separation. The other curves

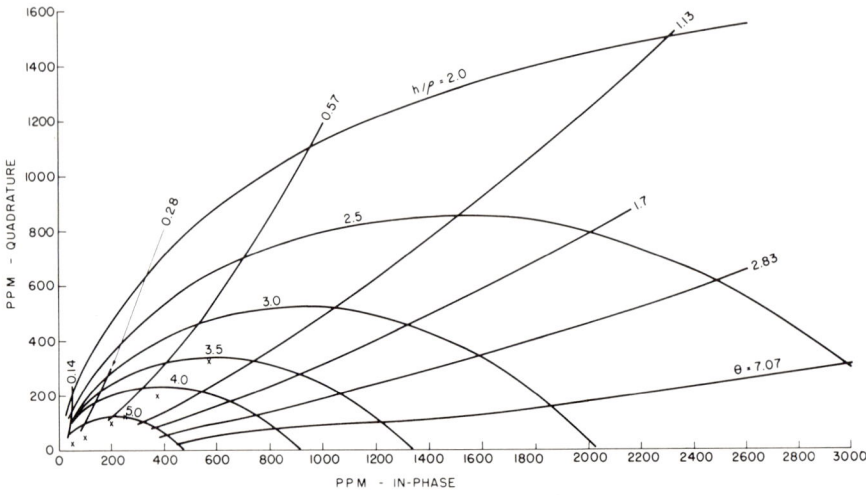

FIG. 25. Characteristic diagram for a vertical coaxial coil pair over a homogeneous earth. Im($\Delta Z/Z$) vs. Re($\Delta Z/Z$), each plotted in parts per million, for constant $\theta = (\sigma\mu\omega)^{1/2}\rho$ and constant h/ρ where h is system height and ρ is coil separation.

pertain to various values of the induction number $\theta = (\sigma\mu\omega)^{1/2}\rho$, where σ and μ are the electrical conductivity and magnetic permeability of the subsurface, respectively, while ω is the angular frequency at which the device is operated. Figure 26 contains the in-phase and quadrature anomalies recorded by an in-line vertical coaxial coil pair when flown at elevations ranging from 175 to 325 ft over a body of brackish water. The in-phase and quadrature anomalies from a number of points on this record have been plotted as crosses on Fig. 25, where they constitute a curve of constant induction number $\theta = (\sigma\mu\omega)^{1/2}\rho = 0.92$. Since the coil separation was 50 ft and the frequency of operation was 400 Hz, we deduce that the conductivity of the half-space is 1.2 mhos/meter. This value falls within the range for brackish water since sea water exhibits a conductivity of about 4 mhos/meter. The depth of the half-space beneath the coil pair may also be determined from experimental data plotted on the Argand diagram, and in this instance the value so deduced is consistently within about 5% of that measured independently by a radio-altimeter and is also plotted on Fig. 26.

When an electromagnetic system of this type is used in the search for massive sulfides, it is customary to select ω, ρ, and h such that the in-phase and quadrature anomalies of country rock and overburden are negligible, compared to the anomalies recorded from an ore deposit. When one considers that rock and soil typically exhibit conductivities in the range 10^{-2} to 10^{-4} mhos/meter, rather than the high value of brackish water used above, then it may be understood that the response of overburden is readily subdued. In most applications such responses constitute noise, although the discussion of Fig. 26 demonstrates that the conductivity of the subsurface can be determined by such measurements and this could be important for geologic mapping.

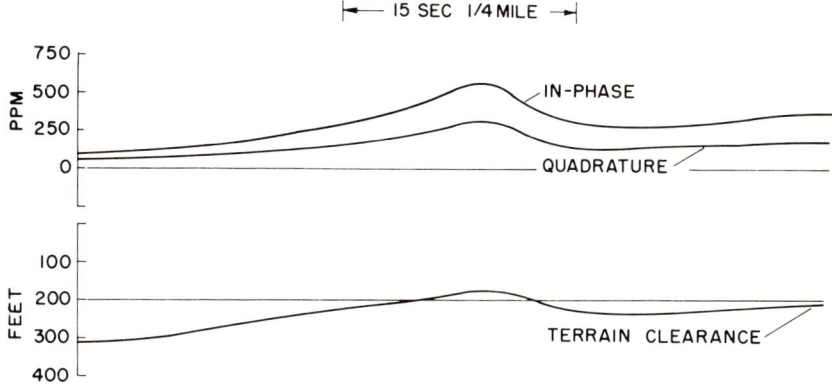

Fig. 26. In-phase and quadrature components as functions of elevation of a vertical coaxial AEM system over brackish water.

3.4.3.2. The response of inhomogeneities in the subsurface. In ore prospection, the target sought usually constitutes an inhomogeneity of high conductivity in an otherwise poorly conducting half-space. If we select the system parameters ρ, h, and ω correctly, we can frequently obtain substantial response from the inhomogeneity and negligible response from the half-space. Figures 19 and 20 provide examples of this type. If then we wish to study the theoretical behavior of ore deposits in an electromagnetic field, we will typically select elementary models to simulate the various shapes of features to be encountered. For example, the following geometrical form—geologic unit association is frequently used:

1. sphere   ore lens
2. vertical cylinder   ore pipe
3. horizontal cylinder   manto
4. dipping half-plane   large vein or tabular replacement deposit
5. dipping disk   small vein or tabular replacement deposit
6. horizontal disk   limited patch of conductive overburden
7. horizontal whole-plane   extensive cover of conductive overburden
8. whole space   country rock

Then, ideally, one would like to have available, for each AEM system, scale model or computed profiles, which would permit comparisons of the systems in terms of the responses to the above targets. With this information, the user could make his own objective decision on the sensitivity of each system for each of the types of targets he expects to encounter. Unfortunately, very few of the purveyors of AEM surveys have deemed it necessary to make this information available. However, there are limited catalogs of anomaly profile types for some of the available AEM systems. The most common model employed is the half-plane. Figure 27 portrays the in-phase α and quadrature

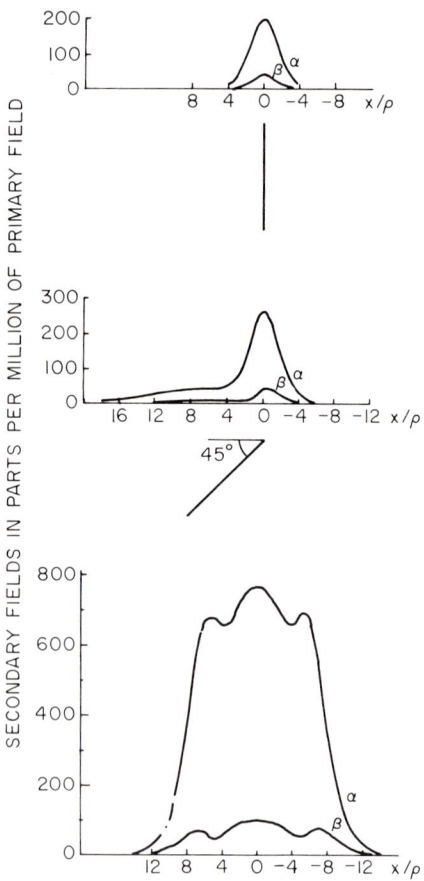

FIG. 27. In-phase α and quadrature β anomalies, in parts per million of primary field, obtained over a dipping half-plane with a model of the Varian–TGS system. Dips are as shown in the figure; x is lateral distance and ρ is coil separation. Traverse normal to strike.

β anomalies obtained with a scaled model of the Varian Associates–Texas Gulf Sulphur Company helicopter electromagnetic system over a thin metallic sheet sufficiently large to be a suitable approximation to a half-plane. The anomalies are expressed in parts per million, as functions of lateral distance measured normal to the long axis of the metal sheet. The lateral distance is expressed in dimensionless units x/ρ of fractions of the coil separation ρ. As mentioned earlier, this particular helicopter electromagnetic system employs vertical coaxial coils separated by 50 ft and flown in-line at an altitude h of about 150 ft. The frequency of operation is 400 Hz. From catalogs of curves of this nature, characteristic curve diagrams similar to that shown in Fig. 28 may be constructed for each model. This particular diagram pertains

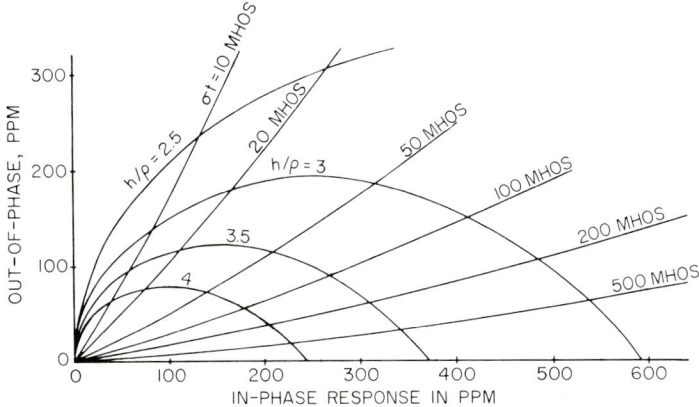

FIG. 28. Argand diagram for the Varian–TGS AEM system applicable to the interpretation of data over a vertical half-plane model.

to a vertical half-plane model. The convex-up curves are lines of constant h/ρ while the other lines represent constant conductivity-thickness product. These diagrams are employed in interpreting AEM anomalies as will be discussed later.

3.4.4. Coil Configurations. From Figs. 21 and 22, and from Table IV, we observe that many different coil configurations have been employed in AEM systems. Some of the factors which in the past have led engineers to particular configurations are treated in the next few paragraphs.

3.4.4.1. Overburden response. There has been a desire to minimize those responses, of country rock, overburden, swamps, lakes, etc., which constitute much of the geologic noise when searching for massive sulfides. Certain coil configurations are better than others in this respect as may be illustrated by computing the response of a double dipole electromagnetic system over a homogeneous half-space. There are six possible in-line dipole pairs which are compared in Table V.

A vertical coplanar coil pair flown broadside would respond the same way as a T_Y-R_Y pair. These six transmitter-receiver configurations have been arranged in order of increasing response to a homogeneous half-space. Any

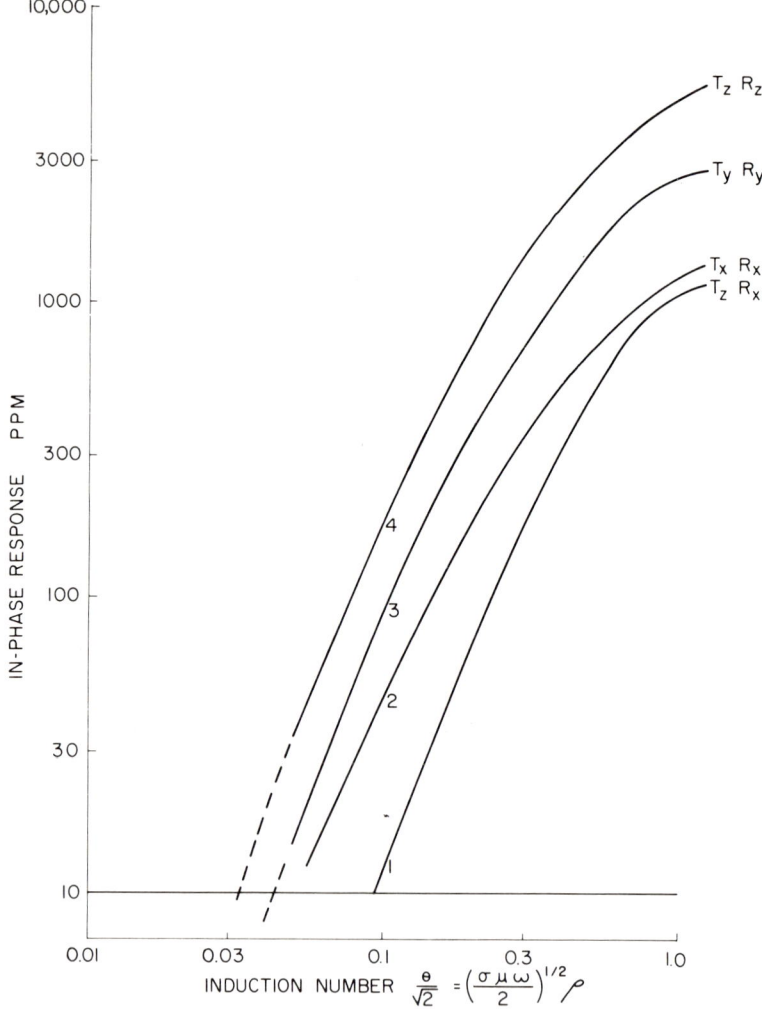

FIG. 29. In phase response of a homogeneous half-space for each of four coil configurations. System elevation is 150 ft.

other coil pair configuration would, through the principle of reciprocity, be a duplicate of one of the six listed. The curves of Figs. 29 and 30 were computed from tables calculated by Frischknecht [13] and pertain to an in-line helicopter system of 50 ft coil separation flown at 150 ft terrain clearance. Thus if minimizing the response of overburden or of country rock is the only consideration, then the first two coil configurations of Table V are superior, since

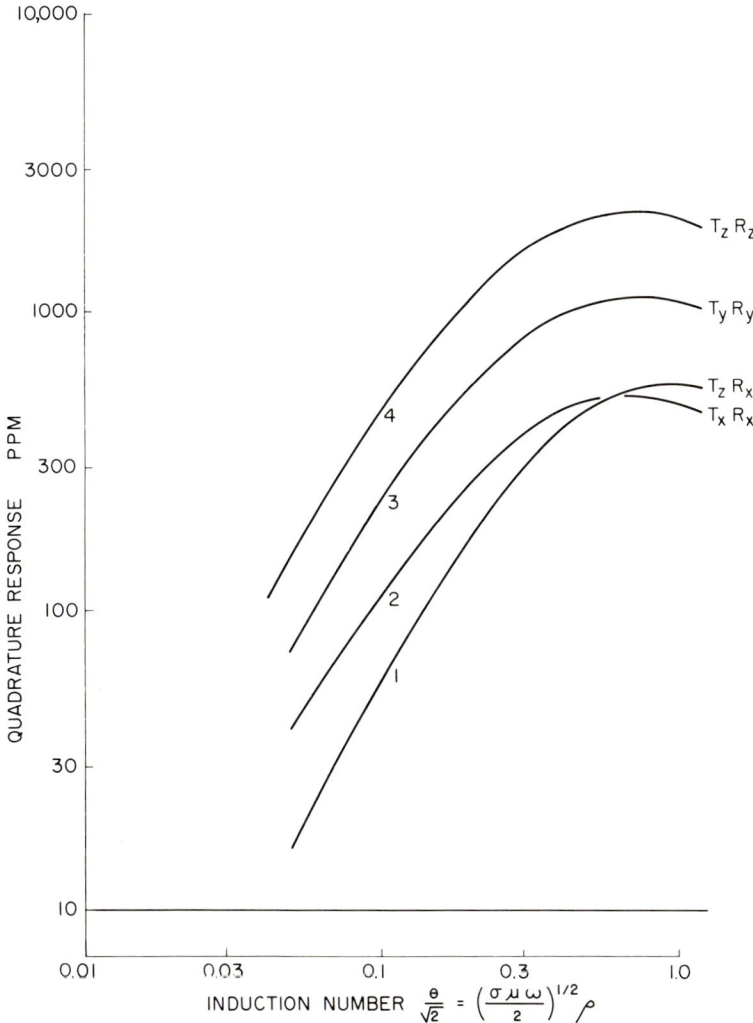

Fig. 30. Quadrature response of a homogeneous half-space for each of four coil configurations. System elevation is 150 ft.

they produce zero response from a homogeneous half-space. Note that the response of all systems to a homogeneous half-space can be lowered below the nominal 10 ppm operating noise level expected for systems of this type, by limiting the induction number to small values. In practice this is accomplished by selecting the frequency of operation to be some low audio frequency.

While our illustration of the response of a half-space has utilized only in-line configurations likely to be used in a helicopter, similar reasoning applies to all other systems. The response of any system to a homogeneous half-space will be controlled by the numbers [13]

(3.13) $$\mathscr{A} = (\sigma\mu\omega)^{1/2}(z+h), \qquad \theta = (\sigma\mu\omega)^{1/2}\rho$$

where Z is the elevation of the receiving coil, h is the height of the transmitting coil, ρ is the coil separation, ω is the operating angular frequency, and σ and μ are the electrical conductivity and magnetic permeability of the half-space. The amplitude of the response decreases with increasing values of $\mathscr{A}$ and increases with increasing values of θ.

TABLE V. A comparison of in-line dipole pairs.

	Transmitting dipole orientation	Receiving dipole orientation	Response to homogeneous earth
1.	T_Y	R_X	Zero
2.	T_Y	R_Z	Zero
3.	T_X	R_Z	Curve 1, Figs. 29, 30
4.	T_X	R_X	Curve 2, Figs. 29, 30
5.	T_Y	R_Y	Curve 3, Figs. 29, 30
6.	T_Z	R_Z	Curve 4, Figs. 29, 30

Some towed bird systems trail the bird, which houses the receiving coil, about 420 ft behind and 220 ft below the aircraft as illustrated in Fig. 31. For this coil separation flown at an aircraft height of 450 ft, the in-phase and

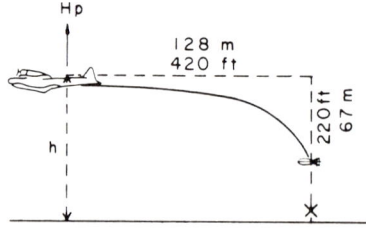

FIG. 31. Diagrammatic representation of a towed bird AEM system illustrating typical position of bird relative to aircraft.

quadrature responses of a homogeneous half-space are as illustrated in Figs. 32 and 33. An acceptable noise level for towed bird systems is about 1000 ppm. Above this ambient level, the quadrature response would indicate the same preferred order of systems as listed in Table V. However, if the in-phase component is to be recorded, then the orders of T_X-R_X and T_Z-R_X are reversed. For most towed bird systems, only the quadrature component is recorded, so that Table V is usually accepted as the preferred order of all airborne systems as far as rejection of the response to country rock or overburden is concerned.

3.4.4.2. Variation of aircraft height. In the design of some AEM systems, it is not possible to ensure that the response of the overburden or country rock is negligible. It then becomes important to learn of the change in country-rock response as the system changes altitude. If this response is only a slowly varying function of variations about the normal aircraft altitude, then it merely adds a constant level to the anomalies from, for example, massive sulfide orebodies. This constant level can be subtracted out and quantitative interpretation of the anomalies caused by sulfide orebodies is still valid to a first approximation.

A computation of the gradients of the mutual coupling expressions for each of the six coil pairs leads to the order of decreasing preference for towed bird and rigid boom systems operating at flying heights of 450 and 150 ft, respectively, over a homogeneous half-space shown in Table VI.

TABLE VI.

	Transmitting dipole orientation	Receiving dipole orientation	Vertical gradient of response to homogeneous earth
1.	T_Y	R_X	zero
2.	T_Y	R_Z	zero
3.	T_X	R_X	weak
4.	T_Z	R_Z	modest
5.	T_Y	R_Y	modest
6.	T_Z	R_X	large

Of course, the variation of anomaly with aircraft height, for all types of anomalies, requires assessment also. Such information is not generally available for all systems. The anomalies over a vertical half-plane decrease approximately as the inverse fourth power of the height of the aircraft above the half-plane for both rigid boom and towed bird types of systems flown at their normal operating terrain clearances.

3.4.4.3. Direction of flight. With the T_Y-R_Y in-line coil configuration, it is necessary to traverse a conductor at approximately 45° to strike to ensure maximum response. When this system is flown at 90° to the strike of a thin sheet conductor, no anomaly results.

With the above exception, all other AEM systems are flown normal to geologic strike, but many of the systems produce anomalies of different shape as the traverse direction is varied relative to strike. The Varian Associates–Texas Gulf Sulphur Company helicopter system and the Hunting Canso

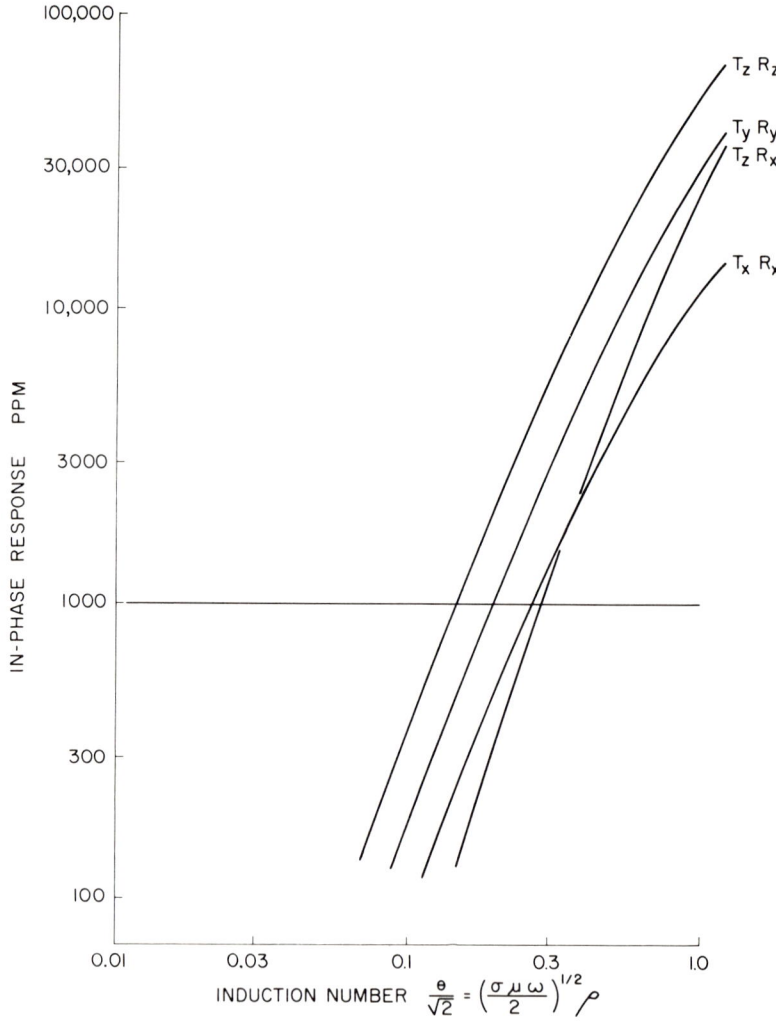

FIG. 32. In-phase response of a homogeneous half-space for each of four coil configurations with coils deployed as in Fig. 31. Aircraft elevation is 450 ft.

towed bird system (T_Z-R_X) are relatively unaffected by deviations from a flight direction normal to strike. Scale model experiments are usually conducted to evaluate this factor prior to design.

3.4.4.4. Relative rotation between transmitting and receiving coils. Coil pairs whose axes are orthogonal to one another and orthogonal to the line joining

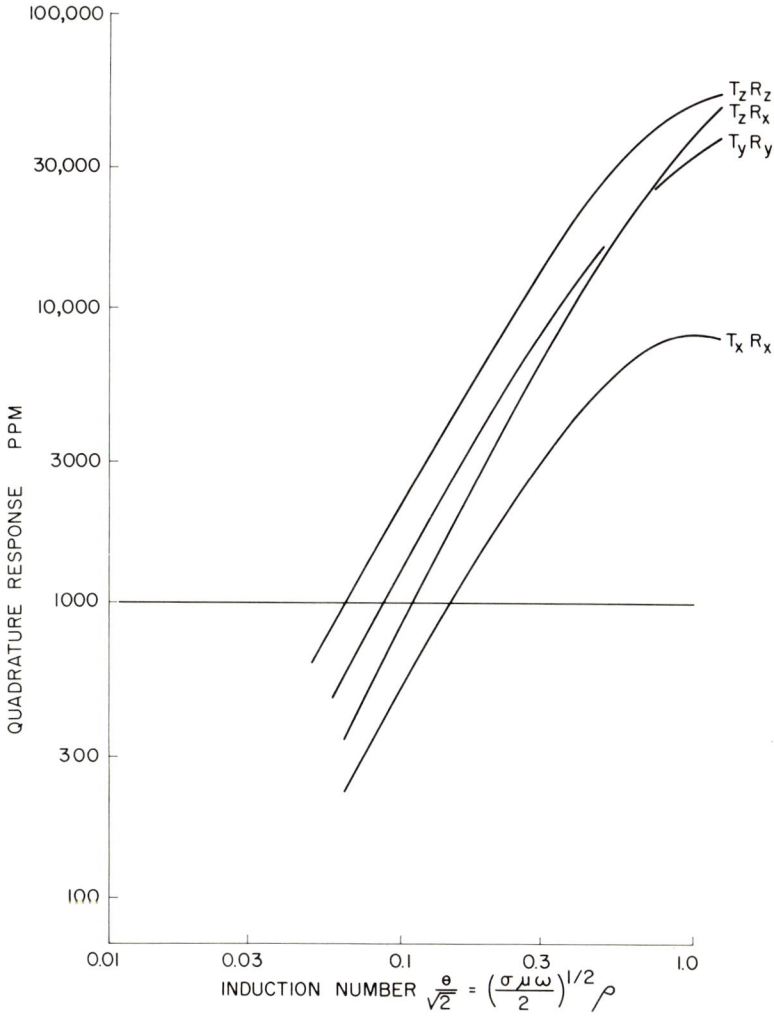

Fig. 33. Quadrature response of a homogeneous half-space for each of four coil configurations with coils deployed as in Fig. 31. Aircraft elevation is 450 ft.

them are referred to as minimum coupled systems; the output of the receiving coil is zero in the absence of adjacent conductors. Any rotation of one coil relative to another will produce a voltage in the receiving coil which is proportional to the sine of the angle of rotation. Thus a minimum coupled coil pair mounted on a rigid 50 ft boom, for which the noise level is to be 10 ppm, require a rotational stability of 1 part in 10^5 or about 2 sec of arc. If the time scale of rotational motion is of the order of the time scale of an anomaly, then relative rotations produce fictitious anomalies which may be mistaken for true anomalies. Until very recently, minimum coupled AEM systems were avoided because of this constraint.

3.4.4.5. Variation of coil separation. Minimum coupled coil pairs may be translated relative to one another, along their separation axis, without deleterious effect. This is an advantage of minimum coupled systems.

Coaxial and coplanar coil pairs are referred to as maximum coupled systems. The coil system T_Z-R_X (or T_X-R_Z) may also be maximum coupled, provided R_X is at a different elevation to T_Z as in Fig. 31. Rotational changes now introduce spurious receiving voltages only as the cosine of the angle of rotation. However, the separation between the coils of maximum coupled systems will introduce error signals given by

(3.14) $$\Delta H/H = 3\,\Delta\rho/\rho$$

For a 50-ft coil separation, separation should remain constant to within 10^{-3} in. to assure a 10-ppm noise level. Of course, such separation changes are only of importance if their spectra seriously overlap the spectrum of the expected anomalies.

3.4.4.6. Aircraft constraints. If a very large transmitting dipole moment is required, then it usually is only practical to obtain this by making the coil area large. To achieve this, it is customary to string cable from nose to wingtip to tail to wingtip and return to the nose of the aircraft. This disposition of wire automatically dictates a vertical transmitting dipole. The INPUT and Hunting Canso systems were designed on this basis.

Metal wings and bodies of aircraft have currents induced in them by the transmitting dipole. These currents produce a secondary field which is detected at the receiving coil. Usually it is necessary to null out these effects, and any changes in them resulting from aircraft flexing in air turbulence, by means of compensating coil systems. Some coil configurations are superior to others in this regard, but the choice of a particular coil system depends, to some extent, upon the aircraft to be used.

3.4.5. Ratio of Signal to Noise. The incremental voltage Δe induced in the receiving coil by currents flowing in a subsurface conductor is given by Eqs.

(3.3) and (3.8) as

(3.15) $$\Delta e = K(x, \omega) N_T A_T I_T N_R A_R(-i\mu\omega/4\pi)$$

where

$$K(x, \omega) = \left(\frac{-i\mu_0 \omega}{4\pi}\right)\left(\frac{1}{R + i\omega L}\right) N_C^2 A_C^2 f_C f_{CR}$$

describes the dependence of the anomaly Δe on location and configuration of the AEM system, the frequency, and the size and conductivity of the subsurface inhomogeneity. In principle, either the transmitting or receiving dipole may be a corefree or an iron-core coil and for this reason the products $N_T A_T$ and $N_R A_R$ may be considered as the equivalent turns-area products of the transmitting and receiving coils, respectively. The magnitude of the signal voltage $|\Delta e| = e_s$ may be placed in the alternate form

(3.16) $$e_s = (WR)^{1/2}(\sigma/\delta)^{1/2} \mu_0 a f \frac{N_T A_T I_T}{4\pi} K(x, \omega)$$

where W, R, δ, and σ are the total weight, total resistance, density, and conductivity, respectively, of the material of the wire of the receiving coil, a is the radius of the coil, and f is the frequency.

We wish now to compare this signal voltage with various noise voltages.

3.4.5.1. Johnson noise. The Johnson or thermal agitation noise e_{Tn} in the receiving coil is usually designed to be larger than the Johnson noise e_{a_n} in the input of the first amplifier. In other words, the received signal voltage is proportional to the number of turns in the receiving coil, so that we might as well continue to increase the number of turns until the coil itself is the primary noise generator. The Johnson noise in the receiving coil is

(3.17) $$e_{Tn} = [4kTR(\Delta f)]^{1/2}$$

where k is Boltzmann's contant, T is absolute temperature (°K), R is the resistance of the coil, and Δf is the bandwidth of the parallel tuned circuit of which the receiving coil is a component. Then the ratio of signal to Johnson noise is

(3.18) $$\frac{e_s}{e_{Tn}} = K(x, \omega) \frac{N_T A_T I_T}{4\pi} \mu_0 \left[\frac{W A_R \sigma}{\pi\delta(\Delta f/f)}\right]^{1/2} \left(\frac{f}{4kT}\right)^{1/2}$$

Once the material of the wire is selected, then δ and σ are determined. Note that δ/σ for copper is 5.2×10^8 and for aluminum is 9.5×10^7, so that there is an improvement in the ratio of signal to Johnson noise of about 2.3 if aluminum is used rather than copper. The penalty suffered is increased brittleness and usually increased cost.

If in fact the receiving coil is parallel tuned, then a $Q = f/\Delta f$ of about 25 to 30 is customary; higher values of Q lead to instability if the transmitted frequency varies. On the other hand, the receiving coil need not be tuned, in which case the bandwidth of the device is determined largely by a phase-lock amplifier used at a later stage. Thus, the design of the coil reduces to selection of $(WA_R)^{1/2}$ so as to ensure an adequate value of $e_s/e_{\tau n}$. However, mechanical design factors usually dictate maximum values for both A_R and W. Johnson noise seldom turns out to be the limiting noise in the system.

3.4.5.2. Magnetic field noise. Both natural and man-made electromagnetic fields introduce noise in an AEM system, as was discussed earlier. Narrowband detecting equipment is usually employed so that the noise induced in a corefree coil may be written [42]

$$(3.19) \qquad e_{m_n} = (WR)^{1/2}(\sigma/\delta)^{1/2}\mu_0 a h_m (\Delta f)^{1/2} f$$

where h_m is the magnetic noise intensity for unit bandwidth. The ratio of signal to magnetic noise is then

$$(3.20) \qquad \frac{e_s}{e_{m_n}} = \frac{N_T A_T I_T}{4\pi (\Delta f)^{1/2} h_m} K(x, \omega)$$

A value of h_m of order $10^{-2} \gamma$ per unit bandwidth at 100 Hz can be expected from natural sources [3]. Man-made noise fields originating in transmission lines will contribute at the fundamental and harmonics of the power-line frequency. The use of active notch filters at 60 Hz usually suffices to provide adequate reduction of this noise source. Additionally, switching transients may contribute broadband noise.

The dipole moment of the transmitting coil must be chosen to be sufficiently large to produce a field at the receiving coil which is well above the magnetic field noise. If the noise level is to be 10 ppm of transmitted field strength for a coaxial coil system of ρ ft separation, then we may compute the required moment from

$$(3.21) \qquad \frac{H_T}{h_m} = \frac{e}{e_{m_n}} = 10^5 = \frac{N_T A_T I_T}{\pi (\Delta f)^{1/2} h_m} \frac{1}{\rho^3}$$

where ρ is the coil separation as before. The dipole moment of the transmitting coil, m_T, may be placed in the following convenient design form:

$$(3.22) \qquad m_T = \frac{N_T A_T I_T}{4\pi} = \frac{1}{4\pi} (P_T A_T W_T)^{1/2} \frac{1}{2} \left(\frac{\sigma}{\pi\delta}\right)^{1/2}$$

where P_T is the power to the transmitting coil, and A_T, W_T are the area and weight of the transmitting coil. Once again, selection of aluminum or copper determines the δ/σ product and we are left with a choice in selecting P_T, A_T, and W_T to suit the needs of our particular application. Note that the moment only increases as the square root of any one of these items. A further constraint to be met is that the equivalent resistance of the series inductance-capacitance circuit (which will include substantial losses in the capacitors when high power is used) should be matched to the load requirement of the transmitter power source. Series tuning usually permits this impedance matching consistent with a high value of current I_T.

3.4.5.3. Noise mechanically induced in the receiving coil. Careful selection of materials and mechanical isolation will reduce magnetostrictive noise to acceptable levels.

Oscillations of the receiving coil in the earth's main magnetic field is only important if the spectrum of bird oscillation or sway lies within the passband of the receiver. Careful aerodynamic design is necessary to overcome this noise source and to minimize relative rotation between the coils in a towed bird system. A streamline face and a drag face are necessary for the towed bird and the bird's lift should be minimal. Under such design, the bird will fly with a stability of 1° or better.

It was mentioned earlier that flexing of the airframe gave rise to spurious voltages at the receiver which could only be eliminated by auxiliary cancelling coils. Mechanically induced noise in an AEM system is difficult to eliminate and usually sets the noise level of towed bird AEM systems.

3.4.5.4. Aircraft noise. The electrical and mechanical machinery aboard an aircraft must be shielded and so situated relative to the receiving coil as to lead to negligible noise.

3.4.5.5. Thermal drift. Thermal changes in the parameters of the transmitting or receiving circuits may produce spurious unbalances in the receiver bridge network but these voltages are of a sufficiently long time constant as to be separable from anomalies either by inspection or by a simple RC high-pass network.

3.4.5.6. Mutual impedance changes caused by relative movement between coils. This subject was discussed in Sections 3.4.4.4 and 3.4.4.5. Correct mechanical design can place these noise sources within the 10 and 1000 ppm noise limits considered acceptable for rigid boom and towed bird systems, respectively.

3.4.5.7. Geologic noise. Geologic noise properly should be separated from the anomalies sought in the interpretation process. The techniques of statistical communication theory often are helpful at this stage. The observed mutual impedance profiles may be convoluted with expected anomaly waveforms to obtain the most probable location of anomalies in regions where geologic noise is high. This process has not been used frequently, since the ratio of quadrature response to in-phase response is usually higher for geologic noise than for anomalies caused by massive sulfide ore deposits. Use of AEM systems on geologic problems other than the search for massive sulfides will warrant a reexamination of the geologic noise concept.

Additional sensing units such as magnetometers are also used simultaneously with AEM systems in the search for massive sulfide ore deposits, since such deposits frequently are assumed to be magnetic as well as highly conductive. This concept fails when no magnetic minerals accompany the ore minerals.

3.4.5.8. Cultural noise. The most obvious means of treating this form of noise is direct correlation of observed anomalies with all known cultural objects. Occasionally, anomalies from ore deposits are obscured by cultural noise.

3.5. Interpretation

The interpretation of AEM data at present is based largely on a comparison of the field observations with data acquired in scaled model experiments conducted in the laboratory. The conditions under which a model system may reproduce accurately the geometrical configurations of the lines of force in the full scale system have been developed from the theory of similitude. In the full scale system, the wave equation for magnetic field intensity is

$$[\nabla^2 - \gamma^2]\mathbf{H} = 0 \tag{3.23}$$

where the propagation constant γ is given by

$$\gamma = [i\mu\omega(\sigma + i\varepsilon\omega)]^{1/2} \tag{3.24}$$

In the model system, the wave equation is

$$[(\nabla')^2 - (\gamma')^2]\mathbf{H}' = 0 \tag{3.25}$$

We shall now assume linear transformations between full scale and model systems as follows:

$$x = px', \quad y = py', \quad z = pz' \quad t = qt'$$

$$\mathbf{E}(x, y, z, t) = r\mathbf{E}'(x', y', z', t') \tag{3.26}$$

$$\mathbf{H}(x, y, z, t) = s\mathbf{H}'(x', y', z', t')$$

where p, q, r, s are the scaling factors for coordinates, time, electric field intensity, and magnetic field intensity, respectively. If we now transform (3.25) via (3.26), we obtain

(3.27) $$[p^2 \nabla^2 - (\gamma')^2](\mathbf{H}/s) = 0$$

In general there is no requirement to scale magnetic field intensity so that Eqs. (3.23) and (3.27) may be equated to yield the wave number transformation

(3.28) $$(\gamma')^2 = p^2 \gamma^2$$

and for those problems permitting neglect of displacement currents this relation reduces to

(3.29) $$\sigma' \mu' \omega' = p^2 \sigma \mu \omega$$

It is convenient to use a linear scale reduction factor of 200; Eq. (3.29) then informs us that this scale reduction can be offset by increasing the model conductivities over the field conductivities by 4×10^4 with the frequency and magnetic permeability remaining unchanged. Alternatively, a frequency increase of 4×10^4 from field to model, or any increase in the product $\sigma \omega$ which equals 4×10^4 will assure similitude.

The electromagnetic response of an inhomogeneity is governed by one or more *induction numbers:*

(3.30) $$\Theta = \theta^2 = (\sigma \mu \omega) l^2$$

where l is any significant linear dimension of the problem, or combination of linear dimensions. For example, in the earlier discussion of a pair of coils above a homogeneous earth, we used two induction numbers

(3.31) $$\theta = (\sigma \mu \omega)^{1/2} \rho, \qquad \mathscr{A} = (\sigma \mu \omega)^{1/2}(z + h)$$

Since an induction number is dimensionless, it must remain invariant under transformation, and this is an alternate way of looking at the requirements of similitude.

As mentioned earlier, the basic scale model used in providing the in-phase and quadrature responses as functions of x coordinate, depth to top (z coordinate), dip, strike, conductivity, and thickness, is a thin sheet whose dimensions are sufficiently large that its response almost exactly simulates a half-plane. The anomalies illustrated in Fig. 27 were obtained over such a model. One induction number for this model [16] is

(3.32) $$\Theta = \sigma \mu \omega t \rho$$

where t is the thickness of the sheet, ρ is the coil separation, ω is the operating angular frequency, while σ and μ are the electrical conductivity and magnetic permeability of the material of the thin sheet. Most geologic materials possess permeabilities only fractionally different from μ_0, the free space value, so it is customary to use model materials for which $\mu \sim \mu_0$. Curves of the type shown in Fig. 27 are obtained as functions of σt for various values of h/ρ. Characteristic curves similar to that of Fig. 28 are plotted by noting the peak in-phase and peak quadrature responses on the anomaly profiles. Usually the construction of these characteristic diagrams is carried out only for traverses normal to strike, but one characteristic diagram is usually drawn for each 30° of dip.

3.5.1. Determination of Dip. By comparing an observed full scale anomaly profile with a catalog of profiles obtained in the laboratory, an estimate of dip may be obtained. For example, a comparison of Figs. 34 and 27 would indicate that the field profile of Fig. 34 was caused by a body dipping much more steeply than 45° and probably dipping nearly vertically.

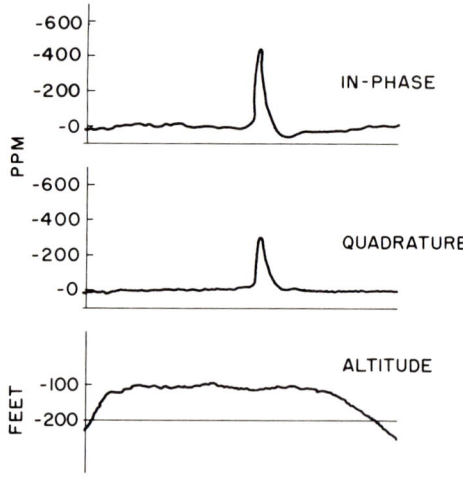

FIG. 34. In-phase and quadrature anomalies recorded with a Newmont Aero helicopter AEM system over a sulfide-ore deposit.

3.5.2. Determination of Depth. For the anomaly of Fig. 34, we have measured the peak in-phase and quadrature anomalies to be 450 and 290 ppm, respectively. We may then enter the characteristic diagram of Fig. 28 and obtain an estimate of h/ρ of 2.6. This particular anomaly was obtained with a vertical coaxial coil system of $\rho = 60$ ft. Hence the depth beneath the aircraft to the top of the sheetlike source was 155 ft. The radio altimeter record of Fig. 34 indicates a terrain clearance of 110 ft, so that the subsurface depth of the conductor is 45 ft.

3.5.3. Determination of Conductivity-Thickness Product. Again the values of 450 and 290 ppm for in-phase and quadrature components of the anomaly of Fig. 34, when entered into the characteristic diagram of Fig. 28, yield a σt product of about 50 mhos. While one might attempt to estimate the thickness t of the body from the width of the anomaly, this is seldom attempted because of the inaccuracies involved. The value of 50 mhos for the σt product is characteristic of massive sulfide deposits [50].

3.5.4. Determination of Strike. When the peaks of the in-phase component of an anomaly are recorded on several flight lines, then the line joining the peaks gives the strike of the anomaly. Figure 35 shows the positions of the

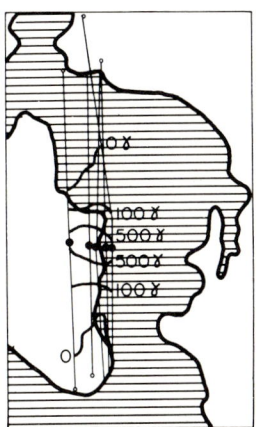

FIG. 35. Correlation of AEM in-phase anomaly peaks (dots) with total magnetic intensity data (contours) recorded simultaneously by the Newmont Aero helicopter AEM system over the New Hosco ore deposit, Quebec, Canada.

in-phase peaks detected on four adjacent lines over the New Hosco ore deposit in Quebec, Canada. Also shown are contours of total field magnetic intensity. Determination of strike and location obviously is important in correlation of AEM anomalies with data from other sources.

3.5.5. Correlation. The example of Fig. 35 shows the usefulness of correlation studies. The causative source in this instance is both magnetic and highly conductive. Given the regional, or better still the local, geologic setting, and given an analysis of the major lineaments and fracture density evident in aerial photographs, the probability of such signatures being caused by massive sulfide mineralization may be assessed. Correlation studies of this nature are essential in assessing the possible economic significance of AEM anomalies.

3.5.6. Effect of Magnetic Rock Types. Induction in weakly conductive but highly magnetic rock units can be of a sign opposite to induction in highly conductive but weakly magnetic ores. Thus "negative" anomalies are sometimes detected by AEM systems. This possibility must be taken into account in interpreting AEM data, as Podolsky [33] has discussed.

Figure 36 contains the in-phase M and quadrature N components of the reflection coefficient of a conductive permeable sphere in a uniform-plane alternating magnetic field. The response of such a sphere is given by

$$(3.33) \qquad \mathbf{H} = -H_0 a^3 (M + iN) \left[\frac{(2x^2 - y^2 - z^2)\mathbf{i} + 3xy\mathbf{j} + 3xz\mathbf{k}}{\rho^5} \right]$$

where H_0 is the magnitude of the x-directed inducing field, a is the radius of the sphere, and the term in brackets describes the dipolar geometry of the secondary or reflected field. The quantities M and N are functions of the induction number

$$(3.34) \qquad \theta = (\sigma \mu \omega)^{1/2} a$$

The curves of Fig. 36 display the transitional behavior of the induced dipole moment of the sphere, from positive at low values of θ to negative at high

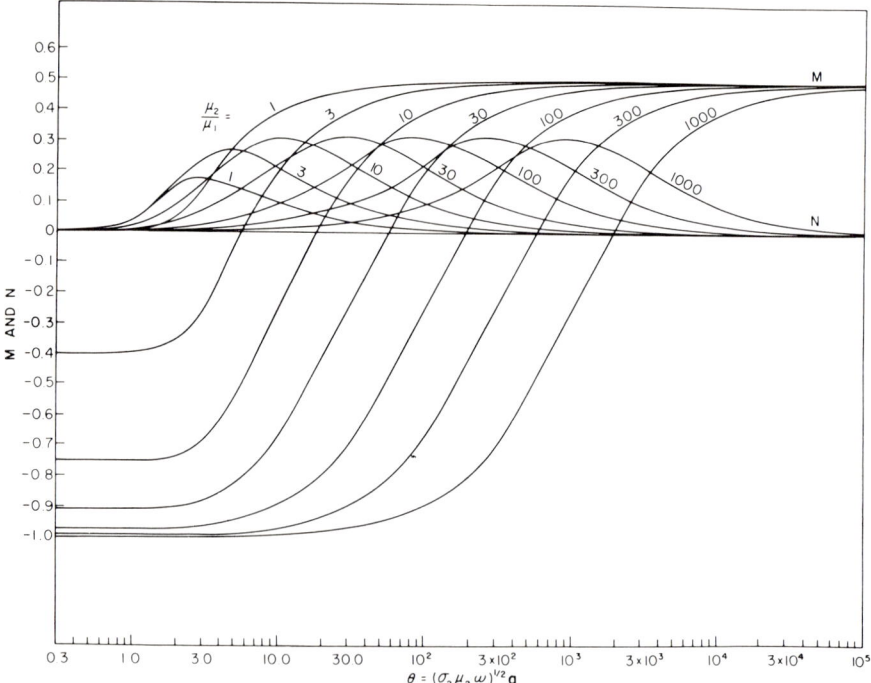

FIG. 36. In phase M and quadrature N components of the reflection coefficient of a sphere in a uniform-plane alternating magnetic field.

values of θ. A similar type of behavior may occur for shapes other than a sphere and for dipolar fields as well as uniform plane-wave fields. However, interpretation of data over thin conductive magnetic dipping sheets can be complicated as Bosschart [5] clearly points out.

A summary of Bosschart's comments follows: Magnetic but weakly conductive rock units (e.g. quartz diorite) frequently produce "negative" AEM anomalies; strongly magnetic and strongly conductive geologic units (e.g. massive magnetite) may produce positive or negative anomalies depending upon the induction number and dip; composite bodies consisting of a strongly magnetic member adjacent to a strongly conductive member may produce some very complicated anomalies. This problem requires considerably more study than it has been given.

3.6. Survey Procedures

3.6.1. Survey Altitude. Helicopter-borne and other rigid boom AEM systems typically are flown at mean terrain clearances of 100 to 200 ft depending upon the system and the terrain, and thereby they facilitate exploration for half-planes (large vein or tabular replacement deposit) to depths of order 150 to 300 ft. Towed bird AEM systems are flown at elevations ranging from 300 to 500 ft depending upon the system and the terrain; depths of exploration to 300-ft subsurface, for half-planes, are thereby achieved. Since the anomaly due to a half-plane decreases approximately as the inverse fourth power of distance between the aircraft and the top of the conductor, and since the anomalies due to spheres or disks (ore lens or small vein) decrease according to the sixth power of distance from aircraft to the center of the disk or sphere, it is essential that rigorous control be maintained over aircraft altitude if a constant depth of exploration is to be secured. Further, it was mentioned earlier that variation of aircraft terrain clearance above a conductive half-plane would in itself lead to false anomalies. On this account, it is seldom feasible to produce a contour map of anomaly intensity, since two adjacent flight lines are usually at sufficiently different altitudes that the two anomalies are of grossly different magnitudes on that account alone.

It is desirable to maintain altitude control to within 10% of the mean terrain clearance, if a reasonably constant depth of exploration is to be assured.

3.6.2. Line Density. As in aeromagnetic surveying, the density of lines in a surveyed area is chosen on the basis of the statistical probability of detecting or delineating most of the anomalies in a region [1]. Seldom is it desirable to attempt to detect more than 80% of the massive sulfide deposits in any given region [51], for to increase these odds would lead to excessive survey costs. Line spacings of $\frac{1}{4}$ mile with towed bird installations and $\frac{1}{8}$ mile with rigid boom systems typically yield the same search statistics.

3.6.3. Line Orientation. The survey lines are oriented normal to the expected geologic strike as in airborne magnetic surveying, except for the T_Y-R_Y AEM system which is flown at 45° to the geologic strike.

3.6.4. Flight Path Recovery. The flight path recovery for AEM surveys meets the same requirements as for aeromagnetic surveys. Hence the reader is referred to Section 2.4.2.1 of this article authored by Peter Hood.

3.7. Possible Future Developments

There is evidence that the following expansions of the concept of an AEM system are being initiated now or are soon to be initiated:

(1) multiple component receiving-coil systems,
(2) multiple frequencies,
(3) data acquisition on magnetic tape,
(4) automatic interpretation via digital computers, and
(5) extension of frequency range beyond current range of 10^2 to 5×10^3 Hz.

The first of the expansion steps listed is highly desirable if we are to make a true record of the subsurface conductivity distribution and then to discriminate between the various sources. This discrimination is materially assisted by the use of two or more frequencies and the measurement of both in-phase and quadrature where necessary. It is important that attention be given to acquisition of all possible information on the subsurface electrical parameter distribution and that discrimination between anomaly sources be left to the interpretation stage. To make this feature possible, it is necessary to acquire the vast amount of electromagnetic and auxiliary data on magnetic tape and to process and interpret these data automatically via high speed digital computers.

The reduction of system noise and the recognition but not elimination of geologic noise are two areas of development to which we may look with some considerable optimism. Each of these developments will lead to an enhanced depth of exploration. To facilitate both of these developments, systems must encompass more frequencies and a broader frequency range, more received components or even more than one transmitting coil, and coherent detection schemes either by analog devices in the aircraft or subsequently by computer.

3.8. Conclusion

AEM systems have so far been designed only for the search for massive sulfide ore deposits. One AEM system, INPUT, has exhibited reasonable adaptability to one hydrological problem. Another system, AFMAG, shows promise in mapping faults, shears, and conductive beds such as carbonaceous shales. There is adequate opportunity for the expansion of the scope of AEM systems to expect that they will in the future be applied not only to ore prospection but also to geological mapping *per se*, hydrological problems, and petroleum exploration.

Acknowledgments

The University of California and Newmont Mining Corporation have provided funds permitting completion of this review.

The companies whose names appear throughout this manuscript have very kindly provided information concerning their respective AEM systems.

List of Symbols[2]

(MKS *units throughout*)

μ	Magnetic inductive capacity (permeability)
μ_0	Magnetic inductive capacity (permeability), free space
ε	Dielectric permittivity
σ	Electrical conductivity
δ	Density
x, y, z	Cartesian coordinates, field system
x', y', z'	Cartesian coordinates, model system
ρ	$(x^2 + y^2 + z^2)^{1/2}$
h	Height
t	Thickness or time
l	Any significant linear dimension
p	Scaling factor for coordinates
q	Scaling factor for time
r	Scaling factor for electric-field intensity
s	Scaling factor for magnetic-field intensity
a	Radius of a sphere, radius of a coil
α, β, i, j, k	Unit vectors
$e^{i\omega t}$	Harmonic time dependency
f	Frequency
$\omega = 2\pi f$	Angular frequency
$\theta \ominus \mathscr{A}$	Induction numbers
M, N	In-phase and quadrature parts, respectively, of the reflection coefficient of a sphere
$\gamma = [i\mu\omega(\sigma + i\varepsilon\omega)]^{1/2}$	Propagation constant
α, β	Inphase and quadrature AEM anomalies
π	3.141596

[2] For Section 3 only.

$f(x, y, z)$	Functions
$f, f(\omega)$	
$g'(t), g(t)$	
ppm	Parts per million
Hz	Hertz
R	Resistance
L	Inductance
Z	Impedance
Δ	Incremental change in a quantity
H, H', H_0	Magnetic-field intensity
h_m	Magnetic-noise intensity
N	Number of turns in a coil
A	Area of a coil
I	Current in a coil
W	Weight of a coil
P	Power dissipated in a coil
k	Boltzmann's constant
T	Absolute temperature
m	Magnetic dipole moment
K	Proportionality factor
i	$(-1)^{1/2}$
$T_X T_Y T_Z$	Transmitting and receiving coils with their axes oriented
$R_X R_Y R_Z$	in the x, y, and z directions
$M_{TR} M_{TC} M_{CR}$	Mutual coupling paths
Q	$f/\Delta f$
∇^2	Laplacian operator
Im	Imaginary part
Re	Real part

Other Subscripts

R	Receiving coil
T	Transmitting coil
C	Subsurface conductor
τ_n	Johnson noise in coil
a_n	Johnson noise in first amplifier
p	Primary
s	Signal
m_n	Magnetic noise

References

1. Agocs, W. B. (1955). Line spacing effect and determination of optimum spacing illustrated by Marmora, Ontario, magnetic anomaly. *Geophysics* **20**, 871–885.
2. Barringer, A. R. (1963). The use of audio and radio frequency pulses for terrain sensing. *Proc. Symp. Remote Sensing of Environment*, 2nd, Univ. Michigan, Ann Arbor, pp. 201–214.
3. Bleil, D. F. (1964). Introductory Talk. *In* "Natural Electromagnetic Phenomena below 30 kc/s" (D. F. Bleil, ed.), pp. 1–4 (*Proc. NATO Advanced Study Inst. Bad Homburg, Germany, July 22–August 2, 1963*). Plenum Press, New York.

4. Boniwell, J. B. (1967). Some recent results with the INPUT airborne EM system. *Trans. Can. Inst. Min. Met.* **52**, 60–67.
5. Bosschart, R. A. (1964). Analytical interpretation of fixed source electromagnetic prospecting data. Delft Univ. Ph.D. Thesis, 103 pp.
6. Boyd, D., and Roberts, B. C. (1961). Model experiments and survey results from a wingtip-mounted electromagnetic prospecting system. *Geophys. Prospecting*, **9**, 411–420.
7. Brant, A. A., Dolan, W. M., and Elliot, C. L. (1966). Coplanar and coaxial EM tests in Bathurst area, New Brunswick, Canada, 1956. *In* "Mining Geophysics," Vol. I, pp. 130–141. Soc. Exploration Geophysicists, Tulsa, Oklahoma.
8. Cheriton, C. G. (1959). Anaconda exploration in the Bathurst District of New Brunswick, Canada. *AIME Trans.* **217**, 278–284.
9. Collett, L. S. (1966). Airborne electromagnetic survey over the Winkler Aquifer, Manitoba. Geol. Surv. Canada Paper 66–2, pp. 6–9.
10. Collett, L. S. (1966). The measurement of the resistivity of surficial deposits by airborne pulsed electromagnetic equipment. Geol. Surv. Canada Paper 66–1, pp.137–138.
11. Collett, L. S. (1966). Airborne resistivity surveys useful for overburden tests. Geol. Surv. Canada Paper 66–42, pp. 24–25.
12. Fleming, H. W., and Brooks, R. R. (1960). Geophysical case history of the Clearwater deposit, Northumberland County, New Brunswick, Canada. *AIME Trans.* **217**, 131–138.
13. Frischknecht, F. C. (1967). Fields about an oscillating magnetic dipole over a two-layer earth, and application to ground and airborne electromagnetic surveys. *Quart. Colo. School Mines* **62**, 1.
14. Gaur, V. K. (1963). Electromagnetic model experiments simulating an airborne method of prospecting. *Bull. Natl. Geophys. Res. Inst.* **1**, 167–174.
15. Geleynse, M., and Barringer, A. R. (1965). Recent progress in remote sensing with audio and radio frequency pulses. *Proc. Symp. Remote Sensing of Environment, 3rd, Univ. of Michigan, Ann. Arbor*, 1964 pp. 469–494.
16. Grant, F. S., and West, G. F. (1965). "Interpretation Theory in Applied Geophysics." McGraw-Hill, New York.
17. Hedstrom, E. H., and Parasnis, D. S. (1958). Some model experiments relating to electromagnetic prospecting with special reference to airborne work. *Geophys. Prospecting* **6**, 322–341.
18. Hedstrom, E. H., and Parasnis, D. S. (1959). Reply to comments by N. R. Paterson. *Geophys. Prospecting* **7**, 448–470.
19. Hood, P. J. (1968). Mineral exploration trends and developments in 1967. *Can. Min. J.* **89** (2), 173–194.
20. Joklik, G. F. (1960). The discovery of a copper-zinc deposit at Garon Lake, Quebec. *Econ. Geol.* **55**, 338–353.
21. Keller, G. V., and Frischknecht, F. C. (1966). "Electrical methods in geophysical prospecting." Pergamon Press, Oxford.
22. Khomenyuk, Yu. V. (1961). The double rotating field method. *Bull. Acad. Sci. USSR, Geophys. Ser. (English Transl.)* No. 12, 1182–1183.
23. MacKay, D. G., and Paterson, N. R. (1900). Geophysical discoveries in the Mattagami District, Quebec. *Can. Inst. Min. Met. Bull.* **53**, No. 581, 703–709.
24. Mizyuk, L. Ya. (1960). Methods and apparatus for airborne electrical prospecting. *Bull. Acad. Sci. USSR, Geophys. Ser. (English Transl.)* No. 6, 522–528.
25. Parasnis, D. S. (1966). "Mining Geophysics." American Elsevier, New York.
26. Paterson, N. R. (1959). Comments on paper by Hedstrom and Parasnis. *Geophys. Prospecting* **7**, 435–447.

27. Paterson, N. R. (1961). Experimental and field data for the dual frequency phase-shift method of airborne electromagnetic prospecting. *Geophysics* **26**, 601–617.
28. Paterson, N. R. (1961). Helicopter EM test, Mobrun orebody, Noranda. *Can. Min. J.* **82**, 53–58.
29. Paterson, N. R. (1966). Mattagami Lake Mines—a discovery by geophysics. *In* "Mining Geophysics," Vol. 1, pp. 185–196. Soc. Exploration Geophysicists, Tulsa, Oklahoma.
30. Pemberton, R. H. (1961). Combined geophysical prospecting system by helicopter. *Min. Eng.* January.
31. Pemberton, R. H. (1961). Target Mattagami. *Trans. Can. Inst. Min. Met.* **64**, 16–23.
32. Pemberton, R. H. (1962). Airborne EM in review. *Geophysics* **27**, 5, 691–713.
33. Podolsky, G. (1966). An evaluation of an airborne electromagnetic anomaly in Northwestern Quebec. *In* "Mining Geophysics," Vol. I, pp. 197–205. Soc. Exploration Geophysicists, Tulsa, Oklahoma.
34. Rattew, A. R. (1962). Helicopterborne electromagnetic, magnetic, and radiometric survey, Coronation Mines. *Trans. Can. Inst. Min. Met.* **65**, 143–150.
35. Schaub, Yu. B. (1961). On a quantitative measure of the information acquired by airborne electromagnetic prospecting methods. *Bull. Acad. Sci. USSR, Geophys. Ser. (English Transl.)* No. 3, 232–236.
36. Schaub, Yu. B. (1961). An experimental test of the rotating magnetic field method. *Bull Acad. Sci. USSR, Geophys. Ser. (English Transl.)* No. 7, 669–673.
37. Schaub, Yu. B. (1962). The influence of the specific resistance of the surrounding medium on the form of anomaly curves obtained in aerial electrical prospecting. *Bull. Acad. Sci. USSR, Geophys. Ser. (English Transl.)* No. 5, 423–426.
38. Schaub, Yu. B. (1962). Prospecting by the rotating magnetic field method using LI-2 and AN-2 airplanes. *Bull. Acad. Sci. USSR, Geophys. Ser. (English Transl.)* No. 7, 591–595.
39. Schaub, Yu. B. (1962). Efficiency of electromagnetic prospecting by helicopter. *Bull. Acad. Sci. USSR, Geophys. Ser. (English Transl.)* No. 8, 685–688.
40. Schaub, Yu. B. (1966). Airborne electromagnetic prospecting by the pulse-bunching method over isometric orebodies. *Izv. Acad. Sci. USSR, Physics of the Solid Earth (English Transl.)* No. 12, 794–797.
41. Schaub, Yu. B. (1966). Resolving power of the pulse-bunching method in mapping rocks. *Izv. Acad. Sci. USSR, Physics of the Solid Earth (English Transl.)* No. 12, 798–800.
42. Slichter, L. B. (1955). Geophysics applied to prospecting for ores. *Econ. Geol.* Fiftieth Anniversary Volume, pp. 885–969.
43. Tikhonov, A. N., and Dmitriev, V. I. (1959). The influence of interference in the induction method of electrical prospecting from the air. *Bull. Acad. Sci. USSR, Geophys. Ser. (English Transl.)* No. 9, 991–992.
44. Tikhonov, A. N., and Dmitriev, V. I. (1959). On the possibility of using the induction method of electrical prospecting from the air for geological map-making. *Bull. Acad. Sci. USSR, Geophys. Ser. (English Transl.)* No. 10, 1053–1055.
45. Tornquist, G. (1958). Some practical results of airborne electromagnetic prospecting in Sweden. *Geophys. Prospecting* **6**, 112–126.
46. Ward, S. H. (1957). Airborne electromagnetic surveying. *In* "Methods and Case Histories in Mining Geophysics," pp. 71–78. Can. Inst. Min. Met., Montreal.
47. Ward, S. H. (1959). AFMAG—airborne and ground. *Geophysics* **24**, 761–789.
48. Ward, S. H. (1959). Unique determination of conductivity, susceptibility, size, and depth in multifrequency electromagnetic exploration. *Geophysics* **24**, 531–546.

49. Ward, S. H. (1960). AFMAG: a new airborne electromagnetic prospecting method. *Trans. AIME* **217**, 333–342.
50. Ward, S. H. (1967). The electromagnetic method. *In* "Mining Geophysics," Vol. II, pp. 224–372. Soc. Exploration Geophysicists, Tulsa, Oklahoma.
51. Ward, S. H. (1967). Airborne electromagnetic methods. Presented at Can. Centennial Conf. Min. Groundwater Geophys. October 22–27, 1967, Niagara Falls, Ontario, Canada.
52. Ward, S. H., and Barker, R. A. (1958). Case history of the Juniper Prospect. *Trans. AIME* **211**, 100–104.
53. Ward, S. H., O'Brien, D. P., Parry, J. R., and McKnight, B. K. (1968). AFMAG—interpretation. *Geophysics*, **33**, 621–644.
54. Ward, S. H., O'Donnell, J., Rivera, R., Ware, G. H., and Fraser, D. C. (1966). AFMAG—applications and limitations. *Geophysics* **31**, 576–605.
55. White, P. S. (1966). Airborne electromagnetic survey and ground follow-up in Northwestern Quebec. *In* "Mining Geophysics," Vol. I, pp. 252–261. Soc. Exploration Geophysicists, Tulsa, Oklahoma.
56. Wieduwilt, W. G. (1962). Interpretation techniques for a single frequency airborne electromagnetic device. *Geophysics* **27**, 493–506.

4. Airborne Radiometric Methods

4.1. Introduction

In 1948 experimental airborne radiometric surveys were first carried out in Canada [17] using banks of Geiger–Müller [13, 14] counters. Pressurized ion chambers connected to vibrating-reed electrometers were also tried in the early experimental work, but their performance proved to be relatively poor in comparison to Geiger counters. Similar experimental work was carried out at about the same time in the United States [11] and the United Kingdom [6], and up to 49 Geiger tubes were used in airborne radiometric surveys [25]. During 1949 about 18,000 mi of reconnaissance traverse were flown in the Northwest Territories using a 7-tube Geiger unit developed by the National Research Council of Canada at Chalk River. The equipment was installed in a RCAF Anson aircraft and survey operations were directed by Eldorado Mining and Refining, Ltd. The U.S. Geological Survey in cooperation with the U.S. Atomic Energy Commission (USAEC) carried out its first large-scale survey (1600 mi^2) during May and June 1950 [31]. The first USAEC contract was flown using six 3×30-in. Geiger tubes mounted in a helicopter [30]. However the successful development of large scintillation crystals, which have a superior efficiency in detecting gamma rays, caused the use of Geiger tubes to be discontinued. Thus although they are still useful in ground investigations because they are cheaper and have adequate sensitivity, airborne radiometric equipment utilizes scintillation counters exclusively at the present time. The theory of the Geiger–Müller tube will therefore not be discussed, although the interested reader may refer to the excellent texts by Friedman [12] and Curtiss [10] for the theory of operation.

4.2. Airborne Scintillation Counters

The sensitive element of a scintillation counter is a substance called a phosphor which has the property of emitting a flash of light when struck by a gamma ray. Phosphors which have been used in scintillation counters include thallium-activated sodium iodide crystals, napthalene [4], anthracene [17], stilbene, and a solution of terphenyl in toluene [2]. The incident gamma ray transfers some or all of its energy to the electrons of the atoms comprising the phosphor. The flash of light occurs when these excited electrons lose their excitation energy and fall back into their original ground state. The intensity of the flash of light is therefore dependent on the energy of the incident gamma ray if the dimensions of the phosphor are large compared with the particle range [5]. This fact is utilized in scintillation spectrometers. The light photons are detected by a photomultiplier tube whose output is fed to a suitable amplifying and counting circuit. The instantaneous value of the intensity of the gamma radiation received is not recorded; instead an average of the measured radioactivity over a short period of time is taken. This is usually referred to as the time constant or integration time. Alternatively the number of gamma rays striking the scintillator may be counted over a short period of time.

During 1950 airborne scintillometer surveys were carried out by Godby et al. [17] using anthracene and thallium-activated sodium iodide crystals. The latter type were found to be more efficient because of their greater density. Three 4×2-in. diameter NaI(Tl) crystals were mounted in a Beechcraft Expeditor aircraft, being separated by lead plates so that one monitored to the left, one monitored to the right, and one monitored the whole path of the aircraft. It was hoped in this way, by separate recording of the three outputs, to show at which side of the aircraft the various gamma ray sources were located and also to give some indication of their geometry. The subsequent field results showed that the intensity indications could be attributed to many different topographical configurations and that consequently the additional complication of three recorders, etc. was not worthwhile.

The airborne scintillation counter (Fig. 37) used by the U.S. Geological Survey was designed by the Health Physics Division of the Oak Ridge National Laboratory [24]. Six thallium-activated sodium iodide crystals connected in parallel are utilized, which are 4 in. in diameter and 2 in. thick and are shielded on the sides by 0.5 in. of lead [11]. The output from the photomultipliers is fed via the mixing preamplifier to the linear amplifier in the main unit. A discriminator allows only pulses above a certain amplitude to be counted by the ratemeter. For normal survey operations, this pulse-height level is usually chosen so that gamma radiation having an energy in excess of 50 keV is measured. The signal from the ratemeter is fed to a vacuum-tube voltmeter whose output is displayed on an Esterline–Angus chart recorder. The altitude of the aircraft is measured with a radar altimeter and a continuous terrain-clearance record is obtained on the same recording chart with a second pen. The effective area sampled by the system at a survey altitude of 150 meters is approximately 300 meters in diameter [28].

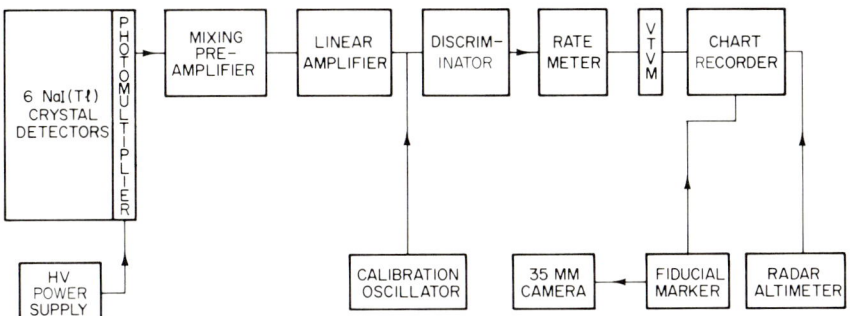

Fig. 37. U.S. Geological Survey airborne scintillation counter.

The U.S. Geological Survey has also used a single crystal scintillometer mounted in a light high-wing plane [7, 8]. In addition, at least a dozen airborne scintillation counters have been produced by companies in the U.S. and Canada.

A number of airborne scintillation counters have been produced by the U.K. Atomic Energy Authority. These include the AERE type 1444A [27] and types 1531A [6, 32] and 1531B [33]. The type 1444A employs a single NaI(Tl) crystal detector 4.625 in. in diameter and 1 in. thick and has a fully transistorized ratemeter having count-rate ranges of 500, 1000, and 2000 cps for full-scale deflection and integration times of 0.5, 1, 2, and 4 sec. The more sensitive types 1531A and B have three Type 1444A NaI(Tl) scintillation detector heads. The transistorized ratemeter has ranges of 1000, 2500, and 5000 cps and the integrating time may be set between 1 and 4 sec.

4.3. Airborne Scintillation Spectrometers

Because uranium is the only radioactive element normally of interest in prospecting surveys, it would be advantageous to be able to discriminate between gamma rays emitted by the three common radioactive elements and their daughter products, namely potassium 40, thorium 232, and uranium

238 [22]. The principal emitters in the uranium series are lead 214 and bismuth 214 [34], which are themselves derived from radium 226. The primary gamma radiators in the thorium series are actinium 228, bismuth 212, and thallium 208.

The amplitude of the voltage pulses from the photomultiplier is proportional to the energy of the gamma rays which produce them. By discriminating between the pulses of different amplitudes, it is possible to use the scintillation counter as a spectrometer [29].

There are two types of discriminator which may be used:

(1) A differential (or window-type) discriminator in which voltage pulses above and below certain levels are eliminated and only pulses in the gate are counted.

Thus potassium (K^{40}) is mono-energetic and has a well-developed peak at 1.46 MeV. By setting the gate to select gamma rays between 1.3 MeV and 1.6 MeV, the presence of potassium can be ascertained.

A differential-type airborne spectrometer, Mark 14, built by Nuclear Enterprises, Ltd. of Winnipeg is shown in Fig. 38. The shock-mounted detector head on the left-hand side of the photograph contains a 4-in. diameter by 5 in. long NaI(Tl) crystal which is optically coupled to a 3 in. diameter photomultiplier. The transistorized electronic unit has four channels whose ranges may be preset to suit specific survey situations. The manufacturers recommended settings for normal operation are listed in Table VII.

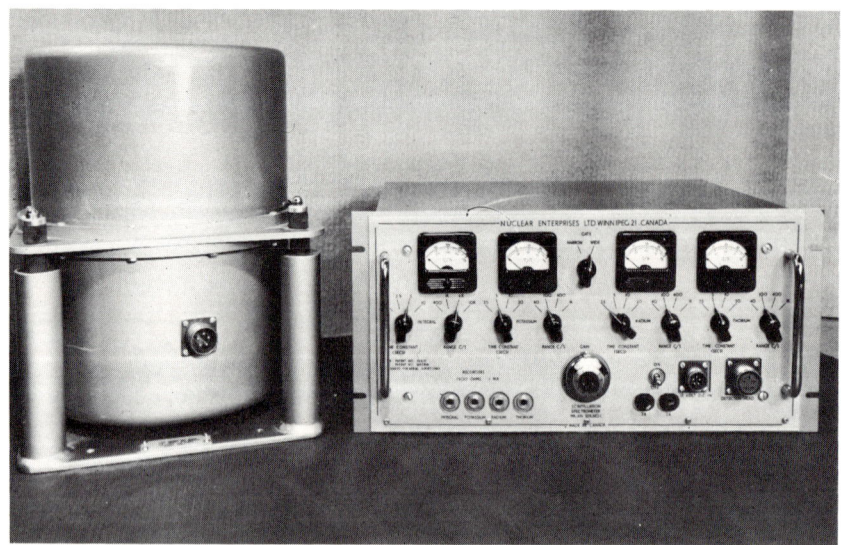

FIG. 38. Differential-type airborne scintillation spectrometer manufactured by Nuclear Enterprises Ltd. of Winnipeg.

TABLE VII.

Channel	Element	Gate width	Isotope detected
1	Total	Threshold set to 100 keV	All
2	Potassium	1.3–1.6 MeV	K40
3	Uranium	1.6–2.2 MeV	Bi214
4	Thorium	2.2–2.9 MeV	Tl208

Each channel has an output for a suitable multipen recorder and the count rate is also monitored by the meters on the front panel. Both the range and the integration time may be individually set for each channel but the total (or integral) channel has a shorter time-constant setting and a higher range.

The Swedish AB Atomenergi Model 1501 is also a differential or window-type airborne spectrometer. The transistorized 127 lb. instrument consists of three units; a scintillation detector, a four-channel analyzer, and an analog recorder. A 5-in. diameter by 6-in. long NaI(Tl) crystal is used as the detector. The window widths and positions of each of the four channels can be varied between 0 and 3.33 MeV and 0.1 to 3.33 MeV, respectively, by means of ten-turn potentiometers. Channel 1, which is normally used for measuring the total counting rate over the whole energy range is fitted with a trip unit which provides an alarm signal at a counting rate level adjustable between 0 and 100% of full range.

(2) An integral or single threshold-type discriminator in which all voltage pulses above a certain level are counted. Members of the French Atomic Energy Commission [3] have developed an integral system with a four-channel recorder. The first channel records all gamma rays with an energy in excess of 0.35 MeV, the second records all above 1.3 MeV, the third all above 1.46 MeV, and the fourth all above 2.1 MeV. Thus the first channel gives the total count. Potassium would be recorded on channels 1 and 2 only. Uranium has few gammas above 2.1 MeV so it is recorded on channels 1, 2, and 3 only. Thorium has a photopeak due to thallium 208 at 2.62 MeV [23] so it registers on all four channels.

Results obtained over known uraniferous occurrences in the Canadian Precambrian Shield using an integral system have been described [26]. A Scintrex SC-1 spectrometer mounted in a Helio Courier light aircraft was used which was equipped with a radio altimeter and a continuously recording 35 mm strip camera having an 18 mm wide-angle (fish-eye type) lens for flight path recovery. The integral spectrometric method was employed so that the count was integrated for all energies in the spectrum above the preset threshold energy value. The three threshold values employed were (1) 0.2 MeV, in order to record the integral count from potassium, uranium, and thorium;

(2) 1.65 MeV, to record uranium and thorium; and (3) 2.5 MeV to record thorium only.

Because the number of 2.62 MeV gamma rays emitted by thorium is rather small, it is advisable to use as large volume scintillator as possible in order to detect the presence of thorium in channel 4 [21]. For instance, the size of the NaI(Tl) crystal used by Berbezier et al. [3] was 18 cm (7.1 in.) in diameter and 15 cm (5.9 in.) in length. Although at distances of a few hundred feet over 50% of the primary radiation is scattered, Gregory and Horwood [20] have shown experimentally that with appropriate counting times the major photopeaks of the radioactive elements can be well resolved at distances of at least 700 ft. Moreover their experiments have shown that gamma radiations from uranium, thorium, and potassium attenuate at about the same rate for air distances from 300 to 1400 ft.

The amplitude of the output peak-voltage pulses from a scintillation counter is dependent on the energy of the incident gamma rays. However peak-voltage instability arises from a number of causes including count-rate hysteresis of photomultipliers, magnetic fields affecting the photomultipliers, high-voltage instability, gain drift of the amplifiers, and temperature effects. One recently developed technique is to ensure that the response of airborne scintillation counters remains constant during survey operations by incorporating a radioactive reference, such as radioactive americium with a peak at 3.5 MeV, into the survey system. This arrangement enables the dynamic gain stabilization of the scintillation counter to be continuously monitored and automatically stabilized.

4.4. Airborne Radiometric Survey Techniques

The factors which determine the gamma radioactivity measured by a given detector are the surface radioactivity of the source material, the effective radiating area of the source, the air attenuation factor, and instrumental parameters [19]. A background count will be recorded which is caused by cosmic radiation, by atmospheric activity owing to radon decay products, and by the radioactivity of the aircraft, its navigation instruments, and the scintillation counter itself. It is the usual practice to assume that a meter reading of two to three times the normal background for an area constitutes a significant anomaly. The average background radioactivity may be obtained by flying over a conveniently located water-covered area close to or in the survey area.

The surface radioactivity is determined by the particular radio elements present and the rock density. The effective radiating area is dependent on the size of the exposed outcrop and its flatness. A partial cover of overburden or water will limit the source area and complete shielding is provided by about 3 to 5 ft of water, 2 ft of overburden, or only 1 ft of rock.

The air attenuation factor depends on the density of the air and the distance between source and detector. For a point source, the intensity of gamma radiation falls off inversely as the square of the distance, for a line source it falls off inversely as the distance, and for an area the signal is porportional to the solid angle subtended at the detector by the area. Thus variations in the terrain clearance will cause fluctuations in the signal strength. Some equipment has used the output from a radio altimeter to correct automatically the recorded radiometric trace, but this procedure is not strictly valid, because the attenuation of the signal strength with height is dependent on the particular geometry of the source.

To summarize, the primary radiations have their origin in a surface layer of rock less than 30 cm thick. The primary emissions are subsequently modified by absorption and scattering in the source material and in the air path between the source and detector. The radiation reaching the detector is thus composed of the original primary radiations with reduced intensity, superimposed on an energy continuum of secondary degraded radiation which predominates at distances greater than a few hundred feet.

The sensitivity of the airborne scintillation counter is controlled by the size, shape, and composition of the detector crystal, the photomultiplier used, and associated electronic circuitry. The time constant is usually obtained from the ratio, width of the radioactive source measured along the aircraft track which is considered of significant interest for subsequent ground investigation divided by the speed of the aircraft [1]. However, for the detection of point sources, a good rule-of-thumb is that the best time constant is given by the height of the aircraft divided by twice its speed. A more sophisticated discussion of the optimum time constant has been given by Peirson and Franklin [25] and by Giret [16], but a typical value for the time constant is about 1 sec for fixed wing surveys.

The following flight specifications are usually recommended for airborne radiometric surveys [9]:

(1) Terrain clearance should be as low and constant as possible commensurate with safety and bearing in mind the rate of area coverage required because the flight paths should be spaced no further apart than about twice the flight elevation.

(2) Flight paths should be across the main strike of the regional geology as with other types of airborne geophysical surveys. Usually tie lines are advisable.

(3) Surveys should be carried out at times of good atmospheric mixing and when the earth and air are free of synthetic radioactive isotopes such as cesium 137, niobium 95, and zirconium 95 [34] produced by fallout from nuclear weapons.

One highly recommended procedure (see for instance [15]) in unexplored terrain is to fly a combined airborne magnetometer/scintillometer survey at 500 ft over the chosen area. The aeromagnetic survey will, of course, aid greatly in any subsequent geological mapping program. Detailed low-level surveys are then carried out over the significant radioactivity anomalies, using a helicopter-borne spectrometer to ascertain the causative radioactive elements.

A special technique called rim or contour flying has been employed in mountainous terrain [2, 7, 8] because mineral occurrences may outcrop along the mountainsides. For instance, in the Colorado Plateau, where the topography consists of buttes, mesas, and deeply dissected plateaus, uranium deposits have been found almost exclusively in nearly flat-lying sediments. Usually light aircraft or helicopters are used and an irregular track is followed along the depressions and escarpments and at a constant distance from them (50 or 100 ft according to Boyle [8]). Random prospecting using a helicopter has also been carried out in the Canadian Shield [18].

There are two main methods of presenting radiometric results. The first is the marking of "hot spots" on topographic maps or air photographs. The radioactivity reading in counts per minute or milliroentgens per hour minus the background count, the number of times above the background count, or perhaps a suitable symbol to indicate that the readings fall in some arbitrary scale, may also be indicated on the airborne radiometric map. Sometimes the area over which the radioactive count exceeds a certain value is delineated. If an aeromagnetic survey had been carried out at the same time as the airborne scintillometer survey, both kinds of geophysical data could be presented on the same map.

The second method of presentation is the drawing of a contour (isorad) map of the radiometric data. The technique is essentially the same as that for the compilation of aeromagnetic maps. Intersection of the radiometric profiles is carried out after removal of the appropriate background count.

REFERENCES

1. Agocs, W. B. (1955). Airborne scintillation counter surveys, *Can. Min. Met. Bull.*, **58**, 109–111.
2. Alekseev, V. V., Grammakov, A. G., Nikonov, A. I., and Tafeev, G. P. (1957). Radiometric methods in the prospecting and exploration of uranium ores, (in translation) *U.S. At. Energy Comm.*, AEC-tr-3738, **2**, 339–640.
3. Berbezier, J., Blangy, B., Guitton, J., and Lallemant, C. (1958). Methods of carborne and airborne prospecting: The techniques of radiation prospecting by energy discrimination. *Second U.N. Int. Conf., Geneva*, **2**, 799–814.
4. Birks, J. B. (1953). "Scintillation Counters," Pergamon Press, Oxford.
5. Bisby, H. (1951). Application of electronics to the detection and measurement of radioactivity, *U.K. At. Energy Res. Est.*, AERE EL/L3, 63 pp.

6. Bowie, S. H. U., Miller, J. M., Pickup, J., and Williams, D. (1958). Airborne radiometric survey of Cornwall, *Second U.N. Int. Conf., Geneva*, **2**, 787–798.
7. Boyle, T. L. (1956). Airborne radiometric surveying, *U.N. Int. Conf., Geneva*, **6**, 744–747.
8. Boyle, T. L. (1963). Airborne radiometric surveying; office and field techniques, *U.S. At. Energy Comm.*, RME-129 (Rev.), 28 pp.
9. Cook, J. C. (1952). An analysis of airborne surveying for surface radioactivity, *Geophysics*, **17**, 687–706.
10. Curtiss, L. F. (1950). The Geiger-Mueller counter, *Nat. Bur. St. (U.S.) Circ.* **490**, 25 pp.
11. Davis, F. J., and Reinhardt, P. W. (1957). Instrumentation in aircraft for radiation measurements, *Nucl. Sci. Eng.*, **2**, 713–727.
12. Friedman, H. (1949). Geiger counter tubes, *Proc. Inst. Rad. Eng.*, **37**, 791–808.
13. Geiger, H., and Müller, W. (1928). Das Elektronenzählrohr, *Phys. Zeit*, **29**, 839–841.
14. Geiger, H., and Müller, W. (1929). Technische Bemerkungen zum Elektronenzählrohr, *Phys. Zeit.*, **30**, 489–493.
15. Giret, R. (1961). La radiométrie aéroportée, outil d'exploration géologique structurale, *Geophys. Prospecting*, **9**, 582–590.
16. Giret, M. R. (1962). Effet de la vitesse et de la constante d'intégration sur la forme des anomalies aéroradiométriques, *Geophys. Prospecting*, **10**, 183–202.
17. Godby, E. A., Connock, S. H. G., Steljes, J. F., Cowper, G., and Carmichael, H., (1952). Aerial prospecting for radioactive materials, *At. Energy Can. Ltd.*, Rep. AECL-13, 90 pp.
18. Gregory, A. F. (1955). Aerial detection of radioactive mineral deposits, *Can. Mining Met. Bull.*, **48**, 479–485.
19. Gregory, A. F. (1960). Geological interpretation of aeroradiometric data, *Geol. Surv. Can. Bull.* **66**, 29pp.
20. Gregory, A. F., and Horwood, J. L. (1963). A spectrometric study of the attenuation in air of gamma rays from mineral sources, Mines Branch, *Dept. Mines Tech. Surv.*, Res. Rept. **110**, 110pp.
21. Hatton, A. (1961). The airborne scintillation spectrometer, *Mining J.* **256**, 9 and 11.
22. Hollander, J. M., Perlman, I., and Seaborg, G. T. (1953). Table of isotopes, *Rev. Mod. Phys.*, **24**, 469–651.
23. Mero, J. L. (1960). Uses of the gamma ray spectrometer in mineral exploration, *Geophysics*, **25**, 1054–1076.
24. Moxham, R. M. (1960). Airborne radioactivity surveys in geologic exploration, *Geophysics*, **25**, 408–432.
25. Peirson, D. H., and Franklin, E. (1951). Aerial prospecting for radioactive minerals, *Brit. J. Appl. Phys.*, **2**, 281–291.
26. Pemberton, R. H., and Seigel, H. O. (1966). Airborne radioactivity tests—Elliott Lake area, Ontario, *Can. Mining J.*, **87**, 10, 81–87.
27. Pickup, J., and Cosgrove, M. E. (1956). A scintillation counter for airborne radiometric surveying, *U.K. At. Energy Res. Est.*, Rep. EL/R 1972, 12pp.
28. Pitkin, J. A., Neuschel, S. K., and Bates, R. G. (1964). "Aeroradioactivity Surveys and Geological Mapping, in Natural Radiation Environment," pp. 723–736, Univ. of Chicago Press, Chicago, Illinois.
29. Pringle, R. W., Roulston, K. I., and Taylor, H. W. (1950). The scintillation counter as a low resolution gamma-ray spectrometer, *Rev. Sci. Instr.*, **21**, 216–218.
30. Rosenzweig, A. (1952). Evaluation of the Texas Company's airborne radioactivity surveying at Blanding, Utah, *U.S. At. Energy Comm.*, RMO-813.

31. Stead, F. W. (1950). Airborne radioactivity surveying speeds uranium prospecting, *Eng. Mining J.*, **151**, 74–77.
32. Williams, D., Cambray, R. S., and Maskell, S. C. (1959). An airborne radiometric survey of the Windscale area October 19th–22nd, 1957, *U.K. At. Energy Res. Est., Rept.* R-2890, 36pp.
33. Williams, D., and Cambray, R. S. (1960). Environmental survey from the air, *U.K. At. Energy Res. Est., Rept.* R-2954, 41pp.
34. Williams, D., and Bisby, H. (1961). The aerial survey of terrestrial radioactivity, *Proc. Inst. Elec. Engrs. (London)*, Pt. B **108**, 403–412.

5. Airborne Gravity Methods

5.1. General

The successful development of an airborne gravimeter has long been a challenge to exploration geophysicists and serious thoughts on the topic probably date from the time of the introduction of the airborne magnetometer, because both gravity and magnetic methods are potential field techniques. The first total-field airborne magnetometer, namely the saturable core variety, was an electronic instrument and did not rely on the balancing of magnetic against mechanical forces. Unfortunately no purely electronic gravimeters have been developed and probably the first airborne prospecting gravimeter will utilize an atomic phenomenon, such as the Mössbauer effect [45], to measure gravity or one of its space derivatives. All the airborne gravimeters described in the following pages rely on the gravitational force exerted by the earth on a suspended mass.

The present generation of airborne gravimeters is capable of measuring the average gravity over a $1 \times 1°$ square to an accuracy of approximately 10 milligals. This is adequate for geodetic purposes but cannot be considered to be an effective tool for exploration surveys except in a regional sense. Actually there is a good case for utilizing a lighter-than-air vehicle for airborne gravity surveys. All the experimental airborne work to date appears to have been carried out either in the U.S.A. [29, 38–40], or in the U.S.S.R. [31, 34].

Airborne gravity measurements are subject to disturbances caused by the accelerations of the aircraft, and even under very good atmospheric conditions, these accelerations amount to the equivalent of thousands of milligals. Because a gravimeter cannot differentiate between motional and gravitational acceleration, the distinction is usually made on the basis of time. Thus the motional accelerations are of short period and consequently can be eliminated by smoothing or averaging over appreciable time periods, whereas the true gravitational effects are not time dependent. This averaging procedure introduces an "uncertainty factor" in that the precision with which average gravity can be measured increases as the length of time of averaging increases, but the finer details of the gravity profile are lost.

The basic equation [29] for the acceleration (g_m) measured by an airborne gravimeter is

(5.1) $$g = g_m + \ddot{h} + E + F + D + H$$

where g = reduced gravity value, h = height above a reference datum as indicated by a sensitive barometer, E = Eötvös correction, F = free-air correction to the datum, D = drift correction of instrument, and H = correction for long-period horizontal accelerations.

Because aircraft motion in general changes the eastward tangential velocity of the gravimeter owing to the earth's rotation, the centrifugal acceleration of the gravimeter is thereby affected. This effect was first noted by Eötvös [7], and the total correction, including the vertical component of the Coriolis acceleration, to be applied to the readings may be obtained from the equation [39]

(5.2) $$E = (R_\varphi + d)[2V_\varphi v_e + v^2]/R_\varphi^2 \quad \text{gals,}$$

where the radius of the earth at latitude φ, $R_\varphi = 637{,}838{,}800(1 - 0.003367 \sin^2 \varphi + 0.0000071 \sin^2 2\varphi)$ cm [15], d = altitude of aircraft above mean sea level, V_φ = tangential velocity of a point on the earth's surface at latitude φ, v = ground speed of aircraft, and v_e is its easterly component. For the Eötvös correction to be in the cgs units of acceleration, i.e., gals, distances must be in centimeters and velocities in cm/sec (1 mph = 44.704 cm/sec). The tangential velocity (V_φ) of the earth's surface at latitude φ is $\omega R_\varphi \cos \varphi$, where ω is the angular velocity of the earth. And $\omega = 2\pi/T = 2\pi/86164.09 = 0.000072921$ sec^{-1} as the mean sidereal day contains 86164.09 sec [1]. At middle latitudes, the Eötvös correction amounts to about 6 milligals per east-west mile per hour [28]. Hence it is by far the largest correction to be made to the gravity data, and necessitates the aircraft speed being known to better than 1 mi/hr and the instrument having a large dynamic range. As an easterly velocity is in the same sense as the earth's rotation (N.B.: earth rotates anticlockwise looking down on the North Pole), the centrifugal (or outward) acceleration will be increased when the aircraft has an easterly component of velocity (which is the reason why satellite-carrying rockets are usually launched from Cape Kennedy in an easterly direction). Thus the measured gravity will be decreased so that the Eötvös correction should be added. Glicken [11] has published nomograms for the determination of errors in the Eötvös correction caused by errors in both the speed and heading and also for errors in latitude determination. Popov [33] has presented similar formulas for calculating the Eötvös correction and has carried out experimental investigations into the errors in determining the Eötvös correction.

The gravitational attraction varies with distance away from the center of the earth—the so-called free-air effect. Thus the acceleration g_φ at a reference datum at height d above mean sea level is

(5.3) $$g_\varphi = [R_\varphi^2/(R_\varphi + d)^2]g_0$$

where R_φ is the radius of the earth at latitude φ, and g_0 is the gravity at the earth's surface at latitude φ. The gravity (g) at height h above the reference datum is

(5.4) $$g = [(R_\varphi + d)^2/(R_\varphi + d + h)^2]g_\varphi.$$

Then free-air correction

(5.5) $$F = g_\varphi - \frac{(R_\varphi + d)^2}{(R_\varphi + d + h)^2} g_\varphi \simeq \frac{2g_\varphi h}{(R_\varphi + d)}$$

because $h \ll R$

$$= \frac{2R_\varphi^2 g_0 h}{(R_\varphi + d)^3}$$

and g_0 may be obtained from the International Gravity Formula, namely $g_0 = 978.049(1 + 0.0052884 \sin^2 \varphi - 0.0000059 \sin^2 2\varphi)$. This formula also indicates the error in the reduced gravity value owing to the uncertainty of locating the exact latitude of the aircraft. The rate of change of gravity along a north-south line is obtained by differentiating the International Gravity Formula, when

$$dg/dx \simeq 1.307 \sin 2\varphi \quad \text{milligals per statute mile,}$$

where x is measured in a north-south direction. Thus at latitude 45°, the error will be approximately 1.3 milligals per north-south statute mile.

For a freely suspended gravimeter subjected to horizontal accelerations, the apparent reading (g_a) is the vector sum of the horizontal acceleration $(\ddot{x})$ and the true vertical gravity (g). If θ is the angle of deflection of the instrument from the vertical, then

(5.6) $$g = g_a \cos \theta \simeq g_a - (\ddot{x}^2/2g)$$

The term $(\ddot{x}^2/2g)$ is called the Browne [4] correction and is obtained by using two horizontal acceleration meters (HAMS) to measure the acceleration along and across the aircraft track. The outputs from the HAMS are squared and added to obtain the Browne correction.

When the beam of a gravimeter is deflected from its zero position, horizontal accelerations produce a couple on the beam. This cross-coupling effect [19] can cause errors in excess of 100 milligals when both the horizontal and vertical accelerations have amplitudes of 50 gals. A coherent phase relationship between leveling errors in the platform and the horizontal accelerations can also cause large errors. This is called the Harrison [14] effect and the periodic stabilization errors should be less than 8″ of arc for an accuracy of 1 milligal when the horizontal accelerations are 50 gals in amplitude.

To summarize, the errors in airborne gravimetric surveys are as follows: (1) those owing to the drift of the instrument, (2) the Eötvös correction which, because of its size, is the limiting accuracy factor, (3) the uncertainty of position, (4) the Browne effect, (5) the Harrison effect, and (6) the cross-coupling effect.

It follows from the preceding discussion that for an accuracy of ± 10 milligals, the Eötvös correction requires that the ground speed be known to one knot or better, the true course be known to $0.5°$ or better, and the latitude be known to $1'$ of arc. Thus navigation is the limiting factor in the accuracy of airborne gravimetric measurements.

5.2. *Airborne Gravimeters*

5.2.1. LaCoste–Romberg Airborne Gravimeter. The LaCoste–Romberg gravimeter [18, 21] appears to be the most extensively tested airborne instrument up to the present time. The basic instrument, which is illustrated in Fig. 39, utilizes the "zero-length" spring principle. Thus the quartz spring is

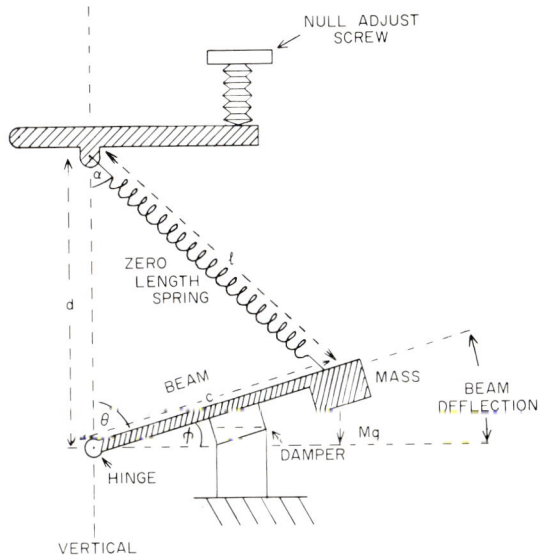

FIG. 39. Basic principle of the LaCoste–Romberg airborne gravimeter.

prestressed in manufacture so that the tension is proportional to the actual length rather than the extension of the spring. Hence for a zero-length spring of length l, the tension is tl, where t is the spring constant.

The airborne instrument differs from the LaCoste–Romberg land gravimeter in a number of important respects. These are the increased damping of the beam, the photoelectric readout system, and the vertical suspension of the instrument. Because of the large Eötvös correction, the range of the airborne instrument is much larger than the ground version, being some 12,000 milligals. A gimbal system was employed in earlier models but a gyroscopically stabilized platform was used in subsequent airborne experiments. Considering Fig. 39, the clockwise torque due to the weight $= Mg \cdot C \cos \varphi$. The anticlockwise torque due to the zero-length spring is $tl \cdot d \sin \alpha = tCd \cos \varphi$ because $l \sin \alpha = C \cos \varphi$. Hence total clockwise torque is $(Mg - td)C \cos \varphi$. If $Mg = td$, the period of oscillation is infinite for all positions of the beam. Therefore the gravitational acceleration g is proportional to the distance d required for equilibrium, which may be varied using the null adjust screw.

The equation of motion of the gravimeter beam when the null adjust screw is set to a value, corresponding to a gravity reading $s = (td/M)$ in the static case, is of the form

(5.7) $$m\ddot{\varphi} + \beta\dot{\varphi} + k\varphi = g + \ddot{z} - s$$

where m, β, and k are instrumental constants, and $\ddot{z}$ is the vertical accleration of the gravimeter as a whole. The term $k\varphi$ is usually referred to as the displaced beam correction [29] in the static case. Because the gravimeter beam is in constant motion during airborne surveys, it is necessary to evaluate the left-hand side of the differential equation using an analog computer; the result is usually referred to as the moving beam correction. Actually the integral of the moving beam correction is obtained and then smoothed. The integral of the spring tension adjustment $(\int s \, dt)$ is subtracted from the resultant smoothed integral together with the vertical velocity, which is the integral of acceleration $(\int \ddot{z} \, dt)$ obtained using a precision altimeter. The values of g are obtained from the slope of the resultant curve. A more elaborate data reduction procedure has been described by LaCoste and Romberg [20].

The first experiments were made in November, 1958 by the Air Force Cambridge Research Center [39] in California using a KC-135 jet aircraft flying at elevations of 20,000 or 30,000 ft. This feasibility test showed that gravity values of better than 10 milligals were obtainable, using a suitable aircraft with adequate navigation instrumentation. A more extensive test was carried out over Imperial Valley in California in May, 1959 by the Fairchild Aerial Surveys, LaCoste and Romberg Co., University of California, and Gravity Meter Exploration Co. [28]. A B-17 aircraft equipped with a precision radar altimeter and aerial mapping cameras for accurate navigation was used. A 20 milligal-contour map was prepared showing the free-air gravity at an elevation of 12,000 ft. The resultant map compared favorably with limited ground gravity data, although rugged topography made a critical comparison difficult.

A careful quantitative test was subsequently made over a triangular test course between Houston, Baton Rouge, and Shreveport in the southern United States in May, 1961 [29]. The instrumentation used is illustrated in Fig. 40 and was mounted in a B-17 aircraft fitted with a Bendix autopilot.

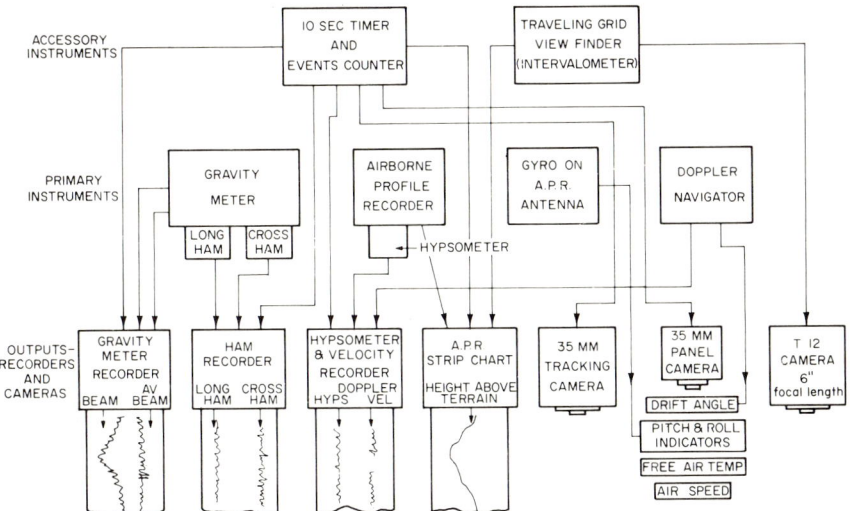

Fig. 40. Block diagram of the LaCoste–Romberg airborne gravimeter (after Nettleton et al. [29]).

The navigation instruments included the following devices:
 (1) Airborne profile recorder and hypsometer for precision altimetry,
 (2) 35 mm track and mapping cameras for aircraft track recovery,
 (3) gyroscopic pitch and roll indicator for correction of information obtained from the navigation instruments above,
 (4) Doppler navigator and computer as a back-up to the primary navigation devices, and
 (5) 35 mm recording camera to photograph the various dials and meters.

The cameras were triggered at 10 sec intervals from a timer which also supplied fiducial marks to the various recorders. The aircraft measurements were compared with the calculated values at 12,000 ft, obtained by upward continuation of ground Bouguer values. The average rms deviation between observed and calculated data for all 6 flights was 6.6 milligals.

5.2.2. Graf–Askania Gravimeter.

Airborne experiments have also been carried out using the Graf–Askania Gss 2 gravimeter [12–14] mounted on a gyrostabilized platform. The basic instrument (see Fig. 41) consists of a

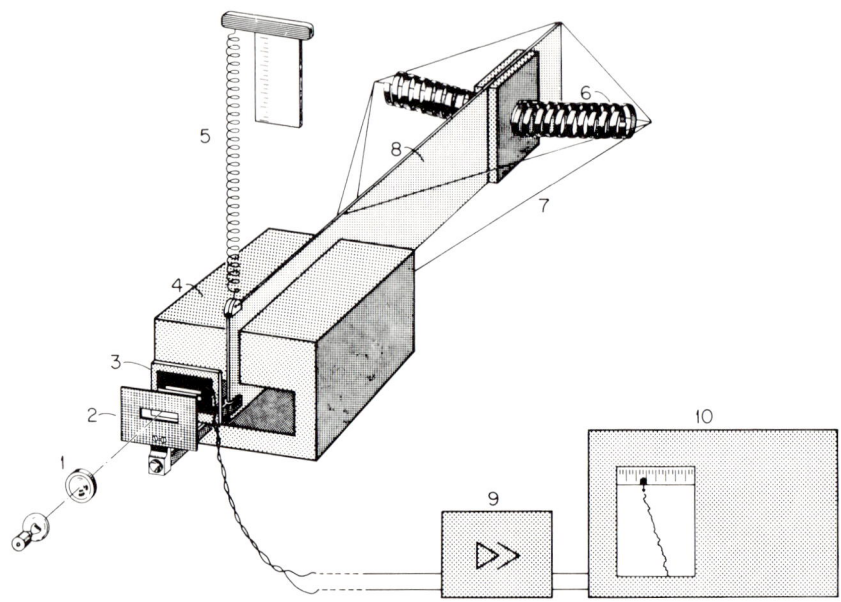

Fig. 41. Graf–Askania Gss 2 gravimeter system. 1. Illumination device (lamp and optics). 2. Diaphragm and slot. 3. Photoelectric cells. 4. Damping magnet. 5. Measuring spring with scale. 6. Torsion springs. 7. Filaments restraining beam motion. 8. Measuring beam. 9. Amplifier. 10. Recording apparatus.

lightweight aluminum beam 30 cm long maintained in a horizontal position by the couple provided by two horizontal cylindrical springs under torsion near one end. Additional horizontal springs running parallel to the beam are provided for fine control and range. Movements other than rotation about the horizontal axis of the main springs are prevented by eight restraining filaments connected between the beam and the case, and heavy eddy-current damping in the vertical plane is caused by the beam being placed between the poles of a strong permanent magnet. The position of the beam is determined by the use of a horizontal slit in the end of the beam through which the light from a lamp impinges on a differential photocell. The output from the photocell is amplified and filtered, and the average over 5 min is recorded.

The static condition of balance (see Fig. 42) is given by $mg\,a = \tau\varphi$ [13] where m is the mass of the beam, a is the distance from the center of mass of the beam to the horizontal axis of the main springs having a torsion constant

τ, and φ is the angle of rotation of the mainsprings. Because of the heavy damping, the instrument has a time constant of 4 to 5 min and short period gravity anomalies are not faithfully reproduced. A servo control system automatically adjusts the measuring spring so that the aluminum beam is kept in the zero position over an average period of time [36]. The position of the end of the measuring spring is determined by means of a circular potentiometer whose central movable arm is connected to a threaded measuring spindle. The number of revolutions of the potentiometer arm is determined using a suitable counter.

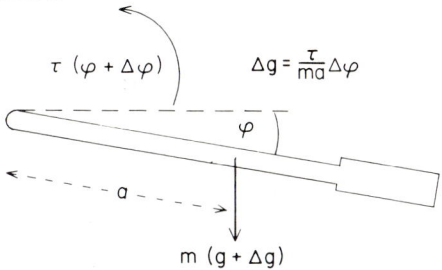

FIG. 42. Basic principle of Graf–Askania gravimeter.

Thompson [40] carried out airborne tests during 1962 in a C-130 aircraft from Edwards Air Force Base in California. A Graf–Askania Gss 2 gravimeter mounted on an ART-25 stabilized platform (Fig. 43) and a LaCoste–Romberg gravimeter in a gimbal suspension were compared over about 6000 line miles of aircraft track. The results from both instruments compared favorably with one another. However, the stabilized platform technique proved to be more advantageous because the operating procedure was simpler and less maintenance was required. The tests also demonstrated that the analog recording system was not the most efficient from the standpoint of data reduction; a digital tape recording system was therefore constructed to accept the outputs of both types of gravimeters [42].

5.2.3. The U.S.S.R. GAL Gravimeter. The GAL sea gravimeter was developed in 1957 by the Aerogravimetric Laboratory of the Institute of Earth Physics of the U.S.S.R. Academy of Sciences from whence it derives its name GAL (Gravimeter Aerogravimetricheskoi Laboratorii). The heavily damped elastic system of these gravimeters consists of two quartz filaments which are twisted in opposite directions by the pendula suspended from them; these pendula remain in an almost horizontal position [31, 32, 34]. Changes in gravity and motional accelerations of the instrument are determined by the changes in the angle between the pendula using an optical system. A permanent record is obtained on moving photographic film.

At least three airborne tests have been made in the U.S.S.R. using the GAL gravimeter [34]. The first was in December 1957, using two GAL gravimeters mounted in Cardan suspensions in an IL-12 aircraft which flew at 3000 meters. Unfortunately less than 10% of the records were legible. This discouraging result caused the second Russian test to be delayed until April 1960. The outputs from the gravimeters together with that from a statoscope (sensitive pressure altimeter) and time fiducials were recorded on a wider film strip with much more satisfactory results. A third test late in 1960 utilized three Cardan-suspended GAL gravimeters mounted in a photosurvey IL-14 aircraft with its more sophisticated nagivational equipment.

Fig. 43. Graf–Askania Gss 2 Gravimeter mounted in a C-130 Aircraft.

5.2.4. Vibrating-String Gravimeter. The vibrating-string gravimeter [10] appears to have a number of advantages as an airborne instrument. It utilizes the frequency (f) of natural vibration of a vertical fiber stretched by a suspended weight (Mg) given by

(5.8) $$f = (1/2l)(Mg/m)^{1/2},$$

where l is the length of the fiber and m is its mass per unit length. The most attractive feature of the vibrating-string instrument is that the measurement is essentially one of force rather than of displacement. The sensitive mass is not displaced with respect to other parts of the instrument and thus cross-coupling effects are not a problem. For the same reason, damping is not required and the response of the vibrating-string gravimeter is essentially instantaneous. However, there are a number of disadvantages to the technique; the main one is that the response of the system is not linear with acceleration, and consequently noise and periodic disturbing accelerations do not average out but instead rectify to produce a time-varying systematic error similar to the Browne error found in pendulum measurements at sea and in measurements made by a freely swinging gravimeter. In Gilbert's original apparatus used on a submarine, the fiber consisted of a flat beryllium-copper strip about 5 cm long which hung vertically between the poles of a permanent magnet. The vibrations of the strip induced an alternating emf between its ends and, using a suitable amplifier and feedback circuit, the metal strip was maintained in continuous oscillation at its resonant frequency. The frequency can readily be determined using any of the standard counting techniques. Subsequently a more sophisticated vibrating-string gravimeter for shipborne use was developed in Japan [43, 44] which used a bifilar suspension. Work has also been carried out in the U.S.S.R. on trifilar suspensions in which the wires are orthogonal [22, 37].

5.3. Airborne Gravity Gradiometers

An instrument that measures the first vertical derivative of the gravitational field would be unaffected by linear motional accelerations, although rotational movement would still produce spurious effects. Consequently there is a distinct advantage to measuring the vertical gravity gradient. Moreover a gravity gradiometer could be used in an orbiting satellite, whereas a gravimeter located at the satellite's center of gravity would give zero reading. Gravity gradiometers also have considerable potential for the control and navigation of space vehicles.

The cgs units of gravity gradient are Eövtös units (E°) and $1~\text{E}° = 10^{-9}$ gal/cm $= 10^{-9}~\text{sec}^{-2}$, so the dimensions of the first vertical derivative of gravity are T^{-2}. Some exploration geophysicists may prefer to use milligals/ft, and 1 milligal/ft = 32,800 E°. The gravity gradient at the earth's surface is approximately 3086 E° = 0.09406 milligals/ft. This varies slightly with latitude [15, p. 54] and a more exact expression for the gravity gradient at height h in kilometers above the earth's surface at latitude φ is

(5.9) $\qquad \partial \Delta g / \partial h = 3087.7 - 4.4 \sin^2 \varphi - 0.73\, h \quad$ E°

Thus the surface gradient value at the poles is 3083.3 E° and at the equator is 3087.7 E°, approximately. The maximum gravity gradient which would be recorded by an airborne gravity gradiometer passing over a 10 million (short) ton excess mass at a vertical separation of 1000 ft would be 42.7 E°. Thus a desirable sensitivity for an airborne prospecting instrument would be of the order of 5 E°.

There is a small Eötvös gradient correction which may be applied to gradiometer results because it follows from Eq. (5.2) that

(5.10) $\qquad \dfrac{\partial E}{\partial h} = \dfrac{(2 V_\varphi V_e + v^2)}{R_\varphi^{\,2}} \qquad$ gals/cm

For a 200 mph aircraft flying east at latitude 45° this amounts to 1.6 E°, so the Eötvös correction for gradient surveys is relatively unimportant.

Chinnery [5] has described a method for the calculation of terrain corrections for airborne gravity gradient measurements. In general these are more important in the reduction of gradient data than in the straightforward gravity case. Paterson [30] has shown how vertical gradient results can be integrated to produce gravity values on any plane above the plane of measurement.

During the late 1950's Lundberg [23–26], experimented with a semiquantitative airborne gravity gradiometer. The device was a balance arrangement [2, 3] in which two masses were suspended vertically above one another, each being attached to an arm of the balance. The displacement of the lower mass was observed by the change in capacity between the mass and a fixed plate which formed a variable capacitor in a bridge circuit. It was claimed that the instrument sensed the sign of the first vertical derivative so that the resultant gradient maps delineated the areas of positive and negative derivative values. To attain this, the masses were periodically clamped and released. A number of test surveys were carried out in North America, Europe, and West Africa but currently (1967) the project appears to be dormant.

During the period 1959–1962 the Hunting (now Lockwood) Survey Corp. of Toronto attempted the development of a gravity gradiometer. The prototype instrument [16, 17] consisted of a sealed glass system, schematically illustrated in Fig. 44, in which two immiscible liquid columns (mercury and an aqueous solution of thallium salts) were enclosed. Because of their different densities, the liquid columns rose to different heights in the enclosed system. The liquids were oscillated at the natural frequency of the system by a mechanical device, and the resultant flow movement was measured by a sensitive transducer placed on the mercury side, which thus enabled any state of imbalance in the oscillating columns to be detected. When a vertical gravity gradient was present, one of the liquid columns would weigh slightly more than the other at the equilibrium balance condition of zero gravity gradient ($\rho_1 h_1 = \rho_2 h_2$ in Fig. 44), which the reader should note is independent of the value of gravity itself. This difference in weight between the columns would thus tend to produce flow movement of the liquid columns. An electronic balance cell was used in the gradiometer system to maintain the zero gradient

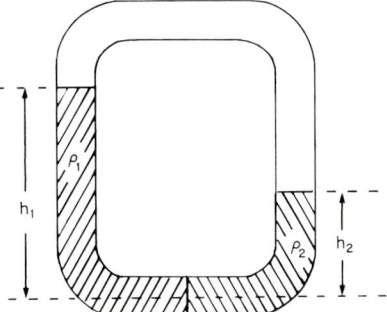

Fig. 44. Schematic representation of the Hunting gravity gradiometer. At balance condition of zero gradient, $\rho_1 h_1 = \rho_2 h_2$; otherwise

$$\frac{dg}{dh} = \frac{2(\rho_1 h_1 - \rho_2 h_2)}{h_2{}^2 - h_1{}^2}.$$

balance condition by applying an appropriate pressure. A servo loop fed the resultant signal from the transducer to the balance cell to maintain the zero gradient balance condition. Because the amplitude of the transducer signal is directly proportional to the gravity gradient, it could be used to record changes in the gradient. Hutchins [17a] has observed that the fully developed gradiometer system would probably have had a sensitivity of about 0.003 milligal/ft, i.e., about 100 Eötvös units, which is a little insensitive for an airborne prospecting instrument.

Hughes Research Laboratories [8, 9] are currently investigating a gradiometer which utilizes the rotational properties of tensors to separate the effects of forces from the effects of the gradients of the forces. It consists of four masses on the ends of four transversely vibrating arms which are spun in a vacuum. The differential torque on the arms caused by a gravity gradient will alternate at twice the rotational frequency of the sensor. Barium titanate strain transducers are used to detect the dynamic strains in the sensor arms, and it is claimed that the system is capable of detecting 1 Eötvös unit.

Thompson et al. [41] have described a gravity gradiometer developed by American Bosch Arma which is a double-string, double-mass vibrating-string gravimeter designed for use in satellites. The difference between the two string frequencies is a measure of the gravity gradient. The instrument has a wide dynamic range and further development should realize a sensitivity of 1 E°.

5.3.1. The Mössbauer Effect.
The effective mass m of a photon of energy E is E/c^2, where c is the velocity of light. Thus if a photon moves a distance d towards the center of the earth, it will lose potential energy $mgd = Egd/c^2$. This loss in energy will appear as an increase in fundamental frequency f, because the energy of a photon is hf, where h is Planck's constant. However, when gamma rays are emitted from radioactive substances, they have a fairly broad spectrum of energies because of the recoil uncertainty in leaving the parent atom, so that it would not be possible to measure the very small frequency shift against such a broad spectrum. Fortunately certain radioactive elements such as Fe^{57} emit gamma rays which are almost monoenergetic. This is because the recoil energy is shared with the crystal as a whole rather than with the emitting atom, and the phenomenon is known as the Mössbauer [27] effect. Because there is very little energy loss, the gamma-ray line widths are extremely narrow [45]. The change in frequency of a photon passing between two points of different gravitational potential has actually been demonstrated [6, 35] but the experiments required many hours of counting so the method is not feasible for gravimetric measurements at the present time. Nevertheless it is an intriguing possibility for a non-mechanical gradiometer system independent of sensor accelerations.

REFERENCES

1. Anonymous (1959). Air Navigation. USAF Manual 51–40. 3.
2. Boitnott, B. D. (1961). Instrument for airborne measuring of derivatives of the vertical component of the earth's gravity field. US Patent 3,011,347.
3. Boitnott, B. D. (1962). Instrument for and method of airborne gravitational geophysical exploration. US Patent 3,038,338.
4. Browne, B. C. (1937). The measurement of gravity at sea. *Roy. Astro. Soc. Mon. Not., Geophys. Supp.*, **4**, 271–279.
5. Chinnery, M. A. (1961). Terrain corrections for airborne gravity gradient measurements. *Geophysics*, **26**, 480–489.
6. Cranshaw, T. E., Schiffer, J. P., and Whitehead, A. B. (1960). Measurement of the gravitational red shift using the Mössbauer effect in Fe^{57}. *Phys. Rev. Letters* **4**, 163–164.
7. Eötvös, R. (1919). Experimenteller Nachweis der Schwereänderung, die ein auf normal geformter Erdoberfläche in östlicher oder westlicher Richtung bewegter Körper durch diese Bewegung erleidet. *Ann. Phys.*, **59**, 743.

8. Forward, R. L. (1964). Rotating tensor sensors. *Am. Phys. Soc. Bull.*, **9**, 711.
9. Forward, R. L. (1965). Rotating gravitational and inertial sensors. A.I.A.A. Unmanned Spacecraft Meeting, Los Angeles. 6 pp.
10. Gilbert, R. L. (1949). A dynamic gravimeter of novel design. *Phys. Soc. Proc.*, **62**, 445–454.
11. Glicken, M. (1962). Eötvös corrections for a moving gravity meter. *Geophysics*, **27**, 531–533.
12. Graf, A. (1958). Das Seegravimeter. *Zeits Instr.*, **60**, 151–162.
13. Graf, A., and Schulze, R. (1961). Improvements on the sea gravimeter Gss 2. *J. Geophys. Res.*, **66**, 1813–1821.
14. Harrison, J. C. (1960). The measurement of gravity at sea. In "Methods and Techniques in Geophysics" (S. K. Runcorn, ed.), pp. 211–229. Wiley (Interscience), New York.
15. Heiskanen, W. A., and Vening Meinesz, F. A. (1958). "The Earth and its Gravity Field," 470 pp. McGraw-Hill, New York.
16. Hutchins, R. W. (1962). Gravity gradiometer system and method. Can. Patent 652, 757.
17. Hutchins, R. W., and D'Arcy, D. F. (1963). Gravity gradient measuring device. U.S. Patent 3,095,744.
17a. Hutchins, R. W. (1966). Personal communication.
18. LaCoste, L. J. B. (1934). A new-type long-period vertical seismograph. *Phys.*, **5**, 178–180.
19. LaCoste, L. J. B., and Harrison, J. C. (1961). Some theoretical considerations in the measurement of gravity at sea. *Geophys. J.*, **5**, 89–103.
20. LaCoste, L. J. B., and Romberg, F. (1963). Instructional manual for LaCoste and Romberg air-sea gravity meter, LaCoste and Romberg, Austin, Texas.
21. LaCoste, L. J. B. (1967). Measurement of gravity at sea and in the air. *Rev. Geophys.*, **5**, 477–526.
22. Lozinskaya, A. M. (1959). The string gravimeter for the measurement of gravity at sea. *Bull. Sci. U.S.S.R., Geophys. Ser. (English Transl.)* 272–278.
23. Lundberg, H. (1957). Airborne gravity surveys. *Can. Oil and Gas Ind.*, **10**, 121–122.
24. Lundberg, H. (1957). Airborne gravity surveys. *Can. Mining J.*, **78**, 121.
25. Lundberg, H. (1957). Airborne gravity surveys. *Can. Mining Met. Bull.*, **60**, 251–259.
26. Lundberg, H. T., and Ratcliffe, J. H. (1959). Airborne gravity meter—description and preliminary results. *Mining Eng.*, **11**, 817–820.
27. Mössbauer, R. L. (1958). Kernresonanzfluoreszenz von Gammastrahlung in Ir^{191}. *Zeit. Phys.*, **151**, 124–143.
28. Nettleton, L. L., LaCoste, L., and Harrison, J. C. (1960). Tests of an airborne gravity meter. *Geophysics*, **25**, 181–202.
29. Nettleton, L. L.. LaCoste, L. J. B., and Glicken, M. (1962). Quantitive evaluation of precision of airborne gravity meter. *J. Geophys. Res.*, **67**, 4395–4410.
30. Paterson, N. R. (1961). An integration technique for airborne gravity gradient measurements. *Geophysics*, **26**, 474–479.
31. Popov, E. I. (1960). Observations with strongly overdamped gravimeters, on airplanes and helicopters. *Bull. Acad. Sci. U.S.S.R., Geophys. Ser. (English Transl.)*, pp. 807–810.
32. Popov, E. I. (1962). The methods of processing the recordings of marine gravity meters and their accuracy. *Bull. Acad. Sci. U.S.S.R., Geophys. Ser. (English Transl.)*, pp. 141–147.

33. Popov, E. I. (1962). Determination of the correction for the Eötvös effect during measurement of the gravity acceleration from flying aircraft. *Bull. Acad. Sci. USSR., Geophys. Ser. (English Transl.)*, pp. 254–256.
34. Popov, E. I. (1963). Results of experimental determinations of the acceleration of gravity in the air from an aircraft. *Bull. Acad. Sci. U.S.S.R., Geophys. Ser. (English Transl.)*, pp. 457–462.
35. Pound, R. V., and Rebka, G. A. (1960). Apparent weight of photons, *Phys. Rev. Letters*, **4**, 337–341.
36. Schulze, R. (1962). Automation of the sea gravimeter Gss 2. *J. Geophys. Res.*, **67**, 3397–3401.
37. Tarakanov, Y. A. (1965). The theory of gravity measurements in an aircraft using orthogonal wire sensing devices. *Phys. Solid Earth, Acad. Sci. U.S.S.R. (English Transl.)* **1**, 300–307.
38. Thompson, L. G. D. (1959). Airborne gravity meter test. *J. Geophys. Res.*, **64**, 488.
39. Thompson, L. G. D., and LaCoste, L. J. B. (1960). Aerial gravity measurements. *J. Geophys. Res.*, **65**, 305–322.
40. Thompson, L. G. D. (1965). Comparison of LaCoste–Romberg and Askania–Graf gravity meters in gimbal and stabilized mounts. *J. Geophys. Res.*, **70**, 5599–5613.
41. Thompson, L. G. D., Bock, R. O., and Savet, P. H. (1965). Gravity gradient sensors and their applications for manned orbital spacecraft. *3rd Goddard Mem. Symp., Amer. Astr. Soc.*, Washington.
42. Thompson, L. G. D., and Hawkins, C. S. (1966). Advances in aerial gravity 1963–1964. Amer. Geophys. Un., *Geophys. Mono. Ser.*, **9**, 28–30.
43. Tomoda, Y., and Kanamori, H. (1962). Tokyo surface-ship gravity meter α-1. *Japan J. Geophys.*, **7**, 116–145.
44. Tsuboi, C., Tomoda, Y., and Kanamori, H. (1961). Continuous measurements of gravity on board a moving surface ship. *Proc. Japan Acad.*, **37**, 571–576.
45. Wertheim, G. K. (1961). The Mössbauer effect: a tool for science. *Nucleonics*, **19**, 52–57.

6. Other Airborne Remote-Sensing Methods

6.1. General

Remote-sensing methods may be defined as those scientific methods capable of detecting an object using sensors, i.e., measuring devices, which are not actually in physical contact with the object sought. The sensors measure a diagnostic physical property of the object which differs from that of its surrounding objects, and they are thus capable of delineating the object of interest from its environment. A practical example of a geological target would be a granite intrusive having higher natural radioactivity and resistivity, lower magnetic susceptibility and density, and a different color from the surrounding country rock. It follows from the above discussion that all airborne geophysical methods are remote-sensing techniques. Remote-sensing methods may be divided into two general groups, namely active and passive sensing systems. An active system is one that actually irradiates the object under investigation with electromagnetic energy in a particular range of wavelengths (see Fig. 45) and then samples the portion retransmitted to the

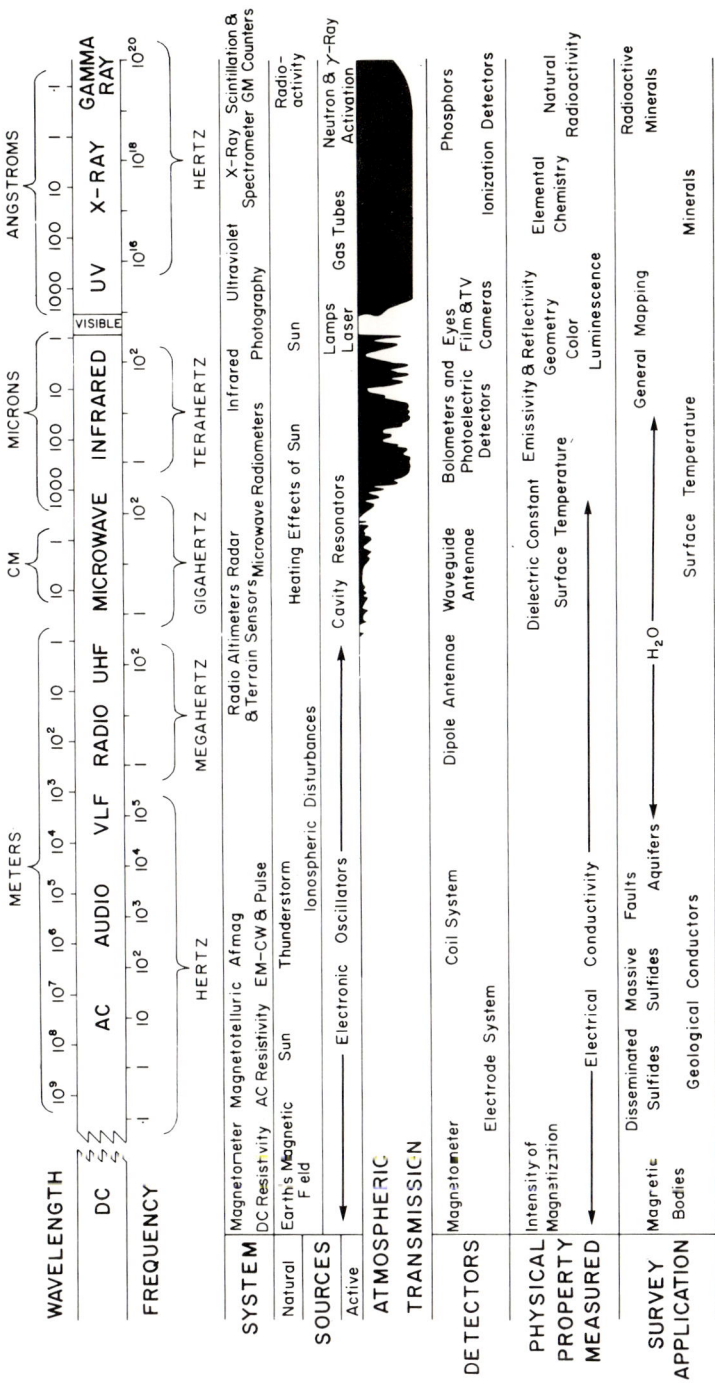

Fig. 45. Use of the electromagnetic spectrum in mineral exploration surveys.

detecting device. A passive system does not use an activating source in the remote-sensing equipment. It should be noted that conventional aerial photography and magnetic surveying methods, although passive systems, use natural energizing sources, namely the sun and the magnetic field produced by the earth's core.

The passive group consists of

(a) Conventional aerial photography and television systems which use the ultraviolet and visible spectrum;
(b) Infrared techniques—photographic and line-scanning methods;
(c) Microwave radiometers; and
(d) Scintillometers.

In addition there are two nonelectromagnetic categories:

(e) Magnetometers; and
(f) Gravimeters.

The active group all use electromagnetic energy; they are:

(a) Radar methods;
(b) Radio-frequency EM—including altimeters; and
(c) Audio-frequency EM.

An excellent review of the basic physics involved in remote-sensing methods has been given by Colwell et al. [10]. A good idea of the present state of the terrain-sensing art may be obtained from the proceedings of the symposia on Remote-Sensing of Environment held at the Institute of Science and Technology of the University of Michigan [2–5].

6.2. Aerial Photography

Conventional aerial photography is historically the oldest, the most useful, and remains the most exploited remote-sensing technique. Consequently it has been developed to a high degree of sophistication. Extensive descriptions of the hardware used and the possibilities of interpretation are contained in the "Manual of Photogrammetry" [1] and "Manual of Photographic Interpretation" [9].

6.3. Ultraviolet Methods

The U.S. Geological Survey is currently carrying out investigations in the ultraviolet part of the spectrum from approximately 2200 to 3700 Å [18].

Initial results suggest that the ultraviolet wavelengths may be the most useful part of the spectrum for the recognition of specific rocks and minerals. For example, many minerals fluoresce at specific wavelengths when excited by UV light. No airborne system using an ultraviolet source appears to have been developed yet, and possibly the method may not be feasible anyway.

6.4. Infrared Methods

Infrared airborne methods in survey use at the present time consist of (a) infrared photography which utilizes that part of the spectrum in the near infrared between 0.7 and 1.35 μ; (b) airborne radiation thermometers, which operate in the range 7.5–13 μ and are used for the measurement of the water surface temperature; and (c) line-scanning systems which operate between 3 and 14 μ.

Infrared photography uses special film which is sensitive in the extreme red and near infrared regions of the spectrum. A deep orange or red filter is normally used because the film is also sensitive in the blue region. The use of infrared photography in geological mapping does not appear as promising as it does in other disciplines such as forestry.

Infrared detectors may be divided into two main classes: (a) thermal types such as thermocouples and bolometers which are based upon measuring the temperature change in resistance of a material owing to the heating effect of absorbed radiation, and (b) photoelectric types which are usually semiconductors capable of detecting photons. Only the latter can be used in airborne devices because the thermal types lack both sensitivity and speed of response.

Airborne radiation thermometers (ART) were first used in 1953 in order to measure water-surface temperatures over large areas of ocean. Actually they measure the temperature of the top 0.1 mm of water, and an accuracy of $\pm 0.5°C$ is readily obtainable. An airborne radiometer receives radiation from three sources, namely (a) reflection from the sky, (b) emission from the atmosphere, and (c) radiation from the ocean, only the last one being of interest to oceanographers.

The atmosphere is opaque to certain wavelengths because of the molecular absorption bands owing mainly to water vapor and carbon dioxide. Fortunately the two gases having the highest concentrations, namely nitrogen (N_2) and oxygen (O_2), possess neither a permanent nor an induced electric moment and hence do not exhibit molecular absorption [23]. The water vapor absorption bands occur at 1.1, 1.38, 1.87, 2.7, and 6 μ and virtually close the atmosphere from 25 μ to the start of the microwave region at about 1000 μ, and CO_2 is observed at 2.7, 4.3, and 14.5 μ. Ozone (O_3), methane (CH_4), nitrous oxide (N_2O), and carbon monoxide (CO) are present to a much lesser extent in the atmosphere and also have characteristic absorption bands which show up over long atmospheric paths.

Water temperatures vary from a minimum of $-3°C$ to a maximum of $37°C$. For infrared wavelengths from 4 to 12.5 μ the ocean surface has an emissivity of 0.98 for radiation normal to the surface [20]; and the reflection of normal incident radiation is only 0.02. The wavelength (λ_p) of the peak energy radiated by a black body at absolute temperature $T°K$ is given by the Wien displacement formula:

$$\lambda_p T = 2893 \quad \text{micron degrees}$$

A water surface therefore has infrared radiation characteristics close to those of a $300°K$ black body, which would emit maximum radiation at a wavelength of 9.6 μ. A typical ART, e.g., Model 14-320 manufactured by Barnes Engineering Company of Stamford, Connecticut, is designed to accept radiation in the range 7.5–13 μ. This region corresponds to a good transmission window through the atmosphere between the ocean surface and the detector (see Fig. 45). A special filter, such as arsenic trisulfide glass coated with a thin antireflection film of indium antimonide, is used to restrict the optical bandpass together with a multielement germanium lens system which has good transmission properties in the desirable range. An optical chopper is utilized so that radiation from the source and then a black body reference are alternately presented to a thermistor-type infrared detector. This enables absolute measurements to be made. Pirart et al. [29] have described an ART, the FRB-2, developed by the Pacific Oceanographic Group at Nanaimo, B.C., which is capable of $0.1°C$ resolution.

HRB–Singer, Inc. has developed a line-scanning system which responds to radiation in the 3–14 μ range, which includes the region of maximum terrain emission and minimum natural reflection [8]. The Reconofax infrared survey system (Fig. 46) scans a line at right angles to and across the path of the aircraft. The output of the amplifying system is fed to a glow tube which illuminates a photographic film whose movement is synchronized with that of the aircraft. Thus a continuous record of the infrared radiation from the underlying terrain is obtained. Detailed description of the system is still classified, but Lattman [24] has given examples of the resultant IR imagery and indicated various geological features such as contacts on the photographs.

Wallace and Moxham [35] have presented examples of infrared imagery (8–13 μ band) of the San Andreas fault system in the Carrizo Plain area of California, which demonstrated that the fault could be clearly traced over most of the 200 mi flown. They concluded that variations in soil moisture caused by the water-barrier characteristics of the fault zone as well as vegetation differences related to soil moisture and microtopography, are factors influencing the visibility of the fault by IR imagery. Presunrise imagery appeared to be more diagnostic than postsunrise imagery because the irregular topography is not heated uniformly by solar radiation and the effects of the thermal inertia of rock units are masked.

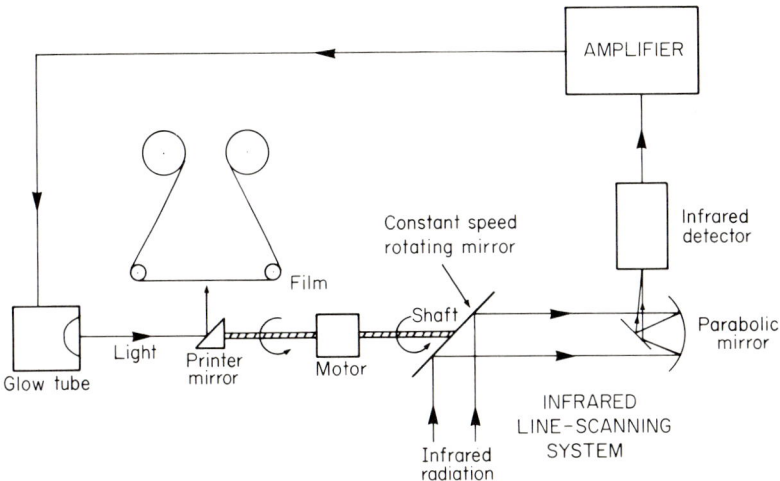

FIG. 46. Infrared line-scanning system (after HRB–Singer Inc.).

6.5. *Microwave Radiometers*

Microwave radiometers are essentially passive high-frequency radar receivers whose development has evolved from their use in radio astronomy. The receivers usually operate in the radiometric or Dicke [12] mode in which the input to the receiver is alternately switched between the antenna and a known thermal reference in order to compensate for drifts in the receiver gain. The instrumentation can be made more sensitive than in the infrared but the thermal energy in the wavelengths used is less by several orders of magnitude. Also the emissivities of surfaces in the microwave region range from about 0.6 to 0.9 that of a perfect black body. The attenuation owing to the atmosphere is much lower for microwave frequency energy and at a wavelength of 1.8 cm there is negligible absorption [27]. The foregoing factors compensate approximately so that present state-of-the-art infrared and microwave systems have comparable performances [11]. However microwave radiation also originates from beneath the surface so that the body and not just the skin temperature may be obtained. This is because as wavelength increases, the electromagnetic energy contribution from the underlying depths becomes significant, thus giving microwave radiometry the capability of detecting radiation from depths of several centimeters in reasonably solid materials and meters for materials having a lower dielectric constant or lower conductivity or both.

An airborne microwave radiometric system, designated the AN/AAR-33, has been built by the Sperry Rand Corporation for iceberg detection. The nominal operating frequency of the system is from 13.5 to 16.5 GHz, and the antenna system scans through 110° at right angles to the aircraft track. Thus the width of the swath of ocean covered is approximately $3\frac{1}{2}$ times the aircraft altitude. The system is capable of measuring the difference in temperature between the background and a target which entirely fills the antenna field of view to an accuracy of 1.75°K under normal operating conditions of 5000 ft flight elevation and aircraft ground speed of 200 knots.

6.6. *Radar Methods*

The principles of radar will be well known to most readers. It is an active system utilizing electromagnetic energy having frequencies between about 10^8 and 10^{11} Hz (see Fig. 45). The various bands are divided as shown in Table VIII, and the frequencies are given in GHz = 10^9 cps.

TABLE VIII.

Band	Frequency (GHz)	Approximate wavelength (cm)
P	0.225–0.39	100
L	0.39–1.55	30
S	1.55–5.2	10
C	3.5–5.85	7.5
X	5.2–10.9	3
K	10.9–36	1
Q	36–46	0.75
V	46–56	0.6

There are two main types of radar which have been used in airborne geophysical surveying: (a) circular scan and (b) side-looking radar. The difference is in antennae and display; the former has a rotating antenna and the resultant radar map is displayed on a plan position indicator (PPI), whose origin may be offset to one side of the cathode-ray tube display. Examples of RAF circular-scan radar photos of Scotland and the Gaspé in eastern Canada have been presented by Cameron [7]. The second transmits energy at right angles to the flight path from antennae on one or both sides of the aircraft [25]. Badgley *et al.* [6], Hackman [19], and Scheps [33] have presented some very striking side-looking radar photographs obtained in the United States which show such features as faults and lineaments and would have an obvious value in any geological mapping program. A permanent record is obtained in both cases on photographic film for subsequent interpretation, and the resultant imagery is similar to conventional photography obtained under low-sun-angle conditions. Thus the density of the photographic image on the survey film is proportional to the amplitude of the reflected radar signal, which is in turn, dependent on (a) electronic parameters of the radar set (including polarization of the radar energy); (b) distance between the set and reflector (the amplitude varies inversely as the distance to the fourth power and there is no appreciable attenuation due to the atmosphere); (c) grazing angle; (d) geometry; (e) surface roughness; and (f) dielectric constant of the reflector. The basic objective of such a survey is, of course, to produce a map giving information on the geology of the region, i.e., a radar geology map [16, 17, 26, 32]. Where the surface roughness and dielectric constants of adjacent outcropping formations are significantly different, they may be differentiated by the radar method. Actually the dielectric constant is increased by moisture content [21, 22], so that the amplitude of the reflected radar return is increased by higher moisture content. Thus faults may be detected by radar techniques owing to moisture differences on either side of the fault. The depth of penetration of the electromagnetic energy is however decreased. The lower frequencies will have the greater penetration, and experimental work at the U.S. Army Corps of Engineering Waterways Experiment Station [30] on soils has shown that a penetration of 18 in. in wet soils was obtained using radar frequencies in the P-band.

Waite and Schmidt [34] and also Evans [13, 15] have described the measurement of ice thickness using modified radio altimeters operating at frequencies between 35 and 440 MHz. This is possible because between a frequency of 1 MHz and the far infrared region there is no absorption in the spectrum of ice [14].

A continuous-wave gas-laser altimeter has been developed jointly by Spectra-Physics, Inc. of California and the Aero Service Corp. of Philadelphia [28, 31]. Flight tests carried out using a Douglas A-26 aircraft have shown that it is capable of remarkable resolution. For instance, ordinary curbstones at the sides of roads could be distinguished, as could the individual seats in an open-air athletic stadium. The laser altimeter should prove particularly useful in an airborne profile recorder system, and in the determination of elevations for gravity survey corrections.

REFERENCES

1. Anonymous (1952). "Manual of Photogrammetry," 2nd ed., p. 876, revised edition, 1966. Amer. Soc. Photogrammetry.
2. Anonymous (1962). *Proc. Symp. Remote Sensing of Environment, 1st, Ann Arbor, Michigan* Rept. 4864–1–X.

3. Anonymous (1963). *Proc. Symp. Remote Sensing of Environment*, 2nd Ann Arbor, Michigan Rept. 4864-3-X.
4. Anonymous (1964). *Proc. Symp. Remote Sensing of Environment*, 3rd Ann Arbor, Michigan, Rept. 4864-9-X.
5. Anonymous (1966). *Proc. Symp. Remote Sensing of Environment*, 4th Ann Arbor, Michigan, Rept. 4864-11-X.
6. Badgley, P. C., Childs, L., and Vest, W. L. (1967). The application of remote sensing instruments in earth resource surveys, *Geophysics*, **32**, 583–601.
7. Cameron, H. L. (1965). Radar as a surveying instrument in hydrology and geology, *Proc. Symp. Remote Sensing of Environment*, 3rd Ann Arbor, Michigan, pp. 441–452.
8. Cantrell, J. L. (1964). Infrared geology, *Photo. Eng.*, **30**, 916–922.
9. Colwell, R. N. (1960). "Manual of Photographic Interpretation," p. 868. Amer. Soc. Photogrammetry.
10. Colwell, R. N., Brewer, W., Landis, G., Langley, P., Morgan, J., Rinker, J., Robinson, J. M., and Sorem, A. L. (1963). Basic matter and energy relationships involved in remote reconnaissance, *Photo. Eng.*, **29**, 761–799.
11. Conway, W. H. and Sakamoto, R. T. (1965). Microwave radiometer measurement program. *Proc. Symp. Remote Sensing of Environment* 3rd, Ann Arbor, Michigan, Rept. 4864-9-X, pp. 339–356.
12. Dicke, R. H. (1946). The measurement of thermal radiation at microwave frequencies, *Rev. Sci. Instr.*, **17**, 268–275.
13. Evans, S. (1963). Radio techniques for the measurement of ice thickness, *Polar Record*, **11**, 406–410.
14. Evans, S. (1965). Dielectric properties of ice and snow—a review, *J, Glaciology*, **5**, 773–792.
15. Evans, S. (1967). Progress report on radio echo sounding, *Polar Record*, **13**, 413–420.
16. Feder, A. M. (1960). Interpreting natural terrain from radar displays, *Photo. Eng.*, **26**, 618–630.
17. Feder, A. M. (1962). Radar geology can aid regional oil exploration, *World Oil*, **155**, 130–138.
18. Fischer, W. A. (1964). Geological interpretation from airphotos, Proc. Seminar on Air Photo Interp., Ottawa, Canada, Pt. 4, 21–31.
19. Hackman, R. J. (1967). Geologic evaluation of radar imagery in Southern Utah, *U.S. Geol. Surv.*, Prof. Paper 575-D, pp. 135–142.
20. Holter, M. R., Nudelman, S., Suits, G. H., Wolfe, W. L., and Zissis, G. J. (1962). "Fundamentals of Infrared Technology," p. 442. Macmillan Co., New York.
21. Howell, B. F., and Licastro, P. H. (1961). Dielectric behavior of rocks and minerals, *Am. Minerolgist*, **46**, 269–288.
22. Keller, G. V., and Licastro, P. H. (1959). Dielectric constant and electrical resistivity of natural-state cores, *U.S. Geol. Surv.*, Bull. 1052-H, 257–285.
23. Kruse, P. W., McGlauchlin, L. D., and McQuistan, R. B. (1962). Elements of Infrared Technology," p. 448. Wiley, New York.
24. Lattman, L. H. (1963). Geologic interpretation of airborne infrared imagery, *Photo. Eng.*, **29**, 83–87.
25. Levine, D. (1960). "Radargrammetry," p. 330. McGraw-Hill, New York.
26. Levine, D., Colbert, C., Graham, L. C., Crane, P. H., and Scheps, B. B. (1966). Combinations of photogrammetric and radargrammetric techniques, *Manual of Photogrammetry*, **2**, 1003–1048.
27. Menon, V. K., and Ragotzkie, R. A. (1967). Remote sensing by infrared and microwave radiometry, Dept. of Meteorol., Univ. Wisconsin, Tech. Rept. 31, p. 59.

28. Miller, B. (1965). Laser altimeter may aid photo mapping, *Av. Wk. and Space Tech.*, March 29.
29. Pirart, M., Carswell, J., Oliver, D., and Bell, W. H., (1965). Airborne radiation thermometer (FRB-2), *Fish. Res. Bd., Canada*, MS Rep. Ser. 188, p. 58.
30. Reid, M. (1962). Airborne soil analysers will help Army plan its campaigns, *Electronics*, **35** (52), 20–22.
31. Rempel, R. C., and Parker, A. K. (1965). An information note on an airborne laser terrain profiler for micro-relief studies, *Symp. Remote Sensing of Environment*, 3rd. *Ann Arbor, Michigan*, pp. 321–337.
32. Rydstrom, H. O. (1967). Interpreting local geology from radar imagery, *Geol. Soc. Am. Bull.*, **78**, 429–436.
33. Scheps, B. B. (1965). Oceanographic applications for radar, in Oceanography from Space, Woods Hole Ocean. Inst., Ref. 65-10, pp. 273–285.
34. Waite, A. H., and Schmidt, S. J. (1962). Gross errors in height indication from pulsed radar altimeters operating over thick ice or snow, *Proc. IRE*, **50**, 1515–1520.
35. Wallace, R. E., and Moxham, R. M. (1967). Use of infrared imagery in study of the San Andreas fault system, California, *U.S. Geol. Surv.*, Prof. Paper 575-D, pp. 147–156.

List of Symbols[3]

Magnetic

γ	Gammas—units of magnetic field intensity
T	Total intensity of earth's magnetic field
Z	Vertical component of earth's magnetic field
ΔT	Residual total intensity
H_e	Maximum value of field applied to fluxgate element
f, F, ν	Frequency in hertz (Hz)
t	Time in seconds
B	Induction
μ	Permeability
γ_p	Gyromagnetic ratio of the proton
θ	Angle
n, N	Counts—number of oscillations
K	Gammas/hertz ratio for nuclear magnetometers or constant of proportionality
S	Sensitivity in gammas
m	Frequency multiplication factor, quantum number of electron
ΔE	Difference in energies between electron sublevels
h	Planck's constant
a	Magnetic activity in gammas
d	Distance, including separation of ground station and survey aircraft in miles
N	Number of readings obtained from the ground monitor during survey flight

Gravity

g	Acceleration due to gravity
h, d	Height

[3] For Sections 1, 2, and 4–6.

D, E, F, H	Gravity corrections
R	Radius of the earth
φ	Latitude in degrees or angle
v, V	Velocity
ω	Angular velocity
θ, α	Angle of deflection
x	Distance measured horizontally
$\ddot{x}$	Horizontal acceleration
m, β, k	Instrumental constants
$\ddot{z}, \hbar$	Vertical acceleration
m, M	Mass
a, c, d, l	Distance or length
t	Zero-length spring constant
τ	Torsional constant of spring
$E°$	Eötvös units
ρ	Density
c	Velocity of light
E	Energy of photo
μ	Micron
λ	Wavelength

LIDAR

R. T. H. Collis

Aerophysics Laboratory, Stanford Research Institute, Menlo Park, California

	Page
1. Introduction	113
2. The Basic Lidar Technique	114
3. Atmospheric Optical Parameters	117
3.1. General	117
3.2. Rayleigh Scattering	117
3.3. Mie Scattering	118
3.4. Backscattering by Atmospheric Turbulence	121
3.5. Absorption	122
4. The Significance of Lidar-Measured Optical Parameters	122
4.1. Meteorological Significance	122
4.2. The Evaluation of Lidar-Measured Optical Parameters	123
5. Application of Lidar Observations to Meteorological Problems and Atmospheric Studies	125
5.1. General	125
5.2. Illustrative Examples	125
6. Lidar Contributions to Atmospheric Studies and Meteorological Problems	133
7. Future Developments	135
Appendix	136
List of Symbols	137
References	138

1. Introduction

The advent, in 1960, of the laser as a source of energy, opened up many possibilities for new techniques of probing the atmosphere or for improving and extending established techniques. The properties of this new form of energy were remarkable even at an early stage of technology. The energy, at optical or near optical frequencies, was monochromatic, coherent, and, with the development of Q-switching techniques, could be generated in very short pulses of very high power. A number of scientists soon recognized the applicability of this device to atmospheric studies and described a variety of ways in which the special characteristics of laser energy could be exploited. These ranged from straightforward radar-type applications to more sophisticated concepts in which the wave nature and coherence of the laser energy were utilized. (See Schotland *et al.* [25], Goyer and Watson [12], for example.)

The first actual use of lasers in atmospheric studies appears to be Fiocco and Smullin's [10] use of a ruby laser "radar" to detect echoes from the atmosphere at heights up to 140 km in June and July 1963. At about the same time however, the late Dr. M. G. H. Ligda [16] had initiated a program at Stanford Research Institute in which a similar pulsed ruby laser "radar" system, or *lidar*,[1] as Ligda called it, was used to probe the lower atmosphere and study meteorological phenomena.

Since that time, such simple "radar" techniques have been applied by a number of workers to map and track concentrations of particulate matter and to study the density profile of the atmosphere by reference to gaseous backscattering. Meanwhile, others have been implementing some of the concepts involving the wave nature and coherence of laser energy. These include the use of multiple wavelength lidars for determining by reference to differential absorption the atmosphere's gaseous composition and also the use of Doppler techniques for determining motion in the atmosphere or, from molecular velocities, its temperature.

This review will consider only the simple "radar" approach and be concerned with the application of determinations of the intensity of backscattering of lidar energy to atmospheric studies and the solution of meteorological problems.

2. The Basic Lidar Technique

Energy generated by giant-pulse (Q-switched) lasers is highly monochromatic, essentially coherent, and concentrated in very short, high-power pulses. This energy is directed by refracting or reflecting lens systems in a beam. Energy backscattered by the atmosphere within the beam is detected by an energy sensitive transducer (normally a photomultiplier tube) after being collected by suitable receiver lens systems. The monochromaticity of the energy makes it possible, by the use of narrowband filters, to limit "noise" in the form of energy of solar origin to a minimum. The coherence of the energy makes it possible to achieve very narrow transmitter beams. A typical lidar system is shown in Fig. 1; its characteristics are given in the Appendix [22].

The essential features of lidar detection of atmospheric targets are described in the equation

$$(2.1) \qquad P_r = \frac{P_t c \tau \beta'_{180} A}{8\pi r^2} \exp -2 \int_0^r \sigma(r)\, dr,$$

---

[1] The word *lidar*, an acronym analogous to radar, from *LI*ght *D*etection *A*nd *R*anging, was earlier used by Middleton and Spilhaus [20] in connection with pulsed-light ceilometers.

using the following notation:

P_r received power
P_t transmitted power
c the velocity of light
τ pulse duration
r range
β'_{180} the volume backscattering coefficient of the atmosphere at range r (having dimensions of area/unit volume). (Following radar practice, β'_{180} is defined as an area that would intercept the same amount of energy as would yield the same return at the lidar if radiated isotropically at range r, as is, in fact, received from unit volume of the atmosphere at that range.)
A the effective receiver aperture
σ the extinction coefficient

FIG. 1. Mark V 1967 neodymium lidar (Stanford Research Institute).

The basic lidar observation consists of an evaluation of received signal power P_r in terms of range and direction. The minimum detectable signal level is determined either by the system noise and that due to solar energy entering the receiver or by the sensitivity of the detector system. At laser wavelengths, even with systems of modest performance, the smallest hydrometeors may be readily detected, as well as the microscopic particles of the "clear" aerosol.

It will be immediately apparent that unless the volume backscattering coefficient, β'_{180}, and the extinction coefficient, σ, are uniquely related, it is not possible to evaluate the intensity information in absolute terms. However, within certain limits, the relationship between these parameters is sufficiently consistent to enable the significance of the variation of received signal with range to be unequivocal and of direct value. This is particularly the case where the lidar beam encounters strongly scattering targets after passing through relatively clear air, as occurs in observing clouds of particulates. Again, minor variations of signal intensity with range are immediately obvious and reveal layers and inhomogeneities in a continuously scattering atmosphere.

In practice, the signal from the photomultiplier is normally displayed on an oscilloscope as a function of range—the familiar A-scope presentation of radar practice. The single transient signal from a single shot may be photographed or magnetically recorded. Polaroid photography allows early inspection of the data in the former case, but the use of magnetic video disc memory makes a continuously viewable oscilloscope display available immediately as well as providing an input for more sophisticated analysis procedures and displays.

Although up to the present, data have largely been converted manually to punched cards or tape for subsequent computer processing and presentation, automatic data input techniques can readily be implemented. In the case of the very weak signals from high altitudes, where the signal is a function of the rate of generation of single photoelectrons, more sophisticated, automatic data processing techniques have already been employed (for example, McCormick *et al.* [15], describe the on-line input of lidar data to a digital computer).

The limited data rate of the early lidar systems (with intervals between pulses measured in seconds if not in minutes) has restricted the resolution of observations in time and has precluded the development of scanning systems capable of developing two-dimensional sections of the type familiar in radar practice. (The lower data rate has perhaps been responsible for an earlier application of quantitative analyses than was the case with weather radar.) Both quantitatively and qualitatively however, lidar has made it possible to study remotely in three dimensions many atmospheric phenomena that hitherto could only be observed grossly or examined piecemeal.

3. Atmospheric Optical Parameters

3.1. General

Electromagnetic energy incident upon a volume of atmospheric gases and the liquid and solid particles suspended therein is scattered and absorbed. The magnitude of these effects is dependent upon the size and number of the particles present and their refractive index (and in this context, gaseous molecules may be considered as particles) and also upon the wavelength of the incident energy. (In the case of laser energy, its highly monochromatic nature is an important consideration, for as shown by Twomey and Howell [26], the effects of critical wavelength/particle-size ratios are not averaged out so readily as is the case with broadband light sources.)

Of the energy scattered, that which is returned in the direction of the lidar is evaluated in terms of the volume backscattering coefficient, β'_{180} (l^{-1}). Energy removed from the direction of propagation, either by scattering or by absorption, can be evaluated most conveniently in terms of the extinction coefficient σ (l^{-1}). This in turn can be considered in terms of the extinction caused by scattering, σ_s, and the extinction caused by absorption, σ_a. The important scattering and absorption mechanisms are now discussed.

3.2. Rayleigh Scattering

Rayleigh scattering from the molecular atmosphere is important for it provides a method by which atmospheric densities may be derived from lidar measurements. In addition, it also provides a convenient datum, to which other scattering and absorption effects may be related, in the upper atmosphere, particularly where layers of purely gaseous composition can be identified.

For wavelengths well separated from the absorption lines of the atmospheric constituents, the Rayleigh scattering cross section C_{RAY} of an individual scattering center is given [29] by

$$(3.1) \qquad C_{RAY} = \frac{8\pi}{3} \left(\frac{2\pi}{\lambda}\right)^4 \alpha^2 \frac{6+3\delta}{6-7\delta},$$

where
- λ wavelength of incident radiation
- δ depolarization factor caused by the anisotropy of the atmosphere
- α molecular polarizability of the scatterer

For the atmospheric gases, the factor δ has a value near 0.035; therefore the fraction $(6+3\delta)/(6-7\delta)$ is about 1.061. The polarizability α is approximately 2×10^{-30} (m^3), and thus:

$$(3.2) \qquad C_{RAY} = 3.96 \times 10^{-56} \lambda^{-4} \qquad (\text{m}^2)$$

and at the ruby wavelength $\lambda = 0.694\,\mu$, for example,

(3.3) $\qquad C_{\text{RAY}}(\lambda = 0.694\,\mu) = 1.71 \times 10^{-31} \qquad (\text{m}^2)$

The total scattering cross section per unit volume of a purely gaseous atmosphere is this elementary cross section multiplied by the number density N of molecular scatterers per unit volume.

(3.4) $\qquad\qquad\qquad \sigma_{\text{RAY}} = NC_{\text{RAY}}$

This quantity σ_{RAY} is also called the Rayleigh attenuation coefficient. It is that quantity which, when multiplied by the incident power density and the effective illuminated volume, gives the total power scattered in all directions from the incident radiation beam.

For pure Rayleigh scattering it can be shown that $3/8\pi$ per steradian of this total will be scattered back toward the source. As a result of the convention used in defining radar cross sections (see Section 2), it follows that for Rayleigh scattering the volume backscattering cross section β'_{180} can be obtained from

(3.5) $\qquad \beta'_{180\ \text{RAY}} = 4\pi(3/8\pi)NC_{\text{RAY}} = 1.5\,\sigma_{\text{RAY}}$

Thus the factor k, which is the ratio of backscattering, β'_{180}, to the extinction coefficient, σ, is for Rayleigh scattering a trusted constant (1.5) and not subject to the fluctuations encountered when the scattering particles become large compared to the wavelength. The significance of the value, $\beta'_{180\ \text{RAY}}$, in determining the density of the upper atmosphere is indicated in Table I.

Table I lists values for N from the "U.S. Standard Atmosphere" [28] and for $\beta'_{180\ \text{RAY}}$ for sea level to 20 km elevation in 5 km increments.

TABLE I. Volume scattering coefficients for Rayleigh component of atmospheric scattering for ruby lidar ($\lambda = 0.6943\,\mu$).[a]

Height (km)	N (m^{-3})	β'_{180} (m^{-1})	σ_{RAY} (m^{-1})
0	2.55×10^{25}	6.55×10^{-6}	4.37×10^{-6}
5	1.52×10^{25}	3.93×10^{-6}	2.62×10^{-6}
10	8.60×10^{24}	2.21×10^{-6}	1.47×10^{-6}
15	4.06×10^{24}	1.04×10^{-6}	$.69 \times 10^{-6}$
20	1.85×10^{24}	4.75×10^{-7}	3.2×10^{-7}

[a] Rayleigh scattering is proportional to λ^{-4}.

3.3. Mie Scattering

Mie scattering is of far greater significance than Rayleigh scattering in the lower atmosphere. It applies to particulate matter having dimensions of magnitude similar to the wavelength of the incident radiation. For large particles the elementary scattering cross section, C_{MIE}, is of the order of twice the geometrical cross section. The scattering pattern in the Mie case does not resemble the symmetrical-dipole pattern of Rayleigh scattering, but can be quite irregular and complicated [7, 19, 29]. The ratio of the backscattered to the total scattered energy is thus highly variable as a function of the particle-size to wavelength ratio and the dielectric characteristics of the particle. This is illustrated in Fig. 2 which shows the relationship between backscattering and total scattering and the size parameter, α, for single spherical particles having a real refractive index of 1.33 (i.e., that of water). (The size parameter α is $2\pi a/\lambda$ where a is the radius.)

It will be seen that neither backscattering nor total scattering show significant *general* dependence on wavelength or particle size. Usually Mie scattering is predominantly forward so that in an assemblage of particles of different sizes, k, in the relation $\beta'_{180} = k\sigma$, is often less than unity. Because the effects of particle size differences tend to average out in such assemblages, useful approximate values can be determined for k and used in evaluating the lidar signal. Stanford Research Institute calculations for water sphere distributions typical of natural water clouds give an average value of $k = 0.625$. This value together with the attenuation coefficients given in Elterman's Clear Standard Atmosphere [8], have been used to compute values for the aerosol contribution to total backscattering for various altitudes as plotted in Fig. 3. (The value $k = 0.625$ is most accurate for water spheres, but is a reasonable approximation for other aerosol components.)

From this figure, it is apparent that even on "clear" days (i.e., those with a horizontal visibility of about 25 km at sea level for the Elterman model) the aerosol backscattering predominates over the molecular backscattering for all elevations below 4 km.

Table II lists a range of typical water-cloud and haze conditions, together

TABLE II. Predicted volume backscatter and extinction coefficients for water clouds and hazes.[a]

Condition	σ_{MIE} (m^{-1})	$\beta'_{180 \text{ MIE}}$ (m^{-1})
Dense water cloud	3.2×10^{-1} to 1.6×10^{-2}	2×10^{-1} to 1×10^{-2}
Light water cloud	1.6×10^{-2} to 4.0×10^{-3}	1×10^{-2} to 2.5×10^{-3}
Thick haze	4.0×10^{-3} to 1.1×10^{-3}	2.5×10^{-3} to 7×10^{-4}
Moderate haze	1.1×10^{-3} to 4.8×10^{-3}	7×10^{-4} to 3×10^{-4}
Light haze	4.8×10^{-3} to 1.6×10^{-4}	3×10^{-4} to 1×10^{-4}

[a] For ruby lidars ($\lambda = 0.6943\ \mu$) ($k = 0.625$).

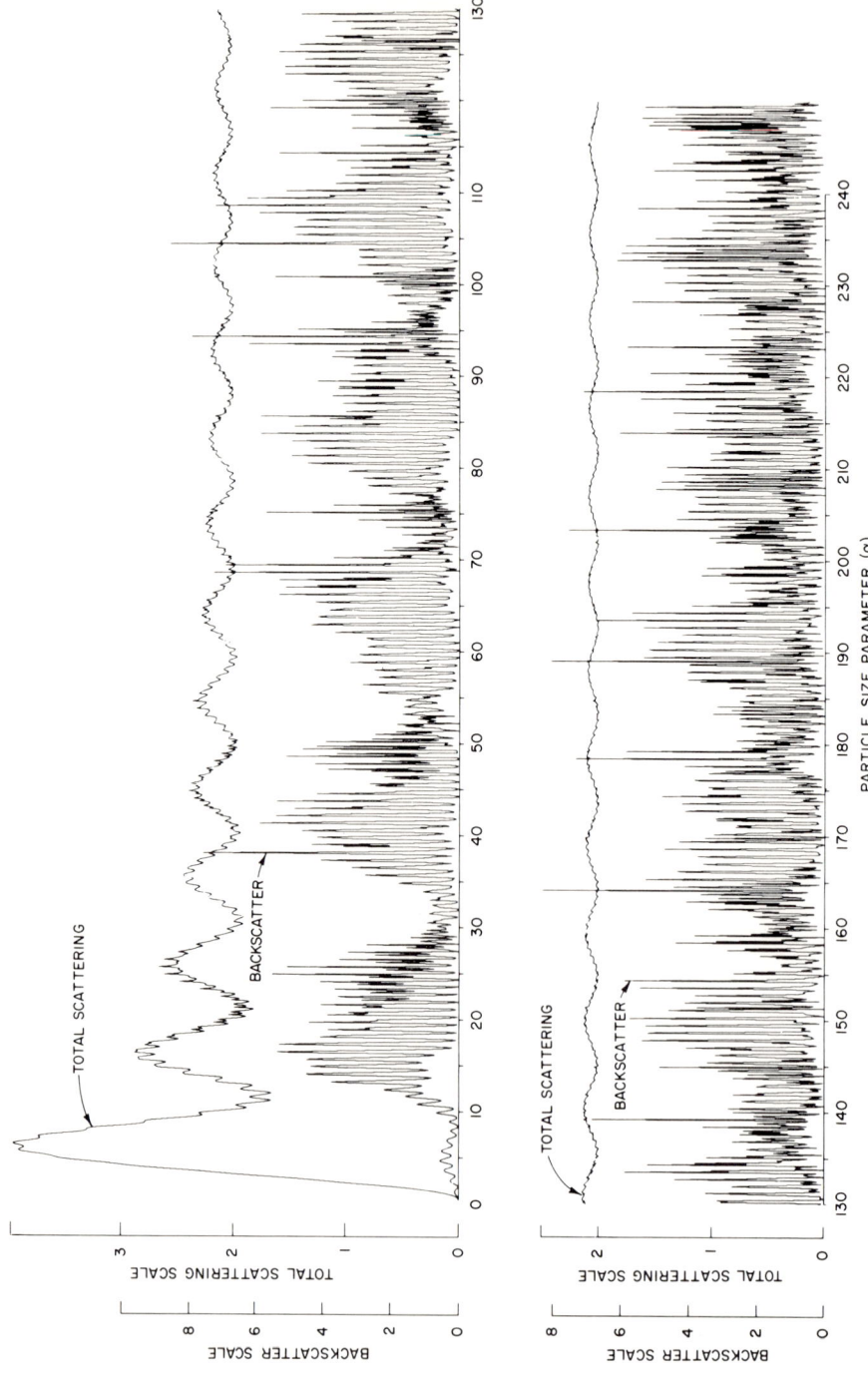

FIG. 2. Scattering efficiency factor as function of particle size parameter. Refractive index: 1.33. Efficiency factor is a ratio of respective scattering cross section to geometric cross section.

with the associated computed aerosol extinction coefficients and anticipated volume backscatter coefficients, β'_{180}, under the assumption that $k = 0.625$.

Note however the generalizations involved in these examples (see Section 4).

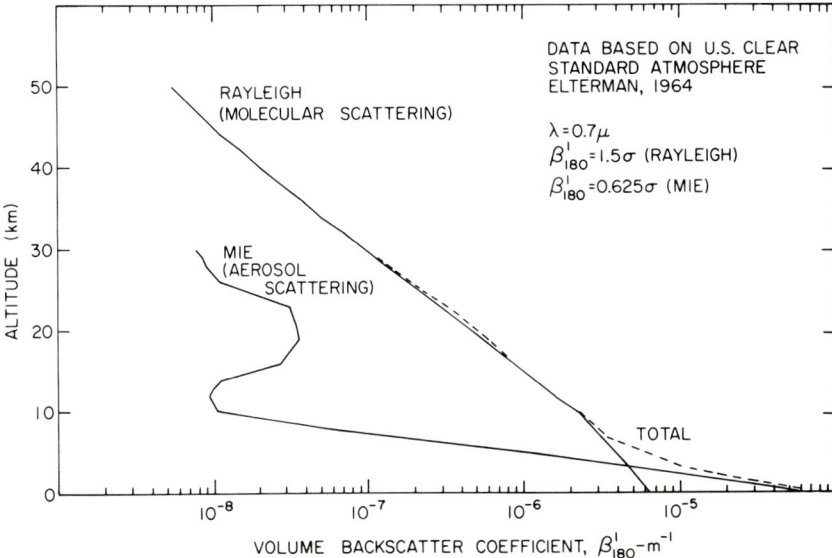

FIG. 3. Volume backscatter coefficients for a clear standard atmosphere (for ruby lidar $\lambda = 0.6943$ μ). Based on a clear standard atmosphere [8]. Note that a recent revision [8a] indicates a substantially larger aerosol content above approximately 4 km. The total backscattering profile based on these data would have the same general characteristics but would have values larger by a factor of approximately two above about 4 to 30 km.

3.4. Backscattering by Atmospheric Turbulence

The possibility of directly detecting atmospheric turbulence by lidar as a function of backscattering by dielectric inhomogeneities has attracted some attention. Among others, Munick [21] has shown, however, that this mechanism is far too feeble to encourage any hopes in this direction. For temperature and molecular number density values typical of altitudes of 10 km and a large temperature structure coefficient (representative of turbulent conditions near the ground), he shows that the backscattering owing to turbulence at ruby wavelengths would be some 7 orders of magnitude *less* than that caused by molecular backscattering!

3.5. Absorption

In addition to scattering, the gaseous atmosphere, and to a certain extent the industrially polluted aerosol, absorbs energy. The attenuation due to this is generally insignificant in comparison to scattering losses, and $\sigma_{\text{total}} \approx \sigma_s$. Absorption is, of course, highly wavelength-dependent (especially at absorption line centers as exploited in spectroscopic lidar techniques) but may be neglected for many purposes in the basic lidar application. Since in the operation of ruby lasers, heating can result in emission at the water vapor line centered on $0.69438\,\mu$ where the attenuation rate is some five times greater, some workers have found it desirable to control the laser operating temperature to avoid this [14].

4. The Significance of Lidar-Measured Optical Parameters

4.1. Meteorological Significance

It is important to recognize that the magnitudes of the coefficients discussed above, and the relationships between the volume backscattering coefficients and attenuation coefficients are by no means absolute. They are given merely to provide general orders of magnitude and illustrate the relationships between the parameters in question.

For many meteorological and atmospheric applications, number and size spectrum of the aerosol particles is all-important. Although in certain cases, e.g., the measurement of visibility, the evaluation of the optical parameter as such, in this case the extinction coefficient, σ, will have direct significance, the quantitative contribution that lidar observations can make to meteorological studies is limited by the degree to which the optical parameters can be interpreted in terms of atmospheric characteristics.

Thus, in the case of the higher atmosphere, i.e., above 30 to 40 km, if the absence of particulate material can reasonably be inferred from the data, an evaluation of the volume backscattering coefficient is essentially a direct method of measuring atmospheric density [14, 24]. In "clear" air in which particulate matter is present, the volume backscatter coefficient and the extinction coefficient can only be related to the particulate loading of the atmosphere within certain limits. Barrett and Ben-Dov [1] discuss these in connection with lidar applications in air pollution measurements. They show that variations in assumed aerosol distribution parameters will produce relatively small errors (less than a factor of 2) in evaluation of particle concentrations from volume backscatter coefficient determinations.

While this degree of accuracy may be acceptable in air pollution studies, for other purposes it is obviously too uncertain. Fenn [9], for example, shows the limitations inherent in the relationship between atmospheric backscattering and the extinction coefficient in connection with the measurement of visual range. Twomey and Howell [26, 27] also discuss the difficulties of deriving information on particle size distribution from optical measurements with special reference in their earlier paper to the monochromatic aspects of laser energy.

4.2. The Evaluation of Lidar-Measured Optical Parameters

The discussion of the meteorological significance of optical parameters in Section 4.1 has been carried on with the tacit assumption that the volume backscattering coefficients and the extinction coefficient *can* be evaluated. As noted in Section 2, the separation and evaluation of these terms cannot readily be accomplished from lidar observations, for unless a unique relationship exists between the volume backscattering coefficient and the extinction coefficient, the lidar equation is unsolvable. The difficulties discussed in Section 4.1 in connection with the interpretation of the significance of optical parameters apply in an especially critical way to the attempt at interpreting the lidar equation. This is particularly true because of the monochromatic nature of the energy [26]. The relationship between backscattering and total scattering in the Mie region (i.e., for the particle sizes commonly involved in atmospheric aerosols and such features as cloud and fog) is then highly variable. For a single scatterer, a diameter variation of, say, 1/100 can change the backscattering coefficient by a factor of 20. Although the averaging that occurs in the case of a volume of multisize particles tends to stabilize the reltionship (k) between the volume backscattering coefficient and extinction coefficient (see Section 3) at single wavelength, uncertainties in the relationship remain. Analysis techniques that rely on assumptions of any specific value of k are consequently apt to be in error. The difficulty lies in the fact that, unlike weather radars (particularly those of wavelength 10 cm or longer), any significant backscattering of lidar energy by atmospheric targets involves *considerable* attenuation.

Various analytical techniques have been proposed. For example, where the atmosphere is homogeneous, the derivative of the logarithm of the range-corrected received signal with respect to range, yields the attenuation coefficient in absolute terms.

(4.1) $$d \log_e P_r r^2 / dr \equiv -2\sigma$$

Barrett and Ben-Dov [1] in the appendix to their paper describe the derivation and solution of an integral equation based on the initial assumption of a specific value of k.

The authors point out the instability inherent in approaches of this type, but show how errors can usually be confined to reasonable limits. At the Stanford Research Institute (SRI) a similar approach has been taken but has been developed in the following form.

The data from the lidar signature is reduced in terms of the atmospheric optical parameters in a form which is called the lidar S-function[2] defined as

$$(4.2) \qquad S(r) \equiv 10 \log \frac{P_r(r) r^2}{P_r(r_0) r_0^2} \equiv 10 \log \frac{\beta'_{180}(r) T_a^2(r)}{\beta'_{180}(r_0) T_a^2(r_0)}$$

where $T_a(r)$ is the one-way atmospheric transmission

$$(4.4) \qquad T_a(r) = \exp\left(-\int_0^r \sigma(r')\, dr'\right)$$

and r_0 is a reference range.

When the backscatter is related to the extinction by

$$(4.5) \qquad \beta'_{180}(r) = k_1\, \sigma(r)^{k_2}$$

the derivative of the expression for $S(r)$ yields a first-order, nonlinear differential equation

$$(4.6) \qquad \frac{d\sigma}{dr} - c_1 \frac{dS}{dr}\, \sigma - c_2\, \sigma^2 = 0$$

where $c_1 = 1/4.34\, k_2$ and $c_2 = 2/k_2$. The transform $\eta \equiv 1/\sigma$ reduces the equation to linear form for which the solution may be written as

$$(4.7) \qquad \sigma(r) = \sigma(r_0) \exp\left(c_1 S(r)\right) \left[1 - c_2\, \sigma(r_0) \int_{r_0}^r \exp\left(c_1 S(r')\right) dr'\right]^{-1}$$

where knowledge of $\sigma(r_0)$ is required for solution.

Even in the absence of complete solutions, it is noteworthy that, unlike much of the work on weather radar, *quantitative* approaches are being developed and utilized in handling and displaying lidar data. This is encouraging for it appears that this will lead to progress both in the analytical technique and in the exploitation of modern data processing and presentation resources.

---

[2] The concept of the S-function was developed from the Spatial Backscatter Function (SBF) previously used at SRI (see Section 5). The SBF was defined as
$$(4.3) \qquad \mathrm{SBF}(r) = 10 \log \beta'_{180}(r) T_a^2(r)$$
where β'_{180} has dimensions of km^{-1}. The S-function has the advantage of being dimensionless.

5. Application of Lidar Observations to Meteorological Problems and Atmospheric Studies

5.1. General

Techniques for remotely probing the atmosphere can be directed towards measuring the temperature, density, or composition (in terms of water vapor, ozone, or carbon dioxide) of its gases, or for delineating and identifying the nature of its particulate content. In addition, the motions of the atmosphere are also of concern—both in terms of wind motion and turbulence.

What can lidar observations accomplish in these areas?

Direct evaluations of the backscattering profile in the upper atmosphere are believed to be capable of providing information on density profiles with sufficient accuracy to show seasonal variations in molecular density, at least in the layer from 50 to 80 km [14, 24]. However the possibility of unexpected particulate intrusions, and the difficulty of making accurate measurements of returns from the tenuous upper atmosphere, make this approach rather uncertain, and in any case, it cannot be used when there are low clouds.

Other direct applications include the detection of the presence, height, shape, and in certain cases, thickness of clouds or haze layers. The evaluation of the atmospheric optical parameters (β'_{180} and σ) can also be considered direct observations which provide descriptive information on the atmosphere and its structure.

Finally from the nature of atmospheric structure, observed in this way, it may be possible to infer the motion of the atmosphere which has given rise to such a structure. Motion, however, is most readily inferred by observing the displacement of recognizable natural features or specifically introduced indicator materials (e.g., smoke).

5.2. Illustrative Examples

The uses of lidar for these purposes can best be appreciated from the following illustrative examples. These are selected from a wide range of applications to demonstrate the salient features of lidar application in this context and show the current state-of-the-art.

5.2.1. Cloud and Cloud Structure.
A good example of the use of lidar in a qualitative role is provided by observations made of cirrus clouds in the Owens Valley, California, early in 1966 [6]. The SRI Mk. I Ruby Lidar (see Appendix for details) was located near Independence and used to make a series of observations in a vertical plane parallel to the direction of air flow. The objective was the observation of the features and dimensions of waves

caused by the Sierra range. Fig. 4 shows an example of the cloud structure observed in this way. The readiness with which the length and amplitude of the waves can be evaluated is obvious. Note that lidar echoes were obtained at slant ranges over 20 km for cirrus cloud in daylight with this relatively modest system. The limited data rate (1 pulse/min) however, restricts the resolution of the cross section both in space and time. Atmospheric structure revealed in this and similar cross sections (even of subvisible inhomogeneities) offers a new capability for studying atmospheric motion with possible

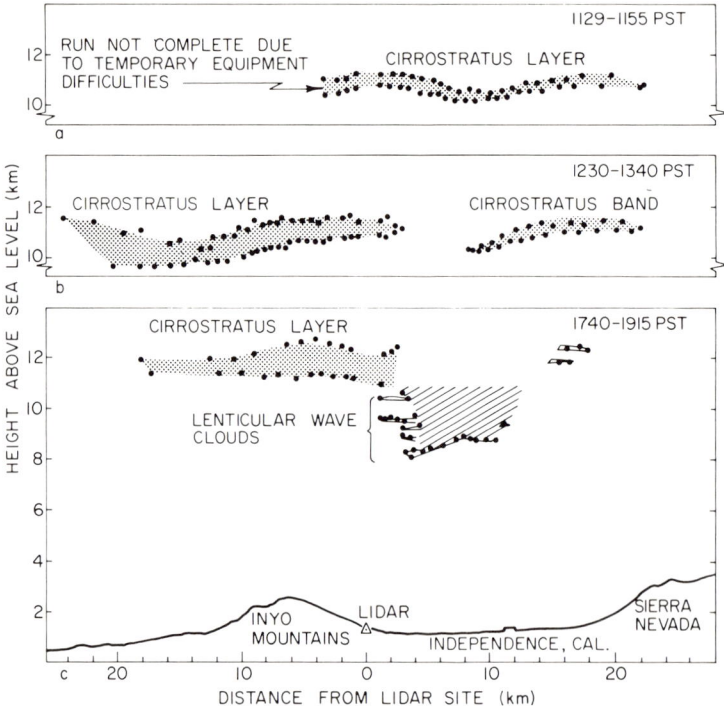

Fig. 4. Lidar observation of wave clouds in the lee of the Sierra Nevada, March 1, 1967. Data were obtained by scanning ruby lidar in the vertical and noting echoes at successive angles of elevation (indicated by data points).

implications in the study of turbulent motion. (See Lawrence et al. [15] for a report of lidar observations associated with turbulence experienced by an aircraft.)

Of course, denser lower clouds can readily be mapped by lidar [2]. More quantitative studies of cirrus clouds are also being carried out at SRI in connection with radiometric measurements such as those made by satellite. An example of quantitative data reduced from lidar observations is shown in Fig. 5. (Manually extracted data are processed and presented by computer and automatic plotter. Quantitative data are available on punched tape for further manipulation.) Data such as these are compared with satellite cloud photographs and upper air soundings. As a subsequent experiment, it is hoped to compare them with radiometric data acquired by the Nimbus satellite.

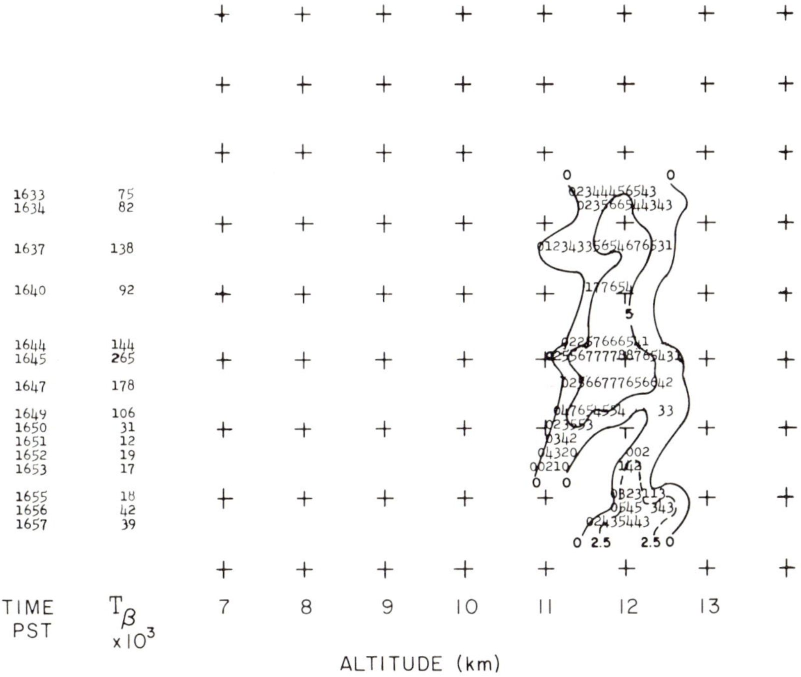

FIG. 5. Graphical/quantitative representation of lidar cloud observations, Menlo Park December 8, 1967. The automatically plotted data show volume backscattering coefficient values (for altitude increments of 100 m) expressed in a logarithmic code. The parameter shown as a number against each time indication similarly describes the transmission measured through the cirrus layer. Input data were manually reduced from Polaroid photographs.

5.2.2. *Inhomogeneity in the "Clear" Air.* Variations in the turbidity of what appears to the eye to be clear air may readily be determined by lidar. Figure 6 shows a time/height cross section made at Menlo Park, California in 1967 by making a series of vertical lidar observations over an extended period. The data in the form SBF values (see Section 4.2) show the stratification clearly. The turbid air in the lower layer is separated from the overlying clean air at

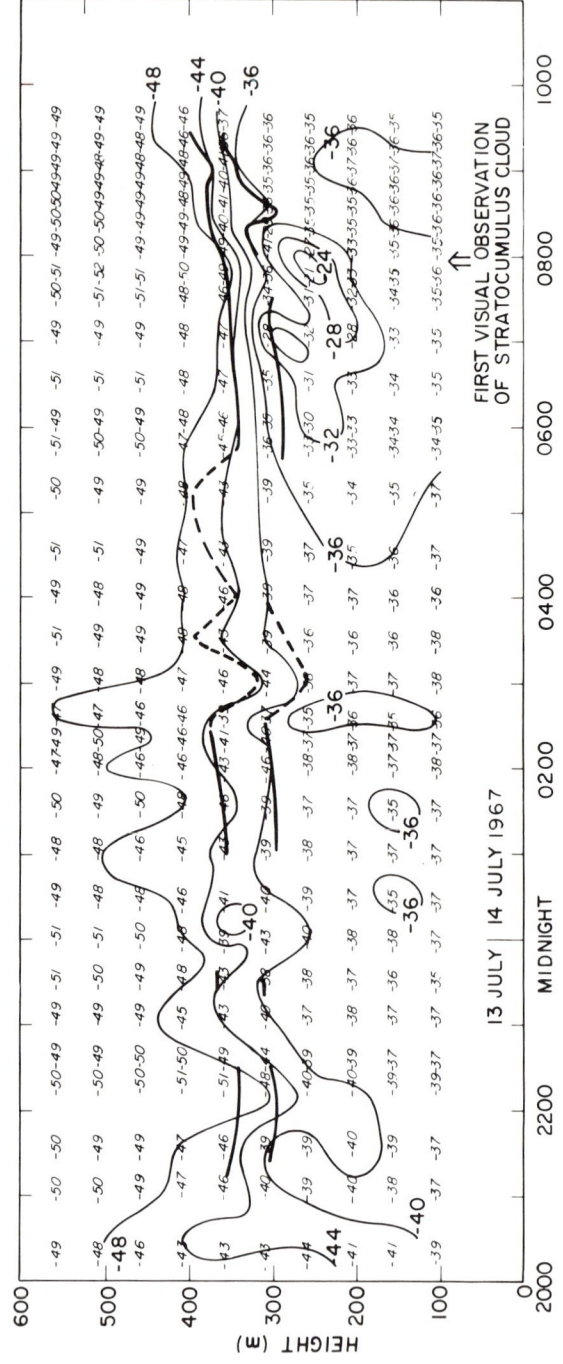

FIG. 6. Time section of the spatial backscatter function from Mk. V neodymium lidar observation at an elevation angle of 30°. Observations were made at intervals throughout the night at a fixed elevation of 30° at Menlo Park.

the level of the temperature inversion base (~300 m). The transition layer between the turbid and clean air is marked in the illustration by the pair of lines roughly parallel to the time axis. This layer was particularly well defined during the period where the lines are heavy. The diurnal effects were apparent as relative humidity increased before sunrise, at which time vestiges of visible stratus cloud were observed to form.

Such lidar observations clearly offer contributions in observing and monitoring the effects of thermal stratification in the atmosphere and possibly changes in its relative humidity. In addition, of course, remote quantitative observation may be made of the density of the particulate pollution loading and its changes with time [1].

At higher levels, i.e., in the stratosphere and mesosphere, a number of workers have reported the detection of particulate layers, some of which are claimed to be associated with noctilucent clouds. (See Fiocco and Smullin [10], Fiocco and Grams [11], McCormick et al. [18], Collis and Ligda [3], Kent et al. [14], and Sandford [24].)

5.2.3. Air Motion. If a suitable indicator or tracer material is injected into the atmosphere, lidar makes it possible to monitor its dispersal quantitatively and conveniently. For example, Fig. 7 shows how a cloud of insecticide released by a low-flying aircraft moved down a wooded hillside, under the influence of air drainage. This example from observations made in Idaho in 1966 in connection with U.S. Forest Service studies of insecticide application shows the position of the cloud (which was quite invisible to the eye) along a fixed line of sight just above the tree tops at successive intervals of time. The velocity of the flow can readily be evaluated. In this case, the cloud remained fairly compact, but in other drops made under different meteorological conditions, the cloud dispersed rapidly. In such cases, especially as studied in a subsequent program conducted in 1967 (Fig. 8), it was possible also to monitor the dispersal in the vertical, and by measuring changes in volume backscattering coefficient, to assess fallout and diffusion [4].

Another example of transport studies is illustrated in Fig. 9 which shows successive horizontal cross sections through a cloud of dust caused by an explosion in Montana in 1967 [23]. These sections were made initially by allowing the cloud to drift through the lidar beam at successive fixed headings and thereafter by scanning in the horizontal plane. Even at the time of the dense first section, the dust suspension was too tenuous to be visible to the eye.

Similar sections have been made of explosion clouds using an airborne lidar [5] and work is continuing at SRI on applications of this type. Hamilton [13] has also described lidar tracking observations of effluent from power station smoke stacks.

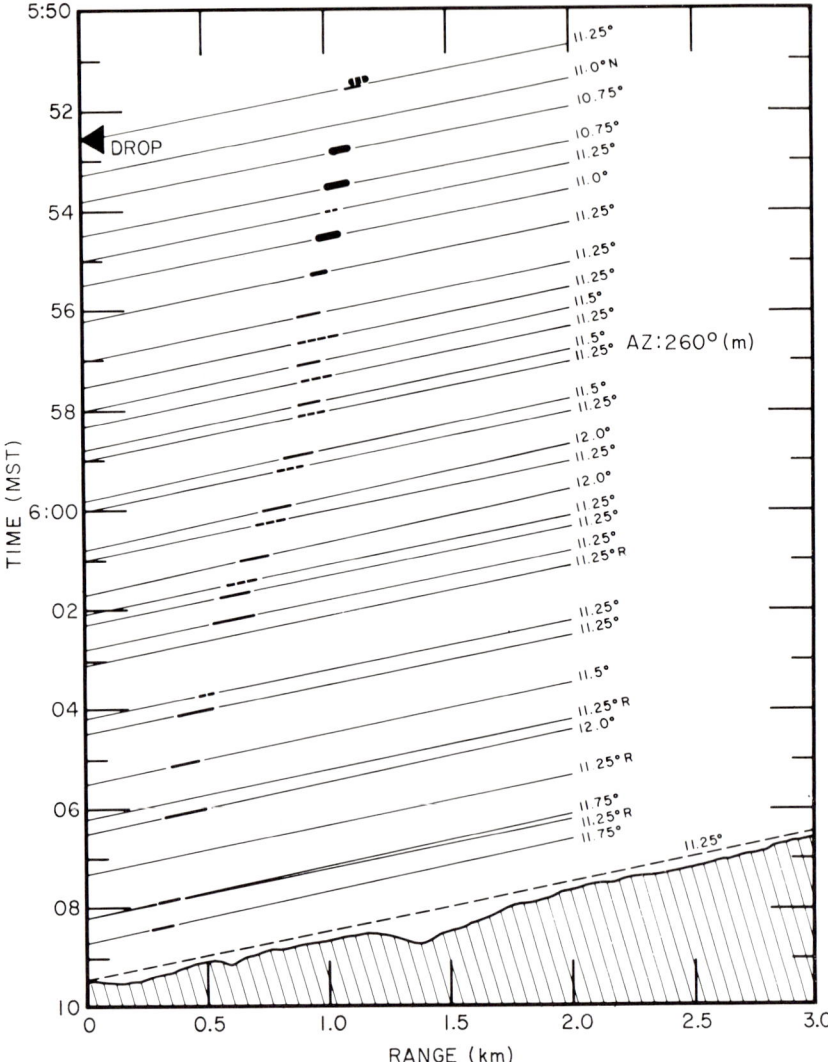

FIG. 7. Lidar observations of insecticide sprayed by aircraft, Idaho, June 1966. Two lidars were used, ruby and neodymium. The echoes (continuous and dashed marks respectively), were detected in the positions shown at about the level of the tree tops (indicated by a dashed line). The insecticide was dropped at approximately 0552 MST (in concentration of 1 qt/acre, droplet size of order 100 μ) from an aircraft flying about 60 m above the surface in a direction normal to the section represented. Note that the ordinate shows *time* and the diagram thus shows the results of lidar observations made at successive intervals as indicated. The insecticide was quite invisible to the eye.

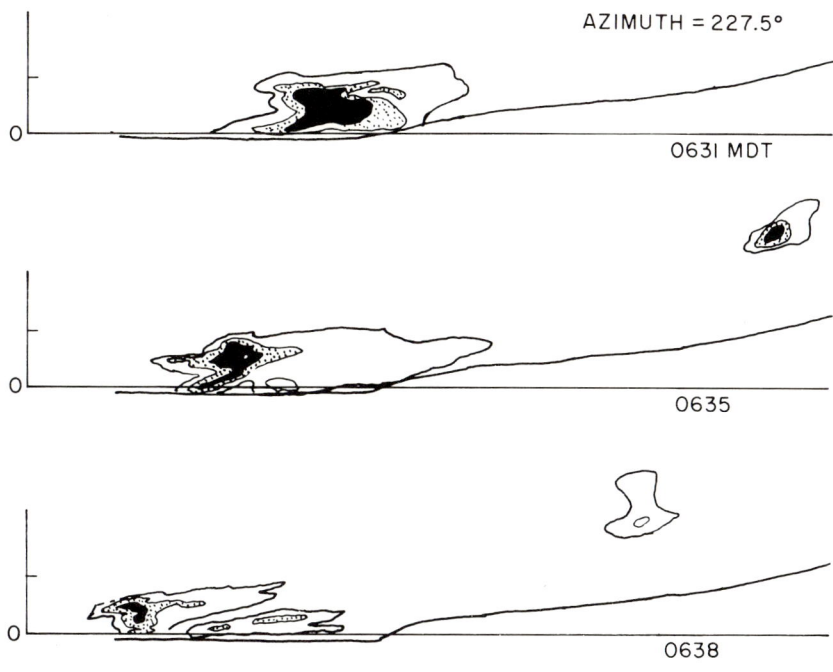

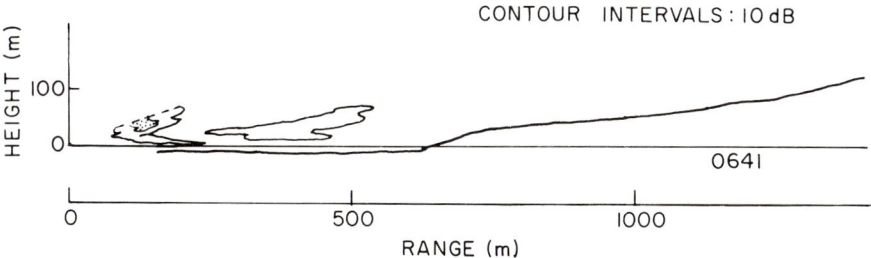

Fig. 8. Lidar observations of insecticide clouds, Idaho, June, 1967. In this case, a neodymium lidar was scanned in the vertical and observations were made at 1° intervals of elevation every 5 sec. (The insecticide was sprayed in a similar manner to that described in Fig. 7, but in this case the concentration was of the order of 0.5 pt/acre, with droplet sizes around 50 μ). (The small cloud on the right of the illustration was smoke, also trailed by an aircraft.) The successive cross sections (in which the internal structure is shown by relative backscattering coefficient isolines at 10 dB intervals) show the motion and rate of dispersal of the insecticide (which was quite invisible to the eye).

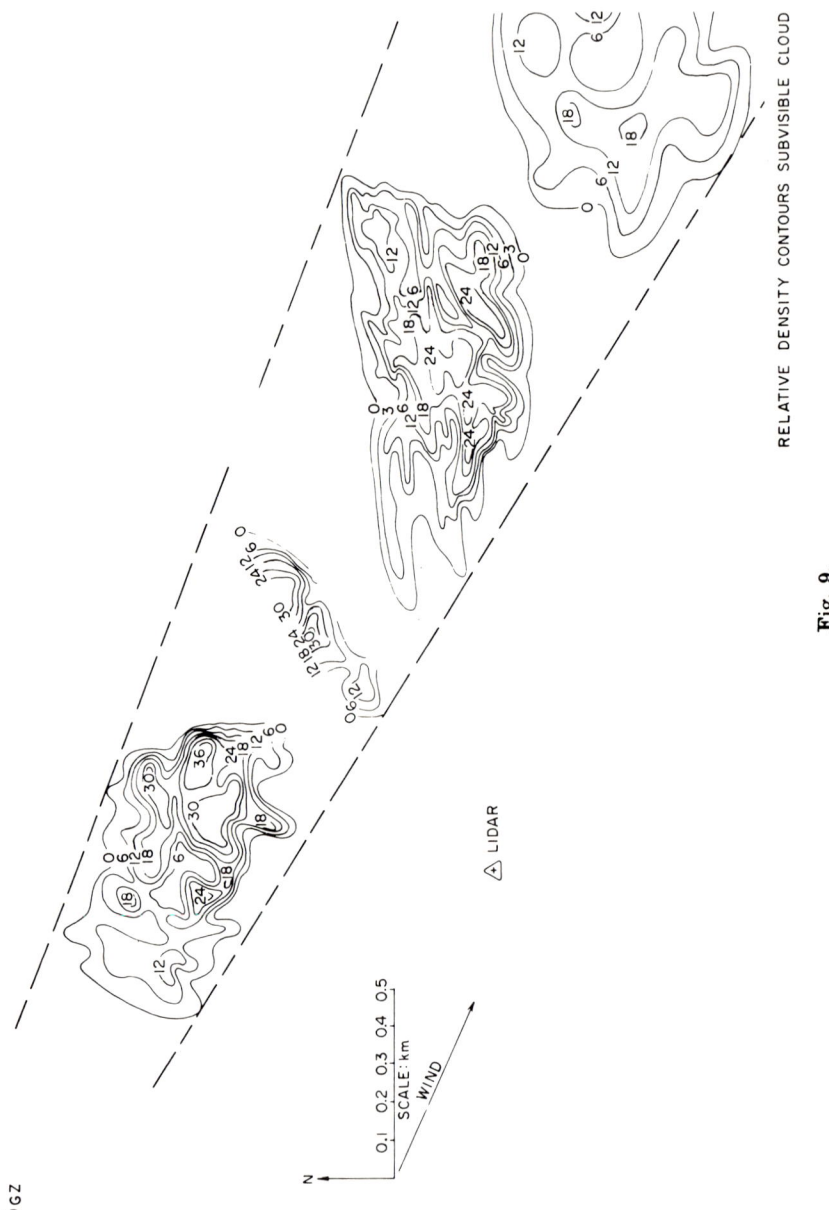

Fig. 9.

5.2.4. Fog and Low Cloud. A recent example of lidar observations of fog and low cloud is illustrated in Fig. 10. It is of particular significance because it demonstrates the important contribution lidar can make in an operational role. At Hamilton AFB, California, the landing approach path on a well used runway lies over the waters of San Francisco Bay and adjoining marshes. The conventional rotating beam ceilometer located near the touch-down point is only capable of monitoring cloud bases immediately overhead. Conditions at this point are frequently not representative of conditions along the approach path. In experimental observations made with an SRI ruby lidar, the nature of the cloud base was monitored out along the approach to distances of up to 2 km in conditions of fog and low ceiling (visibilities of the order of 1000 m). The illustration shows a typical cross section derived from a series of lidar observations scanning in the vertical. In addition to the delineation of the level of the diffuse cloud base (c. 200 m), computations of quantitative parameters related to "visibility" are shown over the section in question. Apart from the ability of lidar to observe cloud bases considerably displaced from its vicinity, this example illustrates the further potential of lidar for evaluating the important, but hitherto inaccessible operational parameter, "slant visibility."

6. Lidar Contributions to Atmospheric Studies and Meteorological Problems

It is important to recognize that the lidar technique applies broadly to a wide range of atmospheric and meteorological studies. We are dealing here with a *generic* rather than a *specific* technique.

The technique may be adapted and applied to a very diverse range of problems and it is hoped that these will be broadly evident from the above discussion and illustrations. The following seem particularly appropriate areas for lidar contributions.

6.1. General Research

(1) Structure of dust layers in upper atmosphere, noctilucent clouds, etc. (20 to 150 km), atmospheric density (50 to 80 km).

Fig. 9. Series of four horizontal cross sections showing approximate relative density distributions of subvisible dust cloud. Cross sections were made with a neodymium lidar. (Observations times are centered at 3.0, 4.0, 6.0, and 8.3 min after the explosion which took place at Ground Zero (GZ) as indicated.) Dust was caused by the explosion of 20 tons of nitromethane at Fort Peck, Montana, November, 1966. Even at the time of the first cross section, however, no dust could be seen by the eye.

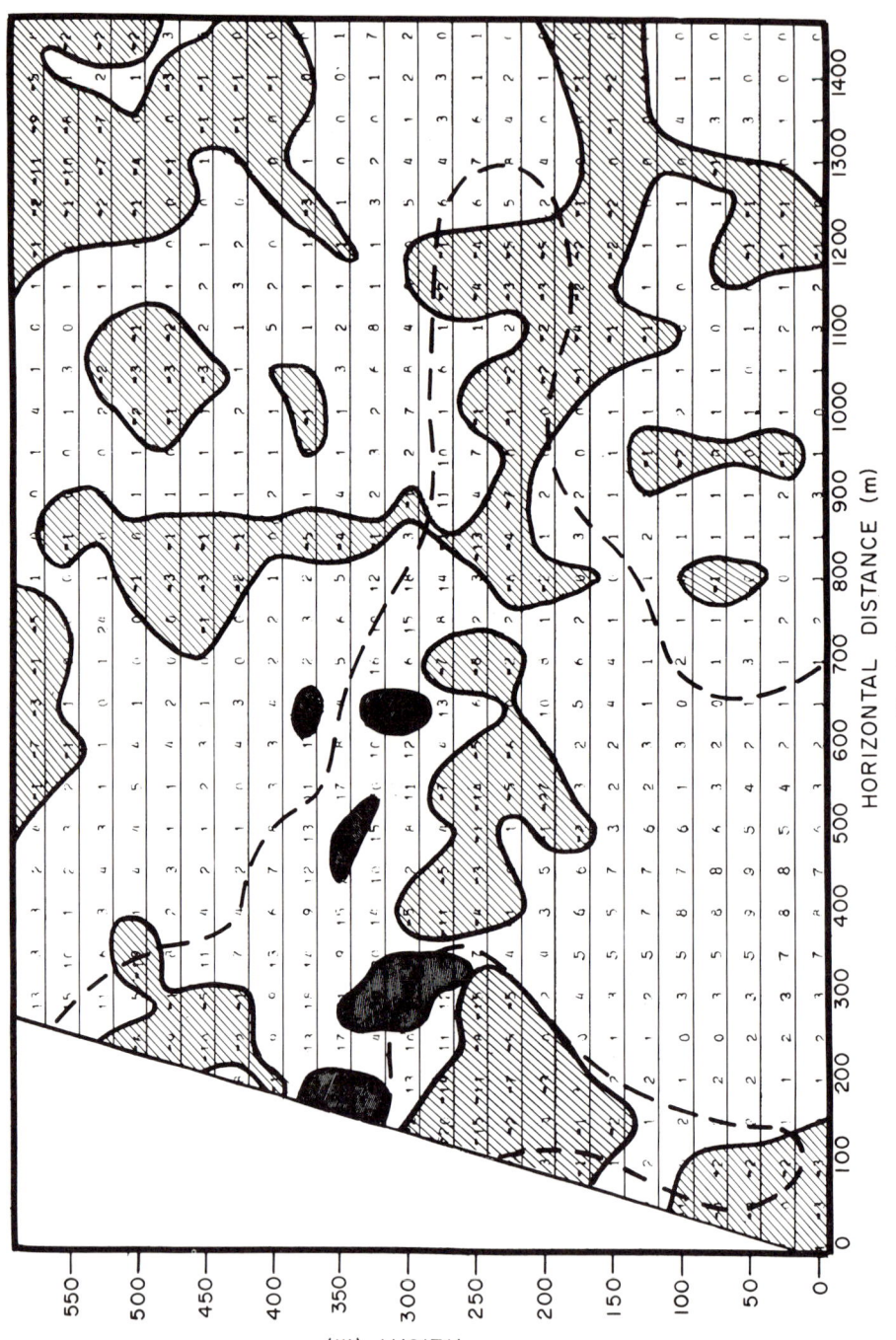

Fig. 10.

(2) Wave motion and turbulent air flow over orographic features and generally, as revealed by clouds and particulate inhomogeneities (at all levels up to say 15 km).

(3) Boundary layer structure (variations in low-level inversion levels, etc.), especially in relation to factors significant to air pollution in urban areas.

(4) Turbulent mixing and diffusion processes, using indicator materials.

(5) The effect of visible and subvisible cirrus clouds and other aerosol layers on radiative transfer of energy within and through the atmosphere.

6.2. Operational Applications

(1) Ceilometry.

(2) "Visibility" measurement, particularly over elevated slant paths for aircraft operations.

(3) Wind profile measurement (using rocket disseminated trails of tracer material).

(4) Tracking atmospheric pollutants from specific sources, e.g., insecticide spraying, nuclear tests, and smoke stacks.

7. FUTURE DEVELOPMENTS

Progress in the atmospheric and meteorological studies (Section 6) could undoubtedly be made with little or no further technological development. There is much room for progress in the technological basis of the lidar technique, however. In certain fundamental aspects, advances in laser energy generation, for example, progress will emerge as a result of new discoveries which can confidently be expected in this burgeoning field. In other aspects, many possibilities for progress are already readily apparent and achievable within the current state-of-the-art. The restrictive factors here are largely economic.

In the area of new developments, there is a need for higher repetition rate lasers providing higher *average* powers and higher data rates than are currently available. In this context, particularly for operational applications, eye safety considerations are important. Fortunately all requirements would be

---

FIG. 10. Lidar observations of low cloud and reduced visibility conditions, Hamilton AFB, January 8, 1968. This is an analysis based on interim evaluations of the extinction coefficient made by computer from manually entered data from a series of lidar observations. The parameter shown is σ (km^{-1}). Negative values show areas of rapidly increasing volume backscattering coefficient (i.e., dense cloud). Dotted line shows limit of area (i.e., within 700 m at the surface) of higher confidence in the data.

well met by a relatively low peak power with a high pulse repetition frequency, but it would be desirable in addition for such lasers to operate at wavelengths which are outside the visual spectrum.

The development of high pulse-rate lasers would lead to the development of systems capable of scanning in two dimensions to obtain nearly instantaneous atmospheric cross sections from stationary viewpoints, or more complete data from moving platforms such as aircraft or satellites. While such high-PRR systems would facilitate the development of graphical displays comparable to those used in weather radars, a more desirable development would be the input of such data to a computer for automatic *quantitative* processing and display. Techniques for handling data in this way are readily adaptable from currently available technology—but further progress must be made in developing techniques for recovering significant data from lidar observations.

Apart from the more obvious advantages of such data handling and presentation techniques, they open up the way to the powerful resources of modern information analysis procedures and the better coordination of lidar observations with other types of observations.

Acknowledgments

In preparing this review, I have drawn heavily upon the contributions of my colleagues at Stanford Research Institute and I am especially indebted to Dr. Warren Johnson, Dr. E. Uthe, and Mr. W. E. Evans for their assistance.

For work of others elsewhere I have relied mainly upon published descriptions of their work; hence my account is necessarily dated by the usual publication lag.

Appendix

Stanford Research Institute—Typical Lidar Characteristics

	Mark V	*Mark I*
	Transmitter	
Laser material	Neodymium-glass	Ruby
Wavelength (μ)	1.06	0.6943
Beamwidth (mrad)	0.2	0.5
Optics	6-in. Newtonian reflector	4-in. refractor
Peak power output (MW)	50	10
Pulse length (nsec)	12	24
Q Switch	Rotating prism	Saturable dye
Max. PRR (pulses/min)	12 (1967) 1–2 (1966)	1–2

	Mark V	*Mark I*
	Receiver	
Optics	6-in. Newtonian reflector	4-in. refractor
Field of view (mrad) (maximum)	3.0	2.0
Predetection filter wavelength interval (μ)	0.01	0.0017
Detector	RCA 7102 (S-1 cathode)	RCA 7265 (S-20 cathode)
Postdetection filter bandwidth (MHz)	30	30

LIST OF SYMBOLS

A	Effective receiver aperture
C	Scattering cross section for a single particle
N	Number density of scatterers per unit volume
P_r	Received power
P_t	Transmitted power
PRR	Pulse repetition frequency
Q	Quality factor in a resonant system
S	S-function (see Eq. (4.2), Section 4.2)
SBF	Spatial backscatter function (See Eq. (4.3), Section 4.2)
T_a	One way atmospheric transmission
MIE	Mie scattering
RAY	Rayleigh scattering
a	Particle or droplet radius
c	Velocity of light (Section 1)
c_1, c_2	Constants (Section 4).
k	Ratio of volume backscattering coefficient to extinction coefficient (may take form $\beta'_{180} = k_1 \sigma^{k_2}$)
r	Range
r_0	Reference range
α	Molecular polarizability of scatterer (Section 3.2); size parameter ($\alpha = 2\pi a/\lambda$) (Section 3.3)
β'_{180}	Volume backscattering coefficient
δ	Depolarization factor
η	Transform $\eta \equiv 1/\sigma$
σ	Extinction coefficient
σ_a	Absorption extinction coefficient
σ_s	Scattering extinction coefficient
τ	Pulse duration
T_β	Parameter describing transmission through cirrus cloud (see Fig. 5)

References

1. Barrett, E. W., and Ben-Dov, O. (1967). Application of lidar to air pollution measurements, *J. Appl. Meteorol.* **6**, 500.
2. Collis, R. T. H. (1965). Lidar observations of clouds. *Science* **144**, 978.
3. Collis, R. T. H., and Ligda, M. G. H. (1966). Note on lidar observations of particulate matter in the stratosphere. *J. Atmospheric Sci.* **23**, 255.
4. Collis, R. T. H., and Oblanas, J. W. (1967). Lidar observations of forest spraying operations. SRI Final Report, Contract 26-120, Forest Service, U.S. Dept. of Agriculture.
5. Collis, R. T. H., and Oblanas, J. W. (1968). Airborne lidar observations—Pre Gondola II. U.S. Army Eng. Nucl. Cratering Group Rept., PNE-1119.
6. Collis, R. T. H., Fernald, F. G., and Alder, J. (1968). Lidar observations of Sierra Wave conditions. *J. Appl. Meteorol.* **7**, 227.
7. Deirmendjian, D. (1964). Scattering and polarization properties of water clouds and hazes in the visible and infrared. *Appl. Opt.* **3**, 187.
8. Elterman, L. (1964). Atmospheric attenuation model, 1964, in the ultraviolet, visible and infrared regions for altitudes of 50 km. Environmental Res. Papers, No. 46. Air Force Cambridge Res. Lab. AFCRL-64-740.
8a. Elterman, L. (1968). UV, visible, and IR attenuation for altitudes to 50 km, 1968. Environmental Res. Papers, No. 285. Air Force Cambridge Res. Lab. AFCL-68-0153.
9. Fenn, R. W., (1966) Correlation between atmospheric backscattering and meteorological visual range. *Appl. Opt.* **5**, 293.
10. Fiocco, G., and Smullin, L. D. (1963). Detection of scattering layers in the upper atmosphere. *Nature* **199**, 1275.
11. Fiocco, G., and Grams, G. (1966). Observations of the upper atmosphere by optical radar in Alaska and Sweden during the summer 1964. *Tellus* **18**, 34.
12. Goyer, G. G., and Watson, R. (1963). The laser and its application to meteorology. *Bull. Am. Meteorol. Soc.* **44**, 564.
13. Hamilton, P. M. (1966). The use of lidar in air pollution studies. *Air and Water Pollut. J.* **10**, 427.
14. Kent, G. S., Clemesha, B. R., and Wright, R. W. (1967). High altitude atmospheric scattering of light from laser beam. *J. Atmospheric Terrest. Phys.* **29**, 169.
15. Lawrence, J. D., McCormick, M. P., Melfi, S. H., and Woodman, D. P. (1968). Laser backscatter correlation with turbulent regions of the atmosphere. *App. Phys. Letters* **12**, 72.
16. Ligda, M. G. H. (1963). Meteorological observations with pulsed laser radar, *Proc. Conf. Laser Technol., 1st. San Diego, 1963*, U.S. Navy, p. 63.
17. Long, R. K. (1963). Atmospheric attenuation of ruby lasers. *Proc. IEEE* **51**, 859.
18. McCormick, P. D., Poultney, S. K., Van Wijk, V., Allen, C. O., Bettinger, R. T., and Perschy, J. A. (1966). Backscattering from the upper atmosphere 75–160 km detected by optical radar. *Nature* **209**, 798.
19. Middleton, W. E. K. (1958). "Vision through the atmosphere." Univ. of Toronto Press, Toronto.
20. Middleton, W. E. K., and Spilhaus, A. F. (1953). "Meteorological Instruments." p. 208. Univ. of Toronto Press, Toronto.
21. Munick, R. J. (1965). Turbulent backscatter of light. *J. Opt. Soc. Am.* **55**, 893.
22. Northend, C. A., Honey, R. C., and Evans, W. E. (1966). Laser radar (lidar) for meteorological observations. *Rev. Sci. Instr.* **37**, 393.
23. Oblanas, J. W., and Collis, R. T. H. (1967). Lidar observations of the Pre Gondola I cloud. U. S. Army Engineer Nuclear Cratering Group, Rept. PNE-1100.

24. Sandford, M. C. W. (1967). Laser scatter measurements in the mesosphere and above. *J. Atmospheric Terrest. Phys.* **29**, 1657.
25. Schotland, R. M., Nathan, A. M., Chermack, E. A., Uthe, E. E. (1962). Optical sounding, Tech. Rept. 2. New York Univ. Rept., Contract DA-36-039-SC87299, U.S. Army E.R.D.L.
26. Twomey, S., and Howell, H. B. (1965). The relative merit of white and monochromatic light for determination of visibility by backscattering measurements. *Appl. Opt.* **4**, 501.
27. Twomey, S., and Howell, H. B. (1967). Some aspects of the optical estimation of microstructure in fog and cloud. *Appl. Opt.* **6**, 2125.
28. U.S. Government Printing Office, Washington, D.C. (1962). "U.S. Standard Atmosphere."
29. Van de Hulst, H. C. (1957). "Light Scattering by Small Particles," p. 82. Wiley, New York.

STRUCTURE OF BALL LIGHTNING*†

James R. Powell

Brookhaven National Laboratory, Upton, New York

and

David Finkelstein

Belfer Graduate School of Science, New York

and

Brookhaven National Laboratory, Upton, New York

	Page
1. Introduction	141
2. Earlier Observations and Theories	143
2.1. Observations of Ball Lightning	143
2.2. Thunderstorm Conditions	148
2.3. Critique of Ball Lightning Theories	149
3. Experimental Evidence	152
3.1. Experiments with Persistent Atmospheric Luminosities	152
3.2. Convective Mixing Experiments	166
4. New Analysis of Problem	167
4.1. A Model of Ball Lightning	167
4.2. Energy Input to Ball	168
4.3. Townsend Multiplication in Ball	174
4.4. Spatial and Temporal Variation of Currents in Ball	181
4.5. Electrohydrodynamic Forces	183
4.6. Stability of the Ball	184
5. Summary and Conclusions	185
List of Symbols	186
References	188

1. Introduction

Over the years there has accumulated an impressive number of reports of encounters with glowing balls of various sizes that drift through the air for seconds or minutes before softly and suddenly vanishing away or exploding. The common name for these events is ball lightning. The first scientist to observe ball lightning appears to have been G. W. Richman, who worked with M. V. Lomonosov at the St. Petersburg Academy (1753) [1]. Richman's experiment was not very different from Benjamin Franklin's. Richman led an

---

* This work was performed under the auspices of the U.S. Atomic Energy Commission and the Air Force Cambridge Research Laboratories.

† Dedicated to Georg-Wilhelm Richman (d. 1753).

ungrounded lightning rod to an electrometer in his laboratory. At the inquest, the engraver I. Sokolov reported that as a thunderstorm was approaching, he saw a pale blue fireball the size of a fist leave the rod and float silently through the air to Richman's face (Fig. 1). There was then a

Fig. 1. Death of Richman by ball lightning.

sound like a small cannon, at which time Sokolov lost consciousness. When he recovered there was an acrid smoke. Richman was not breathing, there was a red spot on Richman's forehead, and two holes in one of his shoes. Lomonosov wrote ". . . I see that Professor Richman was killed by lightning in precisely the same circumstances I was in at the very same time. I was then sitting at the indicator of the aerial electric force (i.e., the electrometer) . . . the sudden shock which brought death to Mr. Richman reduced and soon entirely removed the force from the rod. . . ."

A thorough investigation was ordered. According to Sokolov's deposition, "Outside the house many people saw how the fireball separated from the clouds and fell to the top of the apparatus." L. Euler wrote from Berlin: "This accident has dampened the courage of the local natural scientists who were engaged in the investigation of lightning phenomena, and they have broken off their studies." Publication of Lomonosov's work on atmospheric electricity was forbidden for some time. Richman's apparatus was removed. His death was attributed by the Academy to ordinary lightning, which was supposed to have struck the rod and been conducted through Richman to the ground. Eye-witness reports that contradicted this conclusion were attributed by the investigating committee to fear.

In this review we shall explore a consistent model of ball lightning, explaining the principal circumstances of Richman's death and Sokolov's testimony.

The outline of the work is as follows:

We give some sample ball lightning observations to emphasize the rather fantastic behavior often reported. This behavior has led to considerable doubt about the existence of ball lightning on the part of people who have not seen it; the resemblance to "flying saucers" is more often quoted to discredit the ball-lightning reports than to provide a natural explanation for some flying saucer stories. Some of the average properties of ball lightning inferred from larger numbers of reports are quoted and the question of existence is considered. It seems safe to say that ball lightning not only exists but *occurs about as often as regular lightning*, on the order of 10^7 ball lightnings per day over the entire earth, and has been observed by about 5% of the population.

We critically review some of the published theories and related experiments that have been used to explain ball lightning. These range from quantum mechanical to magnetohydrodynamic, but none seems able to explain in a consistent way the behavior commonly reported.

A critical parameter for any theory is the luminous efficiency of the model or the power input per lumen of visible light output. In many models, the low luminous efficiency of a ball of hot gas is assumed, the radiation being thermal in origin, and an input of thousands of watts being required to account for the bright appearance of the ball in daylight. We describe experiments showing that just oxygen and nitrogen permit discharges of anomalously high luminous efficiency and afterglow persistence electroluminescence, experiments dating back to Rayleigh's studies of "active nitrogen." The luminous efficiency and half-life are estimated. The presence of electroluminescent air in lightning channels is inferred from photographic records. We conclude that the most likely substance for the composition of ball lightning is electroluminescent air and that a power input of several hundred watts would maintain the brightness at observed levels beyond the normal half-life of 0.5 to 1 sec.

We then consider the question of providing this power. We study the electrohydrodynamic and molecular evolution of a ball of electroluminescent air from ordinary lightning or a glow discharge in the atmospheric electric field. It is estimated that above a certain critical field which lies between 1000 and 2000 V/cm, Townsend multiplication in the ball is ample to replenish the decay of metastable excitation, and that when the atmospheric field is downwards, the electrohydrodynamic forces on the ball are such as to account for much of its observed dynamic behavior. We infer that a ball lightning is a positively charged region of electroluminescent air maintained in its dynamic structure and molecular composition by an atmospheric electric field.

It would seem, after all, that ball lightning is a natural laboratory not of magnetohydrodynamics, as was once hoped, but of electrohydrodynamics and molecular physics.

2. Earlier Observations and Theories

2.1. Observations of Ball Lightning

Evidently eye-witness accounts of a single unreproducible observation cannot carry enough weight to establish the existence of a phenomenon so different from familiar gaseous discharges. The recent surveys of McNally [2] and Rayle [3] are very important in dispelling some of the doubt attached to the existence of ball lightning. Interrogating the people at an A.E.C. and a NASA center, respectively, by questionnaire, McNally and later Rayle found enough positive responses (5 to 10% of the persons questioned) to make a very strong case for the existence of ball lightning, and to determine some of its characteristics. Table I and Figs. 2 and 3, taken from Rayle, give a subjective estimate of the frequency of observation, size, and duration of hundreds of ball lightning observations.

No great change in the physical description of the phenomenon was forced by these surveys which are in general in accord with the individual reports of the past several centuries. Think of a glowing ball about the size of a head lasting for about five heartbeats without much change, and then going out, usually quietly but occasionally explosively, with ordinary lightning, either before or after, somewhere in the vicinity.

However, Rayle has pointed out one totally unexpected consequence of these surveys. He has tried to compare the frequency of occurrence of ball lightning and lightning strokes. To do this, he asked his population not only

TABLE I. Statistics of lightning observations [3].

Lightning	Times seen						Total number observing
	No answer or 0	1	2	3	4 to 8	>8	
Ordinary (impact)	1355	179	179		34	17	409
Ball	1584	111	34	6	29		180
Bead	1652	39	30	10	33		112

Observer location	Thunderstorm exposures per year				
	No answer	0 to 1	2 to 3	4 to 6	>6
Outdoors	22	746	710	185	101
Automobile	30	300	889	380	165

whether they had seen a ball lightning, but also whether they had even seen the actual ground point of striking of a lightning stroke. While the entire stroke is, of course, visible for miles, the chance factors determining whether the ground end of a lightning stroke is actually seen by anyone would seem to be similar to those determining whether a short-lived luminous ball near the ground is observed, and the ratio of the positive replies would give an estimate of the ratio of the occurrence frequencies. The surprising outcome of Rayle's survey is that there is not so much difference in frequency as might have been expected. About half (44%) as many reported seeing ball lightning

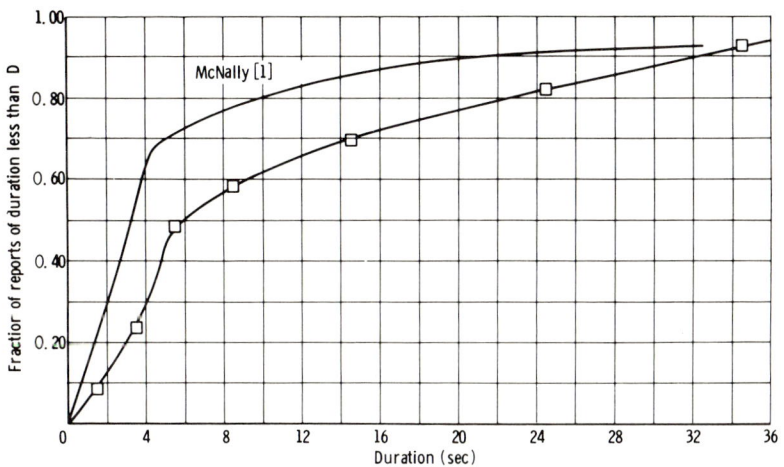

FIG. 2. Distribution of reported ball lightning durations [3].

as lightning stroke impact points, a ratio that might be entirely attributable to the greater brightness and loudness of the latter. We must either regard ordinary lightning strokes as rare or call ball lightning common. Since ordinary lightning strikes at least some 10 million times each day (on Earth), we prefer the second terminology. The point of this conclusion is that there is no longer much use in making up theories in which ball lightning is the product of a rare, freakish kind of lightning stroke. We had already noted that most ball lightning is associated with an ordinary lightning display. We must henceforth be prepared to entertain theories that assert the converse association as well.

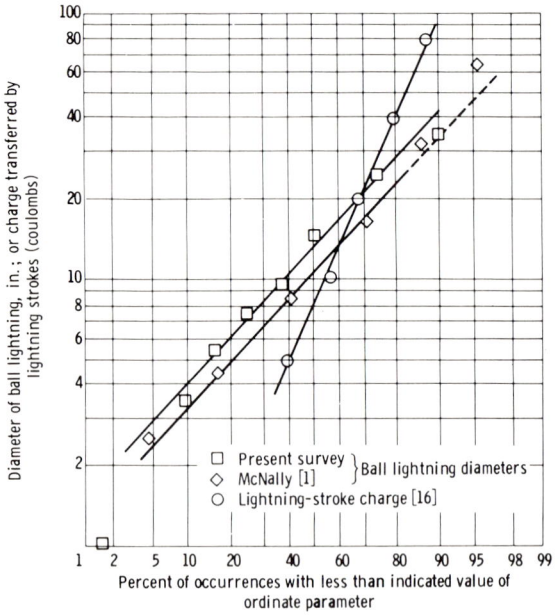

FIG. 3. Distribution of reported ball lightning diameters [3].

Following are some examples from the annals of ball lightning. Silberg [4] quotes W. Kohn's (Meteorological Observatory in Hamburg, 1952) account of a ball lightning witnessed by several people.

"During an electrical storm on the afternoon of July 27, 1952, several witnesses observed accurately the sudden appearance of a ball lightning in a closed room on the ground floor of my house.

"A few seconds after a lightning stroke in the neighborhood, we observed outside the window a brightly gleaming sphere the size of a fist which moved downward in short serpentine lines. Then this luminous ball penetrated through the closed window pane and entered our room. At a depth of about 1 m it performed a sudden turn of 90° parallel to the wall and continued floating another meter further into the room. Thereupon it burst, and the luminous sphere disappeared with a brief deafening explosion.

"This ball lightning was purplish with a reddish cast which persisted during the entire duration of the phenomenon. It lasted approximately 3 sec. No damage whatever was caused either on the inside or outside of the room. After the bursting of the luminous ball, we could perceive the typical odor which occurs in the case of electrical discharge."

Silberg also relates an observation by Mr. Fasold (Saxony, 1925) of a ball lightning which fell from the tip of a funnel attached to the storm cloud toward the ground and exploded when it hit: "The trees looked for a short time like Christmas trees, as if they had candles at the tips of the twigs."

An account related to Manwaring [5] describes the passage of a ball lightning, approximately 0.3 m in diameter, across a room. It appeared on one side of the room (on the second story) emerging through the openings in a wood screen, moved in a horizontal direction across the room at a height of about 1.5 m, and cut an almost perfect circular hole through a closed glass window (the hole measured 28 cm in diameter). It then exploded on the ground below the window. It was bluish-whitish in color with some motion discernible and had a fuzzy outline. The ball appeared a short time after a close stroke; on subsequent investigation, it was found that the stroke had hit a tree near the open window of the room adjacent to the room in which it was seen. To melt a hole in the glass would require tens of thousands of joules.

The last account illustrates that ball lightning can occasionally occur some distance from a stroke [6]. The ball lightning appeared suddenly on the tile floor of a kitchen in the form of a glowing yellow bubble about 10 cm in diameter which exploded 1 sec later, emitting a blinding light and sparks and filling the room with smoke smelling of burned powder. The only trace left by the ball was a white spot on the floor. At the time the ball lightning occurred, lightning struck a house 800 to 1000 m away and a bluish white luminous sphere was observed to travel along an overhead electric line 200 m away.

In Rayle's survey too, about 25% of the ball-lightning observers associated sound and odor with the ball. Considering that only 50% were within 50 ft of the ball, we may conclude that most balls are accompanied by sound and odor. The sound is described by Leonov [1] as a whistling, humming, or sputtering. Leonov comments that after disappearance a smoke is left, light brown in transmitted light and light blue in reflected.

We summarize what seem to us the important points of the above and other observations:

(1) Ball lightning usually occurs near the impact point of a stroke.
(2) The ball often moves downwards.
(3) It can penetrate closed windows or come through cracks.
(4) It is usually accompanied by a hissing sound.
(5) The median diameter is about 20 cm and the ball lasts for several seconds.
(6) Its color is quite variable.
(7) It often ends with an explosion.
(8) An odor or smoke, or both, are noticed after the ball dissipates.

We infer that ball lightning is a high electric-field phenomenon, that the sputtering sounds are electrical in origin, and that the "sulfurous" odors and varicolored smokes seen are those of oxides of nitrogen. These phenomenological conclusions figure importantly in the model we propose later.

2.2. Thunderstorm Conditions

Because ball lightning almost always accompanies a thunderstorm, and usually appears after and near the impact point of a stroke, it is necessary to examine the conditions present in thunderstorms and their relation to ball lightning.

The air temperature is most likely about 70°F and the relative humidity near 100% ($\approx 3\%$ H_2O vapor by volume). Rain may or may not be falling.

The thunderstorm has a bipolar structure with the negative pole closest to ground. There may be a smaller positive region below the negative one, but the predominant effect is to lower electrons to ground. The stroke is preceded by a stepped leader carrying several coulombs of negative charge; the main stroke has up to 200 KA of current in a channel ≈ 0.1 m in diameter and lasts for several tens of microseconds, lowering tens of coulombs.

The predischarge field at the earth may be much smaller than that corresponding to the charge stored in the lower negative pole because positive space charge released by sharp points on the ground cancels the electric field from the cloud. This space charge is contained in a layer several hundred meters thick above the ground and is released for several seconds before the flash. When the negative charge from the cloud is rapidly discharged to ground in the stroke, a positive step change in electric field results; frequently the electric field is then reversed, with the postdischarge field being positive (i.e., a positive ion would move towards the ground) because of this positive ion layer. It takes several seconds for these ions to reach the ground and for the field from them to decay. Wormell [7] noticed many examples of this; for example, one flash 0.8 km from his measuring point had a predischarge field of 60 V/cm negative, and postdischarge field of 400 V/cm positive, exponentially decaying with a time constant of about 1 sec. The field measurement is probably too small because of the long response time in the electrometer (0.5 sec) and possible brush discharge on the measuring plate. From measurements of charge lowered and the cloud electrical moment by Brook *et al.* [8] and their formula,

(2.1) $$\Delta E = 90[(2Q)/H^2]$$

one can compute the field change at the ground, assuming a vertical channel and point source of charge. A value of 500 V/cm is often found. This tends to confirm Wormell's observations, if the predischarge field is small because of almost complete neutralization by a positive space charge layer. The 400 to 500 V/cm average positive postdischarge field might be several times greater near the point of impact if there are inhomogeneities in the space charge layer and field multiplication at conducting structures. Such field multiplication is observed, and indeed is a prerequisite for St. Elmo's fire, where several thousand V/cm is needed. We thus think that postdischarge 1000 to 2000 V/cm positive fields lasting up to several seconds are possible in the region of lightning strokes.

The measurements of Brook *et al.* [8] have further importance. They observed that in 50% of lightning flashes there was at least one period of long continuing luminosity (defined as luminosity interval greater than 40 msec) and in 25% of flashes such a period occurred at the end of the flash. For example, their flash No. 113 had three discrete strokes that lowered a total of 13.8 C and a long continuing luminosity at the end which lowered 31.2 C. This was visible for 0.27 sec with an average current over 100 A. Longer continuing smaller currents are very likely. The authors note that sometimes portions of the channel remain visible long after the rest has disappeared. This is also shown in Young's [9] movies of bead lightning striking a water plume caused by a depth charge. The flash had three strokes, each stroke being of the long continuing luminosity type (average life several hundred msec). As the luminosity in each stroke faded, 10 to 15 ball-shaped luminosities about 30 cm in diameter persisted (some for 200 msec) along the old channel (Fig. 4). The beads reappeared at the same point after each stroke.

2.3. Critique of Ball Lightning Theories

We have abandoned as impractical the attempt to survey all proposed ball lightning theories and will consider here only some of the more prominent proposals. We do not consider those in which new laws are proposed solely to account for ball lightning.

The virial theorem is a consequence of energy-momentum continuity and provides an upper limit on the amount of energy E that can be stored in a ball of gas in the form of random or ordered molecular motion and electromagnetic field [10]:

(2.2) $$E \equiv E_{\text{kin}} + E_{\text{el}} + E_{\text{mag}} \leqslant pV$$

where p is the average pressure at the surface of the volume V and the E's are average energies. For nonrelativistic regimes, a more severe limitation applies. Chemical energy does not appear in this E and is not limited by the virial theorem. The virial theorem is most depressing to the hope, entertained by many including the authors, that ball lightning might be a natural instance

FIG. 4. Photograph of bead lightning [9].

of hydromagnetic self-confinement of a kind pertinent to controlled thermonuclear fusion, limiting hydromagnetically stored energies to levels orders of magnitudes lower than those sometimes inferred from ball lightning descriptions: in 1000 cm^3, $\leqslant$100 J. Moreover, even this modest energy cannot be stored in circulating electric currents in air for more than a few milliseconds, in consequence of ohmic dissipation at lower temperatures, thermal radiation at higher temperatures. We therefore regard hydromagnetic models of ball lightning, such as Stekol'nikov's [11] ring current, as inconsistent with basic laws or too short-lived. Neugebauer's model [12] which proposes exchange forces between air molecules at high density, appears to violate the correspondence principle. Even at room temperature, the density required for electron exchange to be significant would appear to be that of a solid, and at higher temperature still higher densities are called for.

Hill's [13] "miniature thundercloud" postulates separated negative and positive regions in a small ball, analogous to a large thundercloud. The stored energy in this electrohydrodynamic model is subject to the same virial inequality as the magnetohydrodynamic model ($\lesssim 100$ J in 1000 cm^3), and there seems to be no way to account for the continued separation of the charges in moist air.

Radioactive models have been considered, where unstable species of short lifetime are produced by nuclear reactions with a relativistic electron beam. Dauvillier [14] postulated C^{14}, which has too long a lifetime (5000 yr) to be a possible source, but there are positron omitting isotopes like N^{13} (10 min) and O^{15} (2 min) that have shorter lives. However, the necessary electron energy is several tens of MeV's, which is not possible with known electric field and gas density conditions in a stroke, is not consistent with spectroscopic measurements of the channel temperature (24,000°K), and does not account for explosive ball lightning.

Fehr [15] has attributed ball lightning to combustion in a 1 to 2% methane atmosphere, and he has produced fireballs under such conditions. It is very unlikely that such an atmosphere could be present except possibly over some marshes, whereas the vast majority of sightings have been reported over ordinary land, in houses, over water, etc.

There are various decomposition theories of O_3 [16] and NO or oxidation of NO (with the active species formed in the discharge); at the temperatures needed for adequate light emission, the reaction rates are too great for the lifetimes observed.

Kapitza [17] has proposed a focused rf discharge theory, and we report later on experiments that have generated such discharges. Field strengths of ≈ 1000 V/cm are required at several hundred megahertz at the focal point of the reflected rf waves, but in nature Horner [18] shows that only $\approx 10^{-6}$ V/cm (bandwidth 250 c/sec) are found. Silberg [4] attempted to answer this by postulating a discrete frequency spectrum which constructively reinforced at a particular point in space and time. However, this required millions of separate components with a precise regular frequency spacing, something that seems extremely unlikely. It must be concluded that the available energy input is too small by many orders of magnitude.

There remain three theories that seem compatible with basic laws and known energy sources. These are:

(a) Hot air—thermal and molecular dissociation energy
(b) Metastable molecular electronic excitation
(c) Nonlinear dc discharge

The simplest of all theories, that ball lightning is a sphere of hot gas left after a stroke, fits surprisingly well, if the numerous reports of ball lightning prior to any stroke are discounted. O_2 is almost completely dissociated at 4000°K, adding considerable energy (several thousand joules in a 20-cm ball). However, thermal convective mixing would disrupt the ball in less than a second, and it would rise upwards about 120 cm/sec, contravening observation.

The theory of metastable electron excitation of N_2 and O_2 proposed by Powell et al. [19] does not require as high temperatures for energy storage and visible light is emitted at lower temperatures (2000°K in experiments) but it still will dissipate in a second or so, and rise upwards about 120 cm/sec.

The nonlinear dc discharge theory of Finkelstein and Rubinstein [10] proposes a local channeling of cloud-to-ground electron leakage current through the conducting ball. It is heated more than its surroundings and thus is visible. Continuing energy input overcomes the convective dissipation problem but does not provide a force to keep the ball from rising, unless local field peculiarities keep the maximum field at a fixed point in space. If this happens, however, large power inputs (e.g., 10,000 W in a 20-cm ball) are necessary to replenish the energy lost by convective flow through the ball. (These estimates are based on buoyancy experiments described later.) As the calculations of Uman [20] show, this requires local current densities of 10^{-3} A/cm^2 compared with 10^{-9} A/cm^2, the upper limit of the average leakage current under the cloud. Such a current density would cause general heating in the surroundings of >600°C/sec, and a general breakdown of the region from cloud to ground should quickly occur.

It is seen from these cases that the problem of power is a serious one for many ball lightning models, and much hinges on the question of how much power it takes to make a ball of air glow brightly. We turn our attention to this question in the next section.

3. Experimental Evidence

3.1. Experiments with Persistent Atmospheric Luminosities

The phenomenon of "active nitrogen," an afterglow in nitrogen of unusual persistence under certain conditions, was studied by Rayleigh [21], who found that the dark-adapted eye could see this luminescence as much as half an hour after the discharge stopped. This is the recombination light of atomic N and is quite distinct from the molecular emission we describe now.

In 1965 Manwaring [5] produced long-lived luminosities in the open atmosphere using a 75 MHz radio-frequency (rf) arc. A glowing ball could be detached from the electrodes by separating them. The ball would then rise a few feet, lasting for about $\frac{1}{2}$ sec after rf excitation stopped.

Work was then undertaken at Brookhaven by Powell on these luminosities using a 30 kW, 75 MHz rf power source. The optimum configuration for producing the luminosities is found to be that shown in Fig. 5. It is essentially a 15-cm internal diameter glass cylinder (part of a large constant volume

FIG. 5. Apparatus for production of luminosities by rf discharge.

system) with an electrical feedthrough connected to the output terminal of the rf oscillator. Upon separation of part of the electrode an arc forms at the gap. When the movable electrode is fully withdrawn into its well, a long luminous plasma column forms at the tip of the fixed electrode (Fig. 6) and becomes in effect an antenna connected to the rf oscillator. At some point rf excitation is cut off and the glowing plasma column detaches, forming a ball-shaped luminosity at the top and remaining there until it dissipates (Fig. 7). The lifetime for the confined luminosity is about 0.5 to 1.0 sec after rf excitation ceases, and 0.2 to 0.4 sec for the unconfined case, probably because of greater convective mixing. Three parameters seem to have the greatest effect on the behavior of the long-lived luminosity: gas pressure, gas composition, and electrode composition. The effect of each has been studied separately, and the results can be summarized as follows:

(1) *Gas Pressure:* Long-lived electroluminosities are obtained at constant pressures between 0.5 and 3.0 times atmospheric. Below 0.5 atm, the discharges vanish quickly and seem more like conventional glow discharges. Three atm is a practical upper limit: The system cannot contain higher

Fig. 6. Plasma produced in rf discharge.

pressure. However, the luminosities are very similar to those at 1 atm. For experimental convenience, 1 atm was used in most tests.

(2) *Gas Composition:* N_2, air (various N_2/O_2 ratios including 20/1, 10/1, 5/1, 3/1, 1/1, and 1/2), O_2, and N_2O are able to produce these long-lived luminosities (NO was not tried). In argon and CO_2 the discharge is arc-like, not diffuse, and lasts only a few milliseconds after excitation cutoff. In general, the N_2 luminosity tends to be bluish and of relatively low luminosity, while O_2 is white and very luminous. As the percent of O_2 in the discharge increases, the luminosity becomes more like O_2. Ordinary air (20% O_2) tends to be

Fig. 7. Luminosity after end of rf discharge.

yellowish-whitish, of intermediate luminosity. The initial discharges in air are always more whitish, with a gradual change towards yellowish as NO_2 concentration in the discharge gas builds up. The N_2O luminosities are larger, in some cases over 50 cm in diameter, and almost pure orange. They are very easy to excite and last quite long, as much as 2 sec after rf cutoff. Here the chemical energy from N_2O decomposition (1 eV per molecule) is probably responsible for most light emission.

(3) *Electrode Composition:* The luminosities can be obtained with many different electrodes. Among those used successfully are Pt, Au, Ag, Cu, Zn, Cd, C, Sn, W, and Al. Mercury-coated and lead electrodes are not successful. In general, the less easily vaporized electrodes give the longest-lived luminosities, probably because electrode atoms introduced into the arc tend to destroy the energy storing capacity of the luminosity. The spectra of the luminosity depend on the electrode material, but this will be covered later.

Measurements of the output terminal rf voltage with an rf voltmeter give approximately 5000 V (peak) relative to ground; electron density measurements of the afterglow, using a hot cathode-cold anode probe, show about 3×10^{12} electrons/cm³ at 100 msec after rf excitation ceased in N_2. No measurements were obtained on air or O_2.

The lifetime of the ball can be extended by applied electric fields. In a resonant cavity at 75 MHz, Manwaring [5] has succeeded in keeping the ball floating in open air (Fig. 8) for 10 to 20 sec. The cavity is an 8-ft cube made from aluminum plates and is resonant in the TM_{110} mode. The electric field inside the cavity is calculated to be ≈ 1000 V/cm, from the dimensions and power input and assuming ideal resonance.

This "ball" is really rod shaped, ≈ 5 to 8 cm in diameter and ≈ 60 cm long, often sinuous. Higher frequencies would probably produce more of a ball shape. The ball sometimes disappears by explosion.

A calibrated photodiode has been used to obtain total radiant power from the luminosity (≈ 15 cm diameter) as a function of time after rf excitation ceased.

The photodiode[1] has zero sensitivity below 0.4 μ. Sensitivity is almost linear with wavelength between this maximum and the cut-off limits. In most cases the instrument has been used to measure only the total radiant emission in the range 0.4 to 1.1 μ. In a few cases, a rotating shutter with a Kodak 88A filter measured first total radiation, 0.4 to 1.1 μ, and then, the infrared radiation, 0.72 to 1.1 μ. The shutter speed was rapid enough so that a complete measurement was taken every 10 msec. Since the luminosity lasted hundreds of milliseconds, simultaneous curves of infrared power, 0.72 to 1.1 μ, and visible, 0.4 to 0.72 μ, could be derived.

---

[1] Edgerton, Germeshausen, and Grier, Inc., "Lite-Mike."

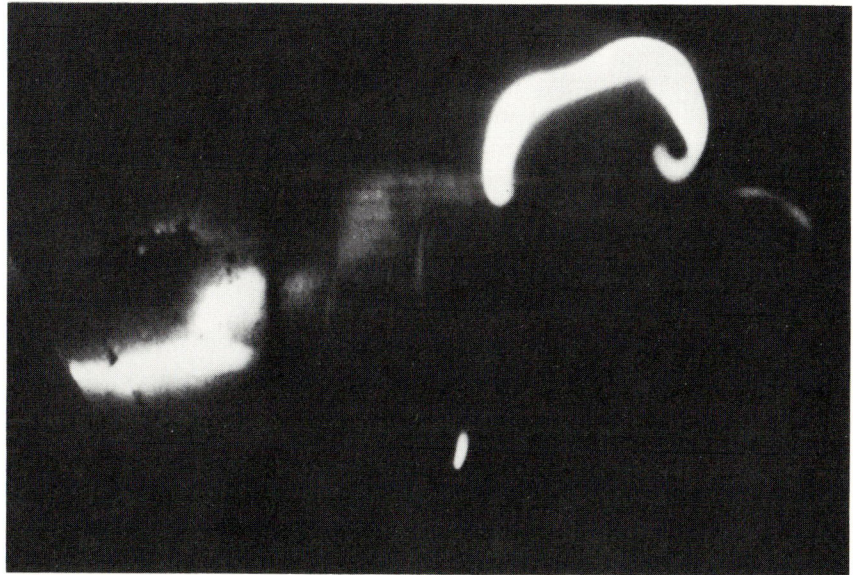

Fig. 8. Floating rf discharge ("ball lightning") in continually excited 75 MHz cavity.

Some results for air and O_2 with platinum electrodes are shown in Figs. 9 and 10. In some cases the decay is not monotonic, perhaps because spatial inhomogeneities exist in the discharge and a more luminous region may arise into the viewing some time after the rf is turned off. However, the power

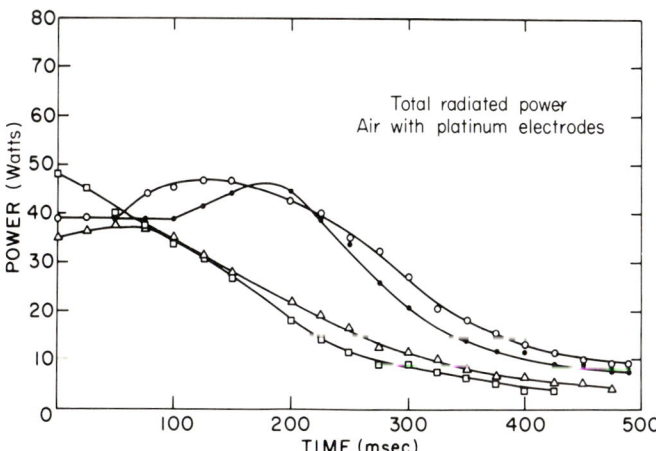

Fig. 9. Total radiant power as a function of time after the end of rf discharge—air luminosity.

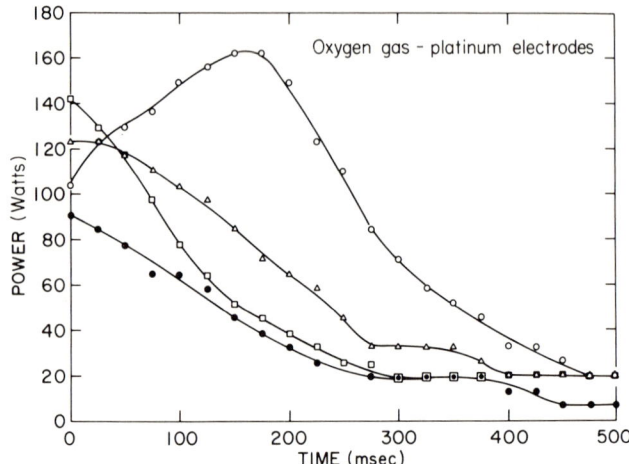

Fig. 10. Total radiant power as a function of time after the end of rf discharge—O_2 luminosity.

decay curve is approximately exponential in many cases, with a time constant of 200 to 300 msec. The O_2 discharge is the brightest, and it hurts the eyes to look at it. It thus appears that only 100 W of radiant power is sufficient for a quite bright lightning ball.

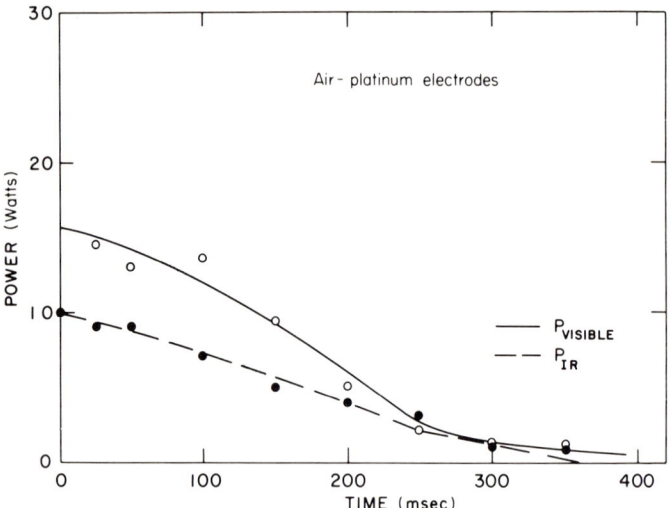

Fig. 11. Visible and infrared radiant power as a function of time after end of rf discharge—air luminosity.

Visible and infrared power for air with platinum electrodes is shown in Fig. 11. Visible radiance (0.4 to 0.72 μ) is greater than infrared (0.72 to 1.1 μ).

The total radiant power is integrated with respect to time to give stored energy at $t = 0$. Some results are given in Table II. Knowing the dimensions of the luminosity and assuming an energy level for the excited species, one can calculate the concentration of such species, which range from 10^{15} to $10^{16}/cm^3$, necessary to supply the radiant energy. Heat transfer calculations indicate that energy loss by convection to the cold wall of the container is approximately 10 times greater than the radiant emission, so that the equivalent excited species concentration is between 10^{16} and $10^{17}/cm^3$. Thus, depending on the gas and electrode, between 1 and 10% of the molecules are excited. In nature, with considerably more input energy available, more energy might be stored in the ball.

TABLE II. Radiant power measurements.[a]

Gas and electrode	Run number	Exponential decay time (msec)	Total energy (joules)	Energy density (J/cm^3)	Metastable concentration (molecules/cm^3)
N$_2$ platinum	4A	50	1.9	1.1×10^{-3}	1.2×10^{15}
	7A	300	5.4	3.1	3.2
	12A	50	1.7	1.0	1.0
	22A	50	3.3	1.9	2.0
Air platinum	5B	300	14.0	7.9×10^{-3}	12.4×10^{15}
	6B	300	15.3	8.7	13.6
	1C	200	8.9	5.0	7.8
	8C	250	9.6	5.4	8.5
O$_2$ platinum	8A	200	15.1	8.5×10^{-3}	1.8×10^{16}
	9A	150	20.4	11.5	2.4
	13A	250	25.8	14.5	3.0
	14A	300	41.9	23.7	5.0
Air copper	6B	200	7.1	4.0×10^{-3}	6.3×10^{15}
	10B	300	5.9	3.3	5.2
	13B	300	5.8	3.3	5.2
	15B	250	4.4	2.5	3.9
N$_2$ copper	2A	250	6.0	3.4×10^{-3}	3.6×10^{15}
	9A	50	2.8	1.6	1.7
	10A	50	3.4	2.0	2.0
	14A	50	2.9	1.6	1.7

[a] Assumptions: plasma equivalent to 6-in.-diameter sphere, average N$_2$* metastable energy = 6 eV, average O$_2$* metastable energy = 3 eV, average air* metastable energy = 4 eV.

Temperatures of the luminosities have been measured at about 2000 to 2500°K by 75-μ tungsten wires (resistance vs. temperature method).

The spectrograph is a 0.5-m Czerny–Turner[2] mount, f 6.2 aperture, and 20 Å/mm dispersion. Kodak 2547 sheet film (5 × 7 in.) is used, with a wavelength coverage of 2500 Å at any given exposure. The wavelength limits are about 2500 Å at the short wavelength end because of air absorption and film sensitivity and 7000 Å at the long wavelength end because of film sensitivity. Single and multiple exposures are taken. The ASA number of the film is approximately 2000, and the spectrograph slit views the luminosity through a 6-in.-diameter quartz window. The luminosity completely fills the accept-

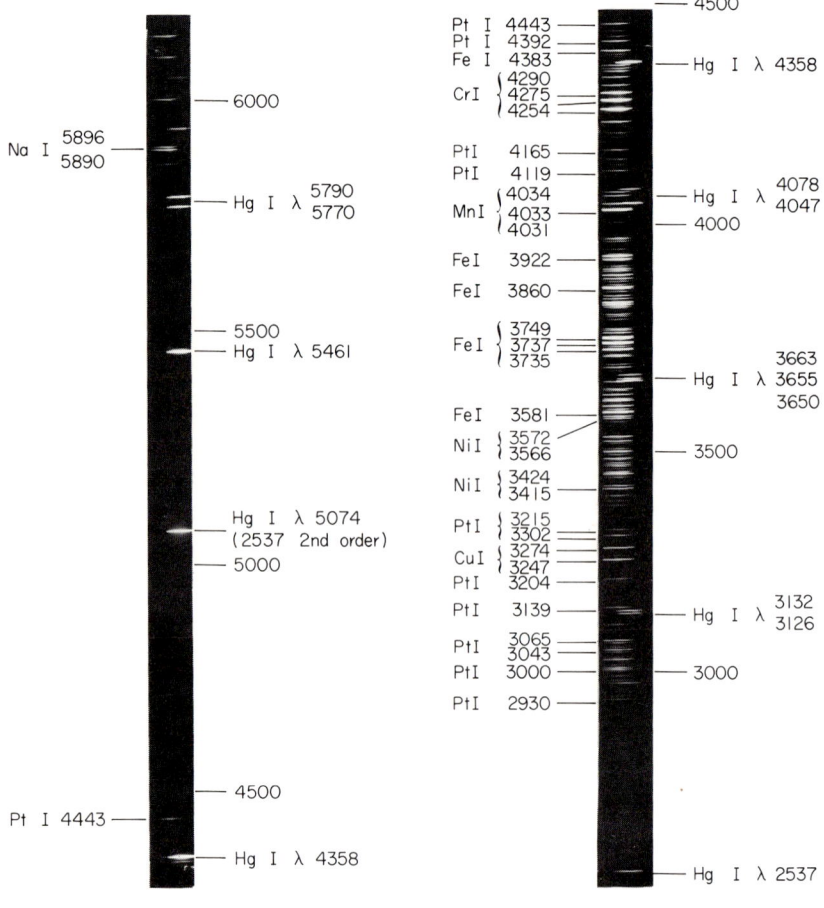

FIG. 12. Spectrum of N_2 afterglow luminosity.

[2] Jarrell–Ash Model 7500.

ance cone of the slit and collimating mirror. Typical spectra for N_2 and air are shown in Figs. 12 and 13. Here the spectrograph shutter opens 85 msec after the rf excitation ceases, and is kept open for 1 sec. This effectively integrates on the plate all light from the luminosity from time of opening until time of extinction. Background light is too faint to affect the spectra.

The spectra from the N_2 luminosity consist solely of atomic lines of impurities introduced from the electrodes. No bands, either from metastable species or the usual first or second positive bands, are observed. In some spectra over 300 lines are identified. Such a spectral analysis is shown in Fig. 14. In general, the average energy of the upper excited level of the emitting species is about 4 eV, and no lines are seen originating from levels above

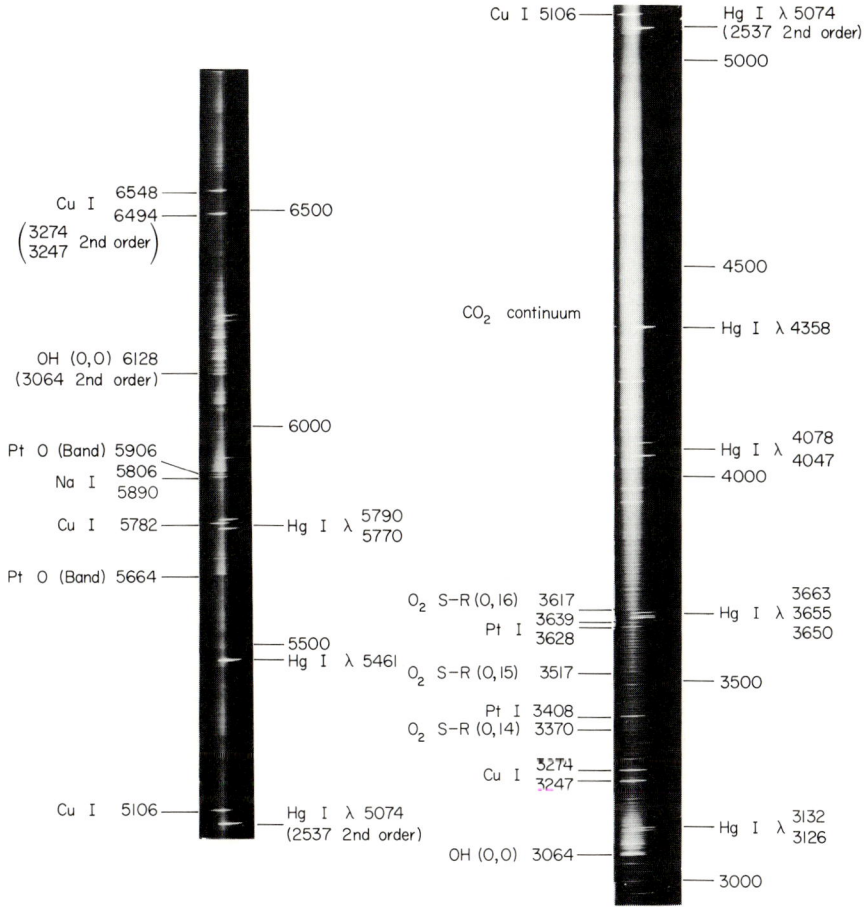

Fig. 13. Spectrum of air afterglow luminosity.

about 5 eV, even though these levels can be populated in the discharge when excitation is present (Fig. 15). CN bands from organic impurities are sometimes observed. Additional sets of spectra are taken, opening the spectrograph slit approximately either 20 or 200 msec after excitation stopped. The spectra are very similar to those at 85 msec, except in intensity. The intensity decreases with time (measured from $t = 0$ at end of excitation).

The spectra from the air luminosity are quite different from those for N_2. They consist mostly of bands with very few lines (Fig. 13), of which none

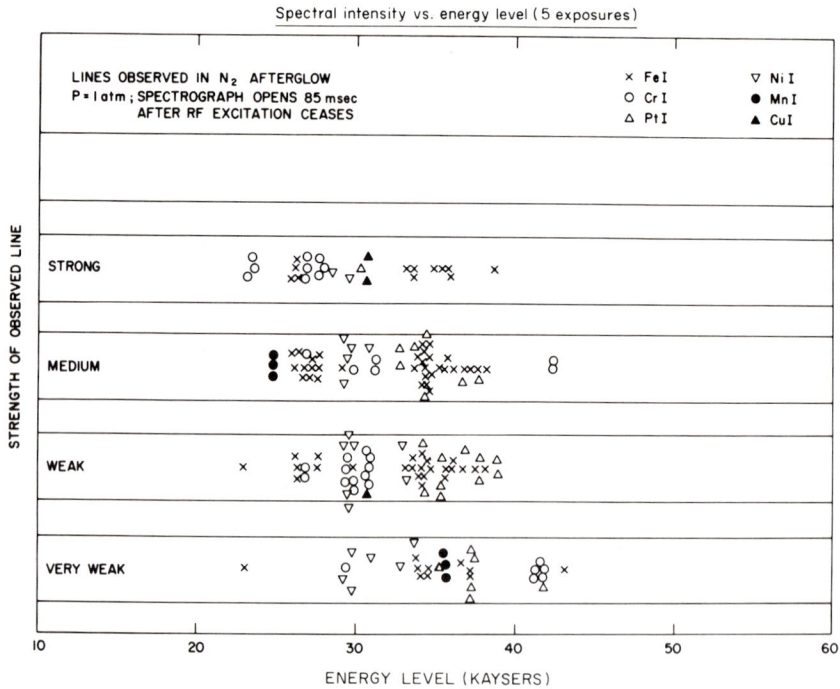

FIG. 14. Spectral Analysis of N_2 afterglow—rf Off.

have an upper excited level greater than 4.0 eV. Prominent bands are: Schumann–Runge of O_2 (0–14; 0–15; 0–16; 0–17); OH bands (0–0); PtO bands. But by far the most emission is from the

$$CO + O \rightarrow CO_2 + h\nu$$

continuum; it remains quite strong down to 3500 Å. With "synthetic" air (80% N_2 + 20% O_2 with no H_2O), no OH bands are observed, but the CO_2 emission and the Schumann–Runge emission are unchanged. The CO_2 probably comes from organic impurities in the system, principally O-rings touched by the discharge. The spectra for pure O_2 are essentially the same as for air. Probably there is some

$$NO + O \rightarrow NO_2 + h\nu$$

emission in air, judging from color photographs, but its spectrum is superimposed on that of CO_2, and identification is very difficult.

The light emission seems to be nonthermal in origin. Our temperature measurements of 2000 to 2500°K are not consistent with the $\approx 4000°K$ re-

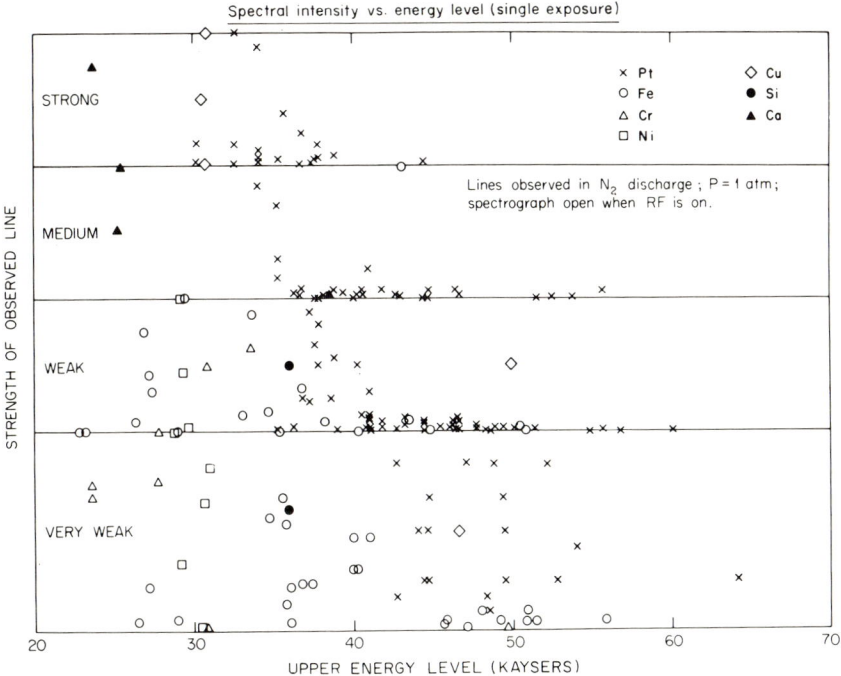

FIG. 15. Spectral analysis of N_2 discharge—rf On.

quired for thermal emission in air [22] (Fig. 16). Also, for the air luminosities, the spectrum indicates that most emission comes from CO_2. Assuming as an upper limit to CO_2 concentration a value of 1% (CO_2 concentrations of this level would be only associated with severe damage to the O-ring, which is not observed), it can be shown [23] that temperatures of 6000°K would be required if the emission were of thermal origin.

Another possible source is chemiluminescence from the recombination reaction

$$CO + O \rightarrow CO_2 + h\nu$$

However, at the pressure and temperature of the experiment, quantum yields are only 0.3% of the total recombinations, requiring impossibly high initial concentrations. Further, the light emission rate constants [24] are too small by several orders of magnitude to account for the observed emission.

The only possible explanation seems to be that a large amount of energy is stored in metastable electronic states of N_2 and O_2 with subsequent exci-

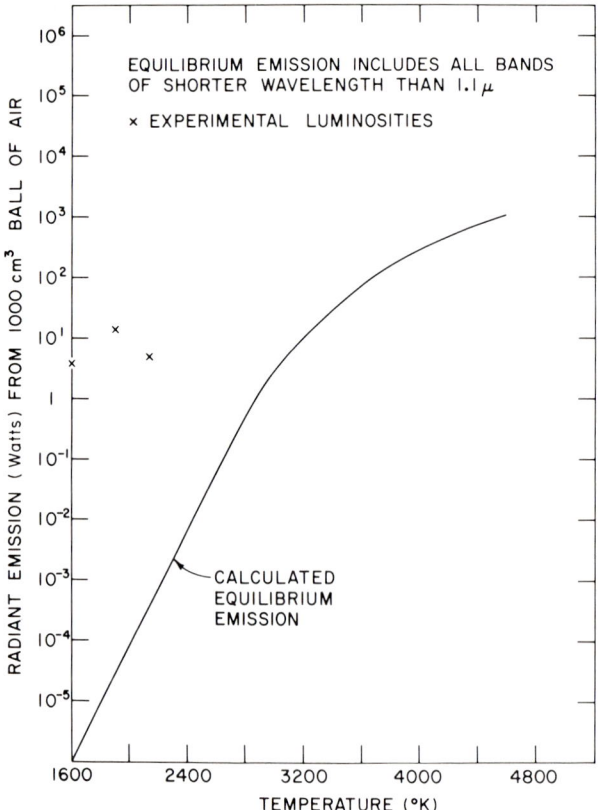

FIG. 16. Calculated thermal radiant power emission from air as a function of temperature.

tation transfer to species that radiate. This may correctly be termed electroluminescence. Table III lists metastable species with their radiative and collisional lifetimes. The results in the N_2 luminosities seem to be explained by the presence of the $A^3\Sigma_u^+$, $w^1\Delta_u$, or $^3\Delta_u$ states or more than one of these. Recombination rates for atoms are too rapid for the observed lifetime of the luminosity, and vibrational excitation of N_2 decays in a few milliseconds.

TABLE III. Characteristics of metastable molecular species.

Species	Energy[a] (eV)	Radiative lifetime (sec)	Collisional lifetime[g] (sec at $T = 2000°K$, $p = 1$ atm)	Quenching partner
$N_2(A^3\Sigma_u^+)$	6.17	13[b]	0.66[h]	N_2
$N_2(A^3\Sigma_u^+)$	6.17	13[b]	2.5×10^{-9} [i]	Air
$N_2(^3\Delta_u)$	7.35	1[c]	Unknown	
$N_2(w^1\Delta_u)$	8.89	1[c]	Unknown	
$O_2(A^3\Sigma_u^+)$	4.43	10[d]	Thermally dissociated	
$O_2(b^1\Sigma_g^+)$	1.63	8[e]	4×10^{-7} [j]	Air
$O_2(a^1\Delta_g)$	0.98	45 min[f]	1[k]	Air

[a] Bond, J. W., Watson, K. M., and Welch, J. A., (1965). "Atomic Theory of Gas Dynamics," Chapter 4. Addison-Wesley, Reading, Massachusetts.
[b] Wentwink, T., and Isaacson, L. (1967). *J. Chem. Phys.* **46**, 822.
[c] Kenty [25].
[d] Herzberg, G. (1950). "Molecular Spectra and Molecular Structure," Vol. I, Spectra of Diatomic Molecules. Van Nostrand, Princeton, New Jersey.
[e] Childs, W. H. J., and Mecke, R. (1931). *Z. Phys.* **68**, 344.
[f] Badger, R. M., Wright, A. C., and Whitlock, R. F. (1965). *J. Chem. Phys.* **43**, 4345.
[g] Collisional lifetimes were calculated using rate constants given in references and molecular densities at $T = 2000°K$ and $p = 1$ atm.
[h] Noxon, J. F. (1962). *J. Chem. Phys.* **36**, 926.
[i] Hunten, D. M., and McElroy, M. B. (1966). *Rev. Geophys.* **4**, 303.
[j] Young and Black [26]. This is not consistent with Hornbeck's [27] observations in CO_2-O_2 explosions.
[k] Vallence Jones, A., and Gattinger, R. (1963). *Planetary Space Sci.* **11**, 961.

Kenty [25] has observed afterglows in N_2–rare gas mixtures that persist for 10 sec and are spectroscopically distinct from the N-atom recombination spectrum (Lewis-Rayleigh). He attributes these to the metastable species $w^1\Delta_u$ and $^3\Delta_u$ because the Vegard–Kaplan bands from $A^3\Sigma_u^+$ are too weak.

In air, the collisional destruction rate of the N_2 states is probably too great to permit any appreciable concentration. Also, increasing the concentration of O_2 stores more energy—pure O_2 afterglows are several times as luminous as air. We conclude that some state or states in O_2 are responsible for the major energy storage. The $A^3\Sigma_u^+$ state has too shallow a potential well (0.6 eV) to prevent rapid dissociation at $2000°K$. The only other possibilities are the $b^1\Sigma_g^+$ and the $a^1\Delta_g$ states. Young [26] has indirectly measured the collisional deactivation rate of $b^1\Sigma_g^+$, which seems fast enough to eliminate it. However, this measurement is not consistent with Hornbeck's [27] spectral observations of very strong $b^1\Sigma_g^+$ bands in CO_2-O_2 explosion flames at several atmospheres. Using Young's collisional deactivation rate coefficients, the lifetime of a nonequilibrium concentration of $b^1\Sigma_g^+$ would be less than a microsecond, far shorter than the time required for appreciable radiation (radiative lifetime ≈ 10 sec). An equilibrium concentration of $b^1\Sigma_g{}^+$ would not produce appreciable radiation. Hornbeck postulates the excitation transfer reaction

(3.1) $$CO_2{}^*({}^3B_2) + O_2({}^3\Sigma_u{}^-) \to CO_2({}^1\Sigma_g{}^+) + O_2{}^*({}^1\Sigma_g{}^+)$$

The triplet CO_2 state is metastable of long lifetime. As Hornbeck's conditions are closest to ours, we accept his results as applicable, and consider $b^1\Sigma_g{}^+$ as an energy carrier. The reverse of reaction (3.1) can keep a high nonthermal concentration of $CO_2{}^*({}^3B_2)$. This occasionally undergoes a radiationless transition to $CO_2{}^*({}^1B_2)$ which can then radiate the CO_2 continuum via an allowed transition. Storage of energy as $b^1\Sigma_g{}^+$ also explains the experiments of Leah et al. [28, 29] with CO_2-O_2 and CO_2-air explosion flames in which they found large (10 to 20%) latent nonthermal energies. This energy persisted for several seconds before being lost. Interestingly, high latent energies were also associated with high O_2 concentrations; when O_2 concentration was low, the latent energy was very small. Also, high latent energies correlated with Na reversal temperatures much higher than the flame temperature. These again seem to favor storage as molecular excitation in O_2.

In effect, the radiant emission is governed not by the thermal temperature of the ball, but by an excitation temperature characteristic of the actual nonequilibrium concentrations of $b^1\Sigma_g{}^+$ and $a^1\Delta_g$ in O_2; that is, a temperature several thousands of degrees hotter than the thermal kinetic temperature.

The $a^1\Delta_g$ can probably play a similar role to $b^1\Sigma_g{}^+$, and it is even more stable collisionally and radiatively, but knowledge of its detailed reactions must await further work.

These metastable states can be formed either by direct electron impact excitation or by recombination of free atoms. It has been previously shown in electrical discharge experiments [30] that more than 10% of the O_2 molecules can be excited by a combination of these mechanisms.

We conclude that the experimental evidence indicates substantial metastable excitation in the luminosities. We suppose that the long-lived luminosity seen in a lightning channel by Young is due to such excitation. Such excitation is present in our model of ball lightning.

3.2. Convective Mixing Experiments

Some experiments have been made to estimate convective-mixing rates in ball lightning using helium filled soap bubbles of approximately the same size as the ball. The soap film is of negligible weight as shown by a very slow descent rate of air bubbles. A helium bubble, therefore, closely approximates a ball of average temperature 2200°K. The terminal upward velocity of such bubbles is found to be ≈ 120 cm/sec for bubble diameters between 20 and 35 cm. Larger bubbles oscillated more from prolate to oblate spheres, thus tending to keep the velocity constant. One interesting experiment has been performed with He filled bubbles containing NH_4Cl smoke. While the film is intact, these rise at the same rate and when they break an expanding smoke ring is left. This ring quickly expands to about 60 cm in diameter in 2 sec. Evidently the air flow around the gas ball induces spherical vortex flow inside the gas ball. Once the gas is free to expand, it quickly does so. If the bubble breaks before it starts to rise, there is no smoke ring, but just a dissipation of the smoke in about a second.

These experiments show that convective mixing of the ball is prohibitively rapid if there is no restraining force; no theory that invokes stored energy could explain lifetimes of several seconds. Further, if the ball is generated at the stationary focus of some dc or rf electric field, the flow velocity of heated air out of the ball is approximately the terminal velocity of the equivalent density He bubbles, i.e., 120 cm/sec. In a 20-cm-diameter ball lightning of temperature 3000°K this would require energy input rates of $\approx 10,000$ W.

We can thus conclude that there must be a mechanism restraining convective mixing. We return to this question later.

4. New Analysis of the Problem

4.1. A Model of Ball Lightning

We now consider the evolution of a ball of electroluminescent air in the terrestrial electrical field, on the basis of the information provided by the experiments we have described, the cited characteristics of the postdischarge field, and the classical laws of gaseous conduction. No *ad hoc* hypotheses are introduced for this study. We suppose the ball is left by a stroke, either cloud-to-ground or cloud-to-cloud, as indicated by the long-continuing luminosity of stroke channels described earlier, or else is created in a glow-discharge (St. Elmo's fire).

We first outline the main physical processes we will consider:

(A) Without external energy input, the ball's lifetime is ≈ 0.5 to 1.0 sec, as found experimentally, and its source of energy is internal (molecular kinetic, dissociation, and metastable electronic excitation energy), formed in the stroke.

(B) With a postdischarge field of 1000 to 2000 V/cm, Townsend multiplication occurs in the ball, and a positive ion and an electron current both leave the ball in opposite directions. Because of the difference in mobilities, the ball becomes positively charged.

(C) Additional electron leakage current is channeled into the ball from a very wide area, many times the size of the ball, by the positive space charge cloud leaving the ball.

(D) As long as the postdischarge field continues, the energy input from (B) and (C) continue, and the lifetime of the ball is greatly extended.

(E) If the postdischarge field is downwards, then an electrohydrodynamical force from the positive-ion current acts downwards on the ball. This force is comparable to the thermal-buoyancy force on the ball and may cause it to move downwards or against a wind. It acts to reduce thermal convective mixing and heat loss from the ball.

(F) The positive-ion space charge cloud is attracted to conductors, and may cause the ball to move towards them, through screens, crevices, etc. A rapid increase in electron leakage current from a conductor when the ball contacts it could cause heating rapid enough to produce an explosion.

This analysis is akin to Finkelstein and Rubenstein's [10] model, with the difference that the positive-ion current from the ball plays a very important role in maintaining it. The short-range dipole field of the earlier model is accompanied by a long-range monopole field in this analysis.

There are many cases where ball lightning is not formed near the impact point of a stroke and thus could not be part of the channel, but issues from structures (rods, stoves, etc.) at which field multiplication is likely. Here the formation process seems to be an initial discharge similar to St. Elmo's fire, with subsequent detachment and independent existence as a ball.

Analytic solution of such a complex phenomenon as ball lightning is impossible; we are attempting numerical solutions on a computer, but this is very difficult.

We examine here a greatly simplified model for which analytic solutions are possible and the effects of various physical parameters can be evaluated. There are five basic questions about this model:

(1) Are adequate energy inputs to the ball possible?

(2) Is Townsend multiplication in the ball possible?

(3) What is the spatial and temporal variation of currents in the ball?

(4) What is the magnitude of electrohydrodynamic forces and can they overcome buoyant forces?

(5) Is the ball stable at a unique size for given conditions?

We analyze only the case where positive ions leave the ball and move towards ground. In principle, balls can be sustained in oppositely directed fields, but such balls would have to rise.

4.2. Energy Input to Ball

Figure 17 shows the basic approximation. There are three regions: *Central*, where Townsend multiplication occurs. This region is the source of visible light and has fairly uniform temperature. *Intermediate*, where temperature decreases with increasing distance from the central region. *Surroundings*, air at ambient conditions (70°F).

Some of the stored excitation energy is used to create ion pairs in the ball, so a current of electrons and negative ions leaves one hemisphere of the ball

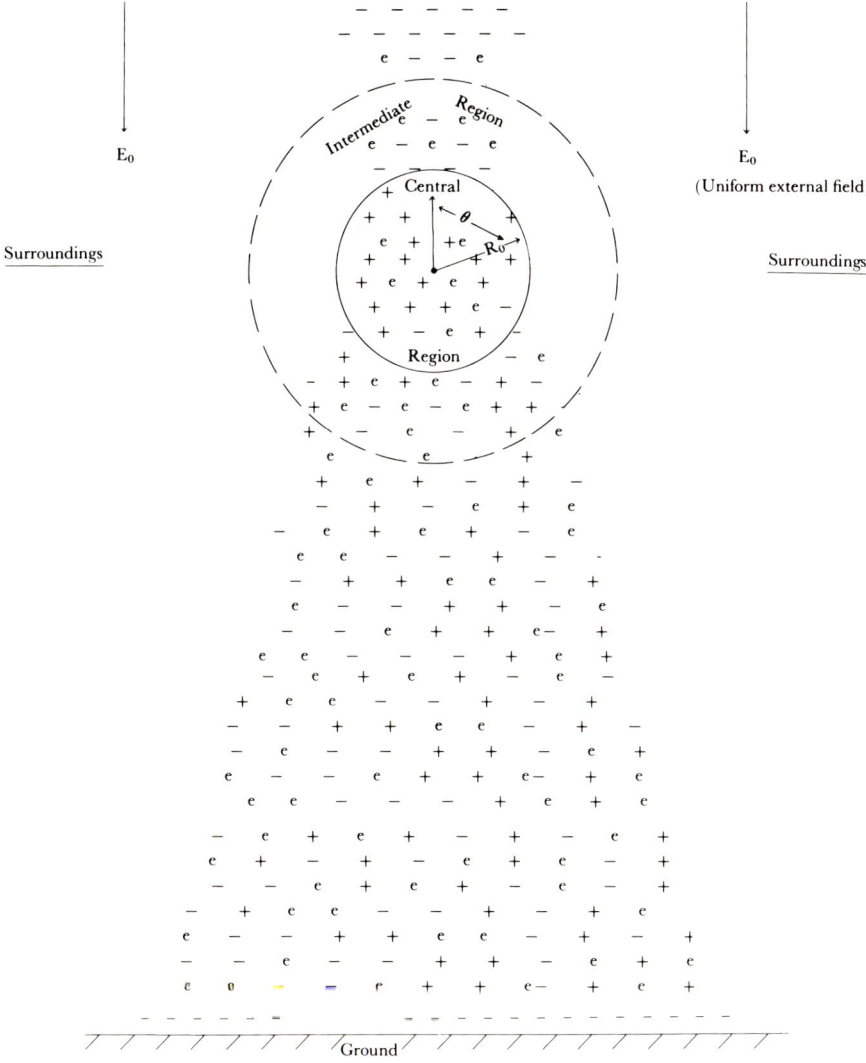

Electron current (e) and Negative Ion current (−) collected from Ground by positive ion channel.

FIG. 17. Schematic of ball lightning model.

and a positive-ion current leaves the other, forming channels of space charge above and below the ball. Additional electrons and negative ions may be attracted into the ball by the space charge of the positive-ion channel.

At first the motion of the gas can be neglected. Then the motion of the electrons and ions in the electric field E is determined by diffusion, with mobilities $\mu^{\pm}$, according to

(4.1a) $$\mathbf{v}^{\pm} = \pm \mu^{\pm} \mathbf{E}$$

This modifies the species distributions $n^{\pm}$ according to continuity:

(4.1b) $$\partial n^{\pm}/\partial t + \nabla \cdot \mathbf{v}^{\pm} n^{\pm} = \sigma^{\pm}$$

where $\sigma^{\pm}$ is the source density constituted by the excited gas, considering both ionization and recombination processes

(4.1c) $$\sigma^{\pm} = \sigma^{\pm}(n^+, n^-)$$

The species distributions modify the field E according to Poisson's law:

(4.1d) $$\nabla \cdot \mathbf{E} = \sum_{\pm} e^{\pm} n^{\pm}$$

We solve this system in the limit where the electron mobility $\mu^- \to \infty$ (so that its concentration is negligible).

Assuming a uniform positive-ion concentration in the ball, and negligible negative ion concentration (all electrons are assumed to be detached from negative ions by the high temperature), the electric field from the space charge in the ball is everywhere radially outwards. At the boundary $r = R_0$, it is

(4.2a) $$\mathbf{E}_{R_0} = 10^2 (R_0 n_0^+ e / 3\epsilon_0)$$

The net field at the surface of the ball is the vector sum of $\mathbf{E}_{R_0}$ and $\mathbf{E}_0$ (the uniform external field from the cloud). At the surface of the ball:

(4.2b) $$E_\perp = E_{R_0} - E_0 \cos\theta$$
$$E_\parallel = E_0 \sin\theta$$

If $K = E_{R_0}/E_0$, then Eq. (4.2b) defines a critical angle θ_1, such that

$$\theta_1 = \cos^{-1} K; \quad 0 \leqslant K \leqslant 1$$
$$0 \leqslant \theta \leqslant \theta_1; \quad \text{electrons leave}$$
$$\theta_1 \leqslant \theta \leqslant \pi; \quad \text{ions leave}$$

Rewriting Eq. (4.1), the positive-ion concentration in the ball is

(4.3) $$n_0^+ = 10^{-2}(3\epsilon_0 K E_0 / R_0 e)$$

and the equal and opposite currents drawn from the ball are

(4.4) $$|I_{iB}^+| = |I_{iB}^-| = n_0{}^+ e\mu^+ \int_{\theta_1}^{\pi} E_\perp(\theta) 2\pi R_0{}^2 \sin\theta\, d\theta$$

$$= (3\pi/100)(\mu^+ \epsilon_0) K(1+K)^2 R_0 E_0{}^2$$

$|I_{iB}^+|$ is bounded by two limits (Fig. 18). The lower limit corresponds to zero negative-ion current coming up the positive-ion channel ($|I_{ic}^-|=0$). Because of the lower ionic mobility in the cooler air below the ball, the space charge effect of the positive ions is greatest there and limits $|I_{iB}^+|$ to a value given by Eq. (4.4) with $K \approx 0.2$.

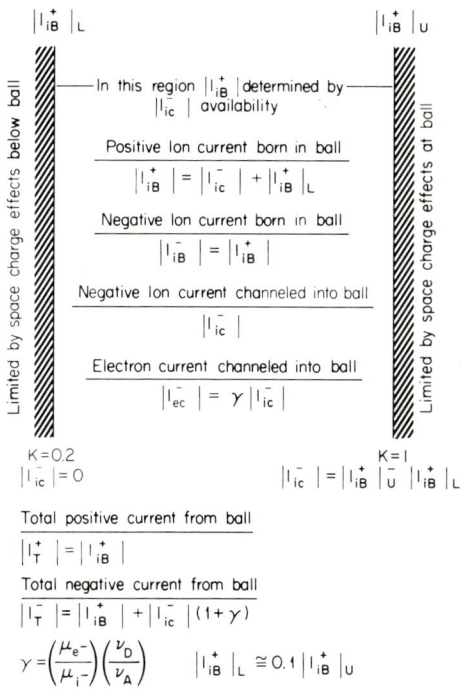

Fig. 18. Schematic of current limits for ball lightning model.

If $|I_{ic}^-| \neq 0$, it neutralizes the space charge effect of an equal magnitude positive-ion current, permitting more positive-ion current to be drawn from the ball. However, an upper limit exists to positive-ion current because for $K>1$ no electrons can be drawn from the ball. (Electrons are the only negative carriers of appreciable concentration in the ball because its high temperature and electric field cause rapid detachment from ions.)

Besides $|I_{ic}^-|$ coming up the channel, there is also an electron current, $|I_{ec}^-|$. The ratio $\gamma = |I_{ec}^-|/|I_{ic}^-|$ is a function of the temperature and metastable concentration in the positive-ion channel.

If $|I_{ic}^-|$ is source limited, then the currents drawn from the ball are determined by $|I_{ic}^-|$ (Fig. 18). If $|I_{ic}^-|$ is not limited, then the $|I_{iB}^+|$ is determined by Eq. (4.4) with $K = 1$, and $|I_{ic}^-|$ is given by

$$|I_{iB}^+|_{K=1} - |I_{iB}^+|_{K=0.2}$$

The total power generated in the ball is given by the product of total current and average voltage drop:

(4.5) $$P_T = [2|I_{iB}^+| + |I_{ic}^-|(1+\gamma)](2R_0 E_0)$$

The power is shown as a function of E_0 (Fig. 19) for the two limits, $K = 0.2$ and $K = 1.0$, with $\gamma = 5$. Since $I \approx (E_0^2)$, $P \approx (E_0^3)$, at 2000 V/cm quite respectable power inputs (20 to 1000 W) are possible. Experiments show that 50 W radiant power gives a quite bright ball, and if convective mixing of the ball is minimized, several hundred watts is enough to maintain its temperature.

The negative-ion current required is small enough (20 mA at 2000 V/cm) that a source should be easily available. For example, Ette [31] experimentally produced 30 μA of corona discharge from a *causarina* plant 18 in. high in a potential gradient of 240 V/cm. In fields of 2000 V/cm, the shrub should yield 10 mA of current. The ion channel will collect over many square feet, so that multiple sources can contribute. In addition, Kamara [32] has shown that in rapid field changes (e.g., caused by a stroke), there is an excess point discharge current caused by the field of the charges that were being emitted from the point before the stroke. Instead of reducing the emission, they now aid it if the potential gradient is reversed.

As to the ability of the negative charge to travel up the positive-ion channel, the ion densities are so low ($< 10^8/cm^3$) that recombination effects are negligible. In the study of stroke channels by Brook et al. [8], they sometimes found currents of 100 A up to 0.3 sec after the stroke. They could not measure smaller currents but presumed that they persisted for some time longer.

The estimation of γ in the positive-ion channel is difficult as it is so strongly dependent on air temperature and metastable concentrations. Phelps [33] gives thermal detachment frequencies for O_2^- ions as a function of T. Attachment frequencies have also been measured [34]—combining these, γ is ≈ 0.2 at 300°K. However, electric fields give O_2^- ions additional energy and increase γ. Metastable molecules can also detach electrons from negative ions. We estimate $\gamma = 5$ as a reasonable approximation for the channel, though it may be considerably higher.

The self-field of the positive ions leaving the ball spreads the diameter of the positive-ion channel. Four feet below the ball, the channel is about 3 ft in diameter; channel diameter increases with distance from the ball less than linearly. In the area bounded by the channel, any negative carrier will be drawn into the ball.

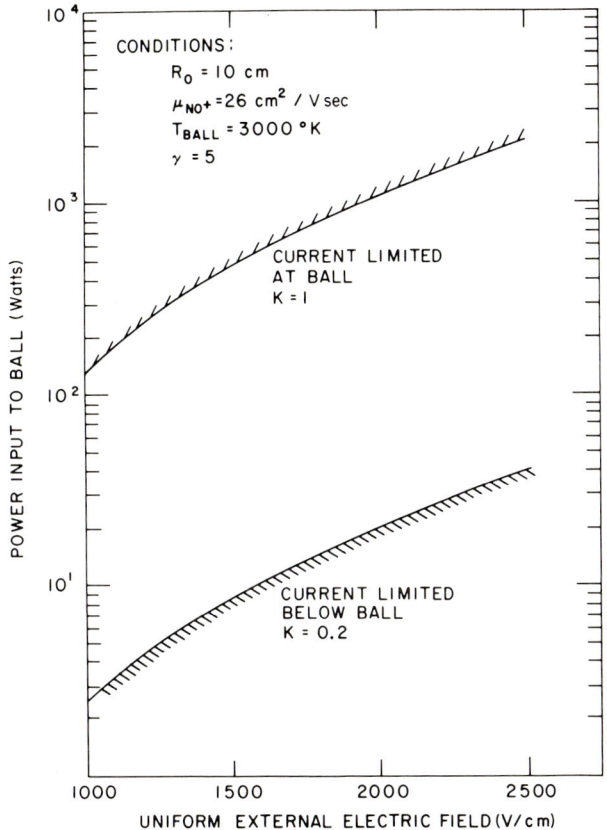

FIG. 19. Power generated in ball lightning as a function of external field.

We infer from this numerical example that, provided the supply of positive ions from the ball is maintained, the channeling of atmospheric electric currents into the ball provides enough power to maintain the losses due to radiation and conduction and extend the lifetime of the ball. Let us now consider the mechanism for maintaining the supply of positive ions.

4.3. Townsend Multiplication in Ball

The ionization processes to be considered in near-equilibrium air at $T \approx 3000°K$, $E \approx 2000$ V/cm, and 1 atm are:

Ionization by electron impact

(4.6) $\quad\quad e + M \to M^+ + 2e \quad\quad [M = \text{NO, O, O}_2, \text{ and N}_2]$

Ionization by excited molecular and atomic states

(4.7a) $\quad \text{N}_2^*$ (electronically excited N_2) $+ \text{NO} \to \text{N}_2 + \text{NO}^+ + e$

$\quad\quad\quad\quad\quad\quad\quad\quad\quad\quad\quad\quad\quad\quad\quad\quad$ Penning-ionization

(4.7b) $\quad\quad \text{N} + \text{N} + \text{NO} \to \text{N}_2 + \text{NO}^+ + e \quad$ chemi-ionization

where N, N_2^* are formed by

(4.7c) $\quad\quad e + \text{N}_2 \to \text{N}$ or $\text{N}_2^* + e \quad\quad$ electron impact excitation

(4.7d) $\quad\quad \text{N} + \text{N} + M \to \text{N}_2^* + M \quad\quad$ recombination excitation

Ionization by thermal collisions

(4.8) $\quad \text{NO} + M + \Delta\epsilon$ (kinetic, vibrational, and rotational) $\to \text{NO}^+ + e + M$

Other species and reactions can be disregarded because of low concentration, insufficient energy, or small rate constants or combinations of these.

Reaction (4.8) cross sections are given as a function of energy by Hasted [35] for NO-N_2 collisions. The less frequent NO-O_2 and NO-NO collisions are assumed to have the same ionization cross sections. For a Maxwellian molecular energy distribution, the calculated ionization rate at 4000°K for reaction (4.8) is only 10^7 ion pairs/cm^3 sec, and is much less at lower temperatures. It is negligible compared with reactions (4.6) and (4.7).

Reaction (4.6) rates can be computed from Townsend multiplication studies. The multiplication factor, α, is defined as the number of new ion pairs created per cm of travel. Figure 20 shows various experimental measurements of α for air. α is expressed as α/p, the multiplication factor divided by the pressure of the given density (at any temperature) referred to a standard temperature (e.g., 15°C). This is a unique function of E/p. For air at 1 atm pressure and temperature T, p is given by

(4.9) $\quad\quad\quad\quad\quad p = 760(288/T) \quad \text{Torr}$

Curves 2 and 3 refer to dc [36] and rf [37] studies of α for dry air with equilibrium composition at 300°K. In the dc studies, a direct measurement of α was made from current multiplication over a defined distance:

(4.10) $\quad\quad\quad\quad\quad I(x_0) = I_0 \exp(\alpha x_0)$

In the rf studies, the time dependent electron density was measured in an rf field

(4.11) $$n(t) = n(t = 0) \exp(\nu t)$$

where the two factors are related through the drift velocity V_D

(4.12) $$\nu = \alpha V_D$$

In the rf measurements, the electron neutral collision frequency was much greater than the rf frequency ($\nu_{coll} \gg \omega$) so that electrons essentially experience a slowly varying dc electric field. The average dc field was assumed equal to the RMS value of the rf field. The validity of the approximation is shown by the close-agreement between curves 2 and 3.

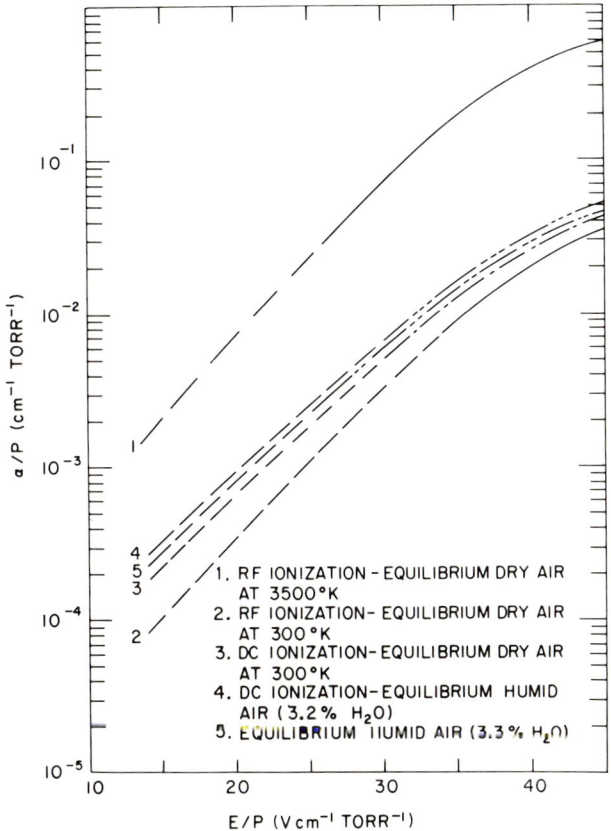

FIG. 20. Experimental Townsend coefficient for air as a function of E/p, T, and H_2O content.

Curves 4 and 5 represent two dc measurements [38, 39] for α with equilibrium air at 300°K containing 3.3% H_2O vapor, which is approximately the concentration at 70°F. This is close to the conditions near thunderstorms. There is a slight increase in α.

A high air temperature causes a much higher α, however, as shown in curve 1 of Fig. 20 and Taylor's [37] original data (Fig. 21). This is a result of change in composition, because the additional thermal energy of N_2 and O_2 is negligible (0.2 eV) compared with ionization energy. In fact, α does not reach its equilibrium value for hot air for some time after a shock front passes

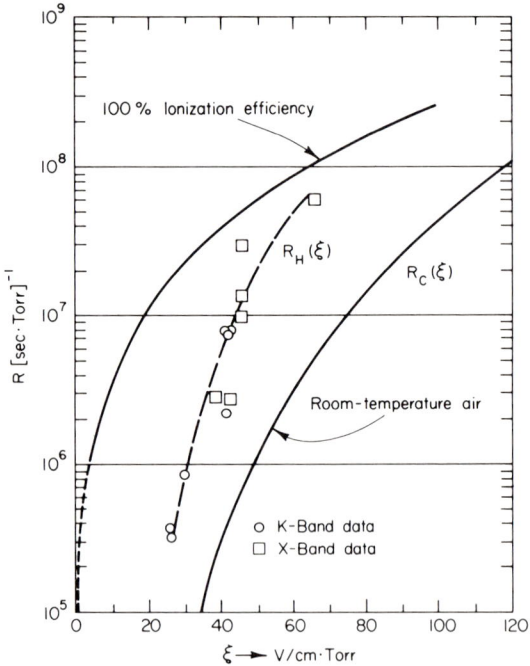

FIG. 21. Taylor's data for rf ionization of high temperature air [37].

the measuring point, because composition change in the air is required. The increase most likely reflects the effect of sizable concentrations (e.g., 5%) of NO which has a much lower ionization potential (9.25 eV) than O_2 (12.15 eV) or N_2 (15.5 eV). Thus, not only does the direct-impact ionization reaction (4.6) become more probable, but reactions (4.7a) and (4.7b) are now possible. This is a well-known phenomenon: For example, Haseltine [40] has observed a 6% reduction in sparking voltage in air after repeated sparking and the consequent formation of oxides of nitrogen. This decrease is probably due primarily to the effect of reaction (4.6).

Kunkel [41] and Young [42] have observed continuing ionization of NO in O_2-N_2 afterglows, with no appreciable electric fields, proof of the presence of reaction (4.7). Kunkel concludes that reaction (4.7a) (Penning effect) is the most likely one in his experiments. Ionization rates of $\approx 10^{12}$ cm^3 sec^{-1} have been observed. The presence of very high NO concentrations will effectively scavenge any N atoms by the very fast reaction

(4.13) $\qquad N + NO \rightarrow N_2(vib) + O, \qquad k = 10^{-11}$ cm^3 sec^{-1}

so that only the (4.7a) and (4.7c) processes need be considered.

The relative importance of the Penning and electron impact processes for ionization of air containing NO can be estimated as follows: α is calculated for electron impact ionization of air containing NO, using equilibrium concentrations at the desired temperature [43] and experimental ionization cross sections for NO, O_2, and N_2 [44] and O [45] and then compared with the measured α. The difference reflected the Penning contribution. The presence of NO is assumed not to change the electron energy distribution function, given by Von Engel [46] for predominantly inelastic energy loss

(4.14) $$f(\epsilon) = C \exp\left[-\frac{(3P_e)^{1/2}\epsilon}{eE\lambda_m} \right]$$

with

(4.15) $$\frac{\alpha_{air, NO}}{\alpha_{air}} = \frac{G(NO) + G(O) + G(O_2) + G(N_2)}{G(O_2) + G(N_2)}$$

and the ionization rate $G(M)$

(4.16) $$G(M) = kN_{iM} \int_{\epsilon_{iM}}^{\infty} \sigma_{iM}(\epsilon) f(\epsilon) \epsilon^{1/2}\, d\epsilon$$

The results are shown in Fig. 22 for air at various temperatures and 2000 V/cm. NO causes about a 3-fold increase in α over ordinary air (curves 2 and 3 vs. curve 1). H_2O was assumed to have the same effect in air containing NO as at 300°K. Curve 4 represents Taylor's experimental rf measurements [37] (extrapolated from data between $E/p = 26$ and $E/p = 60$) on 3500°K air. The difference between 2 and 4 most probably is caused by Penning ionization. The two effects are roughly comparable.

The conclusion is that for fields of ≈ 2000 V/cm, and temperatures of ≈ 3000°K, Townsend multiplication will occur, with a probable value of $\alpha = 0.6$ cm^{-1}. This means that in 10 cm, current would multiply by exp(6) or ≈ 400. Further, the initial electron source to start multiplication can come from the Penning effect on NO, as ionization rates of $> 10^{10}$ cm^{-3} sec^{-1} have been observed due to this mechanism. At temperatures much below ≈ 2000°K, electron multiplication will cease, owing to low E/p and little NO concentration.

Because the Townsend coefficient enters exponentially, it is the physical constant of the dimensions of (reciprocal) length that is most important in determining the steady-state size of the ball.

We have supposed that electrons produced in the ball leave much more rapidly than positive ions, i.e., that electron attachment can be neglected. We now consider the validity of this approximation.

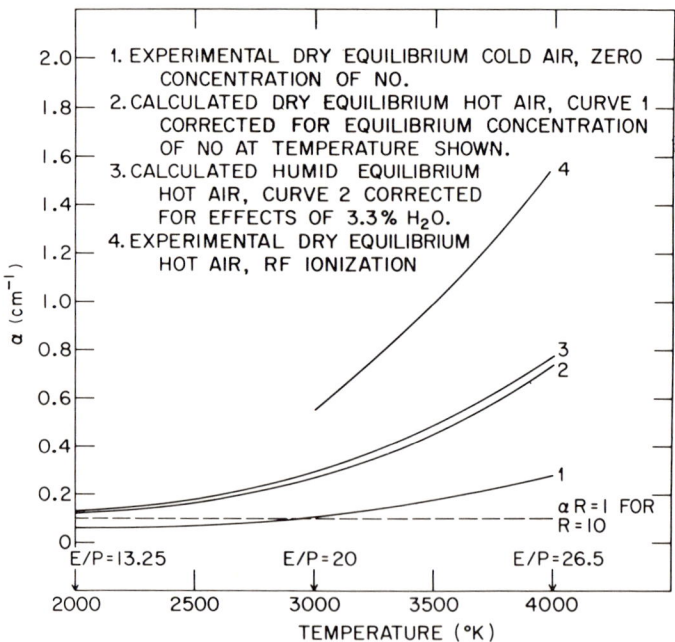

Fig. 22. Townsend multiplication factor for air at fixed electric field and variable temperature.

Electrons attach to air molecules by the following mechanisms:

3-body attachment

(4.17) $\qquad e + X + M \rightarrow X^- + M \qquad (X = O_2, O, OH, H)$

Dissociative attachment

(4.18) $\qquad e + XY \rightarrow X + Y^- \qquad (XY = O_2, NO, H_2O)$

Electrons may be detached from negative ions by the reverse of (4.17) and (4.18), though the latter is very improbable. The reverse of (4.17) can occur either thermally or by the ion gaining energy in the electric field.

Electrons may be lost by:

Recombination

(4.19)
$$e + XY^+ \rightarrow X + Y$$
$$X^- + Y^+ + M \rightarrow XY + M$$

Negative ions have a much larger recombination coefficient with positive ions than free electrons do (typically 10^{-6} vs. 10^{-8} [47]) and if electrons are lost, it will be when they are attached. For a negative ion, the mean lifetime against recombination is

(4.20)
$$\tau_R = (N_{NO^+} \cdot \beta)^{-1} \approx 10^8 \, 10^{-6} \approx 10^{-2} \quad \text{sec}$$

and is long compared with the ion transit time across the ball (10^{-4}) sec. Recombination processes may be neglected therefore.

Electron attachment coefficients, similar in form to the Townsend ionization coefficient, have been measured in air [34]. Figure 23 shows attachment frequencies derived from these measurements, for air at 1500 to 2500 V/cm and temperatures of 2000 to 4000°K. Also plotted are detachment frequencies for the reaction in an electric field,

(4.21)
$$O^- + M \rightarrow O + e + M$$

derived from measurements of Frommhold [48]. O^- is the most important negative ion. Also plotted is the thermal detachment frequency for O^-, neglecting the effect of the electric field. This was calculated for a Maxwellian energy distribution, using Frommhold's detachment cross sections, and is much less than the detachment frequency associated with the electric field. Between 1500 and 2000 V/cm and 2000 and 4000°K, the attachment frequency is no more than 20 times the detachment frequency, and usually much less. At 2000 V/cm and 3000°K, the ratio is 5/1.

Electrons will attach and detach several times on the average before leaving the ball. At any point in the ball, the relative concentration of electrons and ions is

(4.22)
$$\frac{[e^-]}{[X^-]} = \frac{\nu_D}{\nu_A}$$

and the average drift velocity of negative carriers is

(4.23)
$$\bar{V}_D = \frac{V_e}{\nu_A + \nu_D} \left[\nu_D + \frac{V_{ion}}{V_e} \nu_A \right]; \quad \frac{V_{ion}}{V_e} \approx \frac{1}{200}$$

while the ratio of electron/ion travel distance is

(4.24)
$$\frac{D_e}{D_{ion}} = \frac{I_{electron}}{I_{ion}} = \frac{V_e}{V_{ion}} \frac{\nu_D}{\nu_A}$$

Thus for the conditions above, attachment causes electrons to take somewhat longer to leave the ball. At most, the electron transit time is increased an order of magnitude above that for electrons which never attach, and is probably much less. More than 90% of the energy lost by the negative carrier will be in electron-molecular collisions—the remainder in ion-molecular collisions.

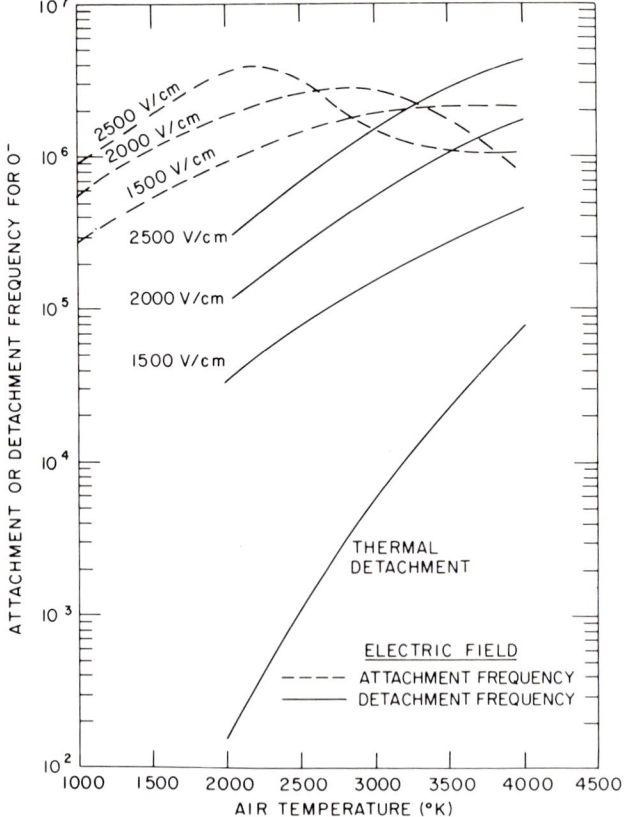

FIG. 23. Attachment and detachment frequencies for O^- as a function of dry air temperature.

This analysis could be applied to other negative ions; however, only OH has a greater electron affinity than O (1.8 vs. 1.5 eV), and its concentration is much less. The other negative ions (O_2^-, H_2O^-, H^-, NO^-) have much smaller electron affinities, and the electron will be very easily detached in thermal collisions [33]. We thus may neglect all other negative ions.

These calculations assume no detachment from negative ions by collision with metastable atomic or molecular species. In fact, there will be considerable detachment by this process, just as there was considerable ionization by metastables. Thus, the calculated detachment frequencies are a lower limit, and attachment is of even less concern to current in the ball.

After electrons leave the ball, they eventually attach permanently because E/p and T decrease. Figure 24 shows the mean distance traveled between attachments at various temperatures and electric fields. When $T \approx 500°\text{K}$, all electrons will quickly attach and then continue as negative ions.

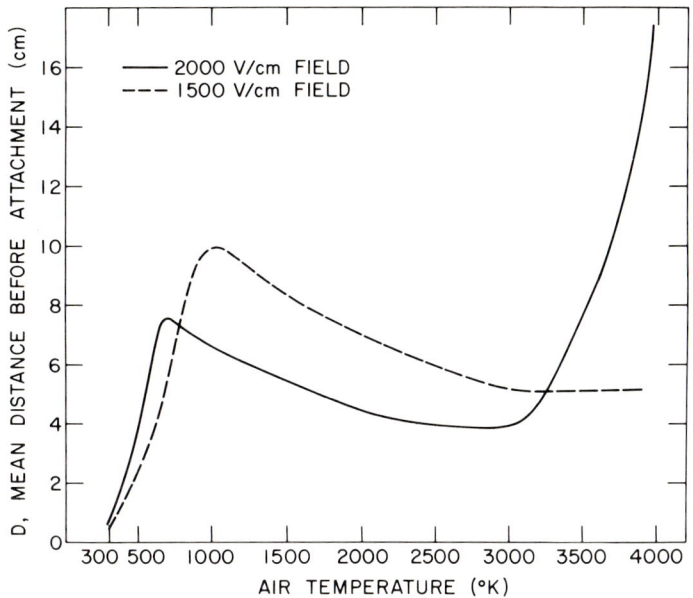

FIG. 24. Mean attachment distances for electrons as a function of temperature.

4.4. Spatial and Temporal Variation of Currents in Ball.

A one-dimensional approximation to the ball is made for simplicity. The thickness of the plasma slab was taken equal to R_0. The following equations govern the behavior of ions and electrons:

(4.25) $\quad \partial n_x^+/\partial t = \alpha_x |J_x^-| - \dfrac{\partial J_x^+}{\partial x} \quad$ positive ion continuity

(4.26) $\quad \partial n_x^-/\partial t = \alpha_x |J_x^-| - \dfrac{\partial J_x^-}{\partial x} \quad$ negative ion continuity

(4.27) $\alpha_x = A \exp(BE_x)$ Townsend multiplication
(4.28) $\partial E_x/\partial x = (n^+ - n^-)/\epsilon_0$ Poisson's equation
(4.29) $J_x^- = -\mu_e n_x^- E_x$ electron flux
(4.30) $J_x^+ = +\mu_i^+ n_x^+ E_x$ positive ion flux

The following boundary conditions are appropriate:

$$J^-(x=0) = 10^{11} \text{ cm}^{-2} \text{ sec}^{-1}; \quad J^+(x=R_0) = 0$$
$$n^+(t=0) = n^-(t=0) = 0$$

Since the ball is charged, in the one-dimensional model its field extends to infinity. The external field E_0 is then represented by the average

$$E_0 \equiv [E(0) + E(R_0)]/2$$

This is also the field at the center of charge of the slab.

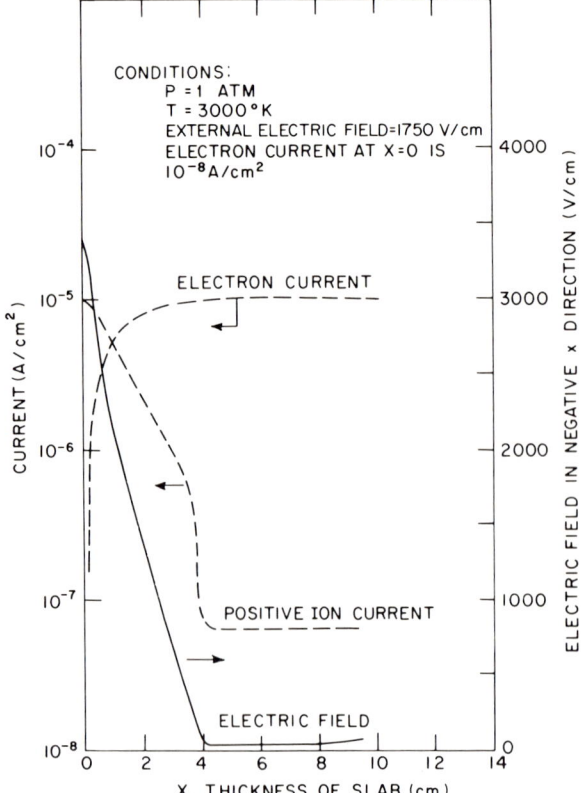

FIG. 25. Current and electric fields as a function of distance for one-dimensional slab approximation to ball lightning.

The above equations were numerically integrated on a CDC-6600 computer. Integration intervals were $10^{-2} R_0$ for Δx and 3×10^{-9} sec for Δt. The latter interval corresponds to an electron transit distance of $\approx 0.3 \Delta x$. For each time step, there was a complete integration over x.

Parameters are: $E_0 = 1750$ V/cm, α determined by Taylor's rf data [37], $\mu^+ = 26$ cm^2/V sec, $\mu_e = 4 \times 10^3$ cm^2/V sec, gas temperature $= 3000°$K.

A steady-state solution (Fig. 25) is reached after $\approx 10^{-4}$ sec (one ion transit time of R_0) and remains constant up to the largest time investigated, 10^{-3} sec.

The rapid decrease in electric field, $0 < x < 4$, is caused by the high positive-ion concentration. For $x > 4$, positive ion and electron concentrations are nearly equal and the net effect on the field is very small. Interestingly, the currents that can be drawn from the slab are ten times greater than those corresponding to a model of the type used in the energy-input analysis, i.e., uniform positive-ion concentration and zero electron concentration. This means that when a more detailed analysis is made of the processes in the ball, the space charge effects of the ions may not be as limiting as thought, and that the upper limit predicted on the power (Fig. 19) may be very conservative. When E_0 was increased to 2000 V/cm, the current increased by about 30%. Of note is the independence of current drawn from the slab on $J^-_{(x=0)}$, remaining almost constant as long as $J^-_{(x=0)}$ was above 10^{11} cm^{-2} sec^{-1}. This low flux can easily be supplied by negative ion-electron current coming up the positive-ion channel or Penning type ionization or both.

4.5. Electrohydrodynamic Forces

If on each volume of gas there is a net downwards force due to ion space-charge (electron concentrations are too low to have any appreciable effect) nearly equal and opposite to the upward thermal buoyancy force,

$$\rho_S g[1 - (T_S/T)]$$

then convective mixing is reduced and heat transfer is that predicted by conduction. The required energy input is then lower.

The average ion force on the central region is

(4.31) $$F_E = 10^7 n_0^+ e E_0 = 3 \times 10^5 \epsilon_0 (K E_0^2 / R_0) \quad \text{dyn/cm}^3$$

and the ratio of ion to thermal buoyancy forces is:

(4.32) $$\frac{F_E}{F_G} = \frac{3 \times 10^5 \epsilon_0 (K E_0^2 / R_0)}{\rho_S g[1 - (T_S/T_0)]}$$

This ratio is shown in Fig. 26 as a function of E_0, as being between the limits $K = 0.2$ and $K = 1$. At the most likely conditions, $E_0 = 2000$ V/cm and $K = 1$, the forces almost balance. In addition, the positive ion channel will also tend to be stabilized by this force in those portions where thermal buoyancy is present. At those points where the channel temperature is equal to the temperature of the surroundings, this will be destabilizing, but the ion density will be low enough there so that it will be only a slight effect.

This gas dynamic force can explain the downwards movement of ball lightning and its ability to penetrate cracks.

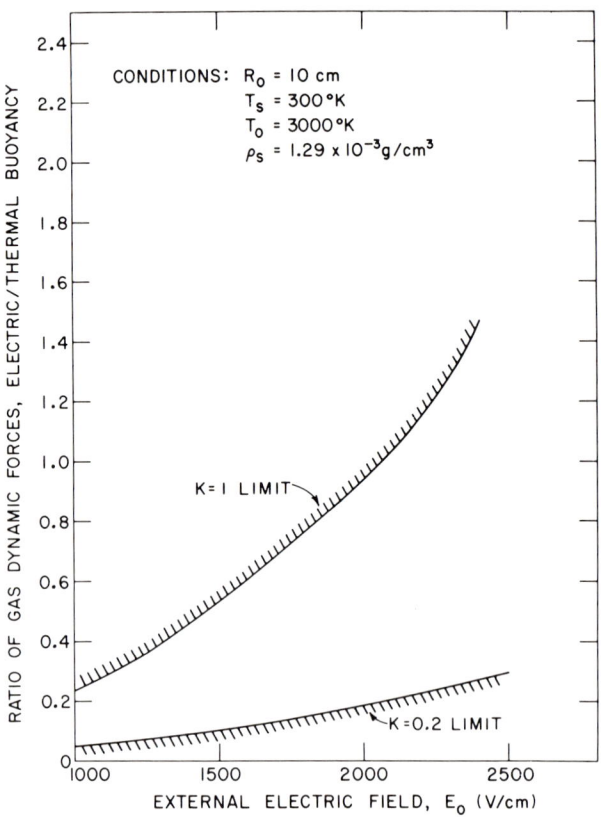

FIG. 26. Ratio of electrohydrodynamic/thermal buoyancy forces as a function of electric field.

4.6. Stability of the Ball

We imagine a small variation δT of the ball temperature. If the processes that determine the temperature are so affected as to restore T to its previous value, the ball is thermally stable; otherwise not. In this calculation we suppose shape and size are fixed and independent of T.

The energy losses from the ball are mainly radiative

$$P_R \approx -T^4 \qquad \partial P_R/\partial T \approx 4P_R/T$$

and conductive (including convection)

$$P_c \approx -T \qquad \partial P_c/\partial T \approx P_c/T$$

The energy input to the ball comes from electron and ion currents in the ball's electric field. As shown previously

$$P_I \approx I_{iB}^+ E_0$$

I_{iB}^+ is an exponential function of the Townsend multiplication coefficient, α, for small α. α is in turn a sensitive function of temperature (Fig. 22, curve 4). However, for large α, I_{iB}^+ is essentially independent of α because positive-ion space charge effects are limiting. This is the case for α comparable to that estimated for the ball, e.g., ≈ 0.5. An increase in temperature thus increases current mainly by its effect on ion mobility, with

$$P_I \approx +T; \qquad \partial P_I/\partial T \approx P_I/T$$

This is seen in the one-dimensional computer solution, where the increase in current is approximately proportional to the change in temperature.

Thus,

$$\frac{\partial}{\partial T}(P_{tot}) = \frac{\partial}{\partial T}(P_I + P_R + P_c) < 0$$

since at equilibrium $P_{tot} = 0$. The ball is stable to variations in temperature of either sign. Radiative loss is the stabilizing mechanism.

A similar analysis shows the ball to be stable against variations in radius. Input and radiative loss are proportional to R^2. Convective loss is proportional to $f(R) \cdot R^2$; the mixing rate, $f(R)$, increases with R because of the decreasing importance of electrogasdynamic stabilization. Convective loss stabilizes radius in the same way as radiative loss stabilizes temperature.

Thus there is a unique radius and temperature for the ball for a given set of conditions. The most important variable is the external electric field, E_0.

5. Summary and Conclusions

To summarize, the rf experiments show that it is possible to make, in an electric discharge, a persistent luminous ball that has many of the characteristics reported of ball lightning. Because of rapid dissipation of its stored energy by convective mixing, the ball will last at most 1 sec after rf cutoff in the absence of other external fields, while the average reported ball lightning lasts several seconds.

However, the dc electric field remaining after a stroke is often positive (40% of flashes [49]) and of high magnitude, because of a positive-ion space charge layer above the ground. In this field, positive ions will leave the ball towards the ground, electrons towards the cloud. Townsend multiplication of the electron current generates positive ions in the ball. These ions leave the ball and move towards the ground. The electrohydrodynamic force on these ions can prevent convective mixing and may even move the ball downwards or against a wind. In addition, the ions collect negative current over an area much larger than the area of the ball and channel it into the ball. Energy inputs from the currents in the electric field of the ball are of sufficient magnitude to sustain the ball or at least greatly extend its lifetime. The theory predicts a stable unique ball size for given conditions. The ball may move towards conductors because of the positive-ion current, and upon contact, may explode because of a rapid increase in power generated in the ball.

It appears that this is a viable explanation of ball lightning. More detailed theoretical and experimental work is needed to further develop this theory; for example, to determine experimentally what external applied dc electric fields are needed to sustain the balls produced in rf discharges, and whether such fields are present in nature. About 2000 V/cm are required by our calculations. This may be too large an estimate. Because of a nonuniform ion distribution in the ball, significantly more current may be channeled into the ball than our present estimate. Also, the ratio of electron-negative ion current may be greater than assumed, giving more power input. If lower fields of ≈ 1000 V/cm are found sufficient by experiment, this would require much less unusual conditions to be present in nature. A more accurate 3-dimensional computer solution for the static properties should be undertaken, and the dynamical behavior has hardly been touched, though there is a clear hint of vortical motion in the results of the helium-bubble observations. It seems possible, notwithstanding, that the basic ingredients of a theory of ball lightning are now in hand.

Acknowledgments

The authors wish to express their deep appreciation to Mr. Josh Manwaring for his discovery of the long-lived rf discharge, to Mr. Roy Domish for his invaluable help in planning and carrying out the experiments, to Mr. Michael Callan for his great help in the experiments, to Mr. Robert Walton for the photographic work, to Mrs. Anna Potter for computer programming, and to Mrs. Bobbe Friedmann, Mrs. Nora McDonald, and Mrs. Jean Reynolds for typing the manuscript.

List of Symbols

A	Constant in Townsend multiplication factor (cm^{-1})
B	Constant in Townsend multiplication factor (cm V^{-1})
D_e	Average travel distance of electron—unattached (cm)

STRUCTURE OF BALL LIGHTNING 187

D_{ion}	Average travel distance of electron—attached (cm)
e	Electron charge (C)
E	Electric field (V cm^{-1})
ΔE	Field change caused by lightning stroke (V cm^{-1})
E_0	External uniform electric field (V cm^{-1})
E_{R_0}	Electric field at R_0 produced by positive charge in central region of ball (V cm^{-1})
$E_\perp$	Component of $(\mathbf{E}_{R_0} + \mathbf{E}_0)$ perpendicular to surface at R_0 (V cm^{-1})
$E_\parallel$	Component of $(\mathbf{E}_{R_0} + \mathbf{E}_0)$ parallel to surface at R_0 (V cm^{-1})
F_E	Electrohydrodynamic force in central region of ball (dyn cm^{-3})
F_G	Gravitational buoyancy force in central region of ball (dyn cm^{-3})
g	Acceleration of gravity (cm sec^{-2})
G	Ionization rate, ion pairs (cm^{-3} sec^{-1})
H	Mean height of cloud negative charge (km)
$|I_{iB}^+|$	Magnitude of positive-ion current leaving ball, produced by ionization in ball (A)
$|I_{iB}^-|$	Magnitude negative-ion current leaving ball, produced by ionization in ball (A)
$|I_{ic}^-|$	Magnitude of negative-ion current attracted into ball along positive-ion channel (A)
$|I_{ec}^-|$	Magnitude of electron current attracted into ball along positive ion channel (A)
J^+	Positive-ion flux in slab approximation, ions (cm^{-2} sec^{-1})
J^-	Electron flux in slab approximation, electrons (cm^{-2} sec^{-1})
K	Space charge parameter, E_{R_0}/E_0
n^+	Positive-ion space charge density (cm^{-3})
n^-	Electron space charge density (cm^{-3})
n_0^+	Positive-ion space charge density in central region of ball (cm^{-3})
N_M	Molecular density (cm^{-3})
p	Pressure of given gas density at 288°K (Torr)
P_e	Probability of electron-molecular inelastic collision
P_T	Total power input to central region of ball (W)
Q	Electrical moment of cloud (C km)
R_0	Radius of central region of ball (cm)
t	Time (sec)
T_0	Temperature of central region of ball (°K)
T_s	Temperature of surroundings (°K)
V_D	Mean drift velocity of negative carriers (cm sec^{-1})
V_e	Electron drift velocity in field (cm sec^{-1})
V_{ion}	Ion drift velocity in field (cm sec^{-1})
x	Distance (cm)
α	Townsend multiplication coefficient, ion pairs (cm^{-1})
γ	Ratio of $|I_{ec}^-|/|I_{ic}^-|$
ϵ	Electron energy (eV)
ϵ_0	Permittivity of free space $[(36\pi \times 10^9)^{-1}$ F m$^{-1}]$
θ	Polar angle
θ_1	Critical polar angle dividing negative and positive surfaces
λ_M	Mean free path for momentum transfer of electrons (cm)
μ^+	Positive-ion mobility (cm^2 sec^{-1} V^{-1})
μ_e	Electron mobility (cm^2 sec^{-1} V^{-1})

ν_{coll} Electron collision frequency (sec^{-1})
ν_A Attachment frequency for electrons to form negative ions (sec^{-1})
ν_D Detachment frequency for electrons from negative ions (sec^{-1})
ρ_S Density of air at NTP (g cm^{-3})
σ Cross section for electron-molecular interaction (cm^{-2})
ω rf frequency (rad sec^{-1})

References

1. Leonov, R. A. (1966). "The Riddle of Ball Lightning." TT 66-33253, Dept. of Commerce Translation.
2. McNally, J. R., Jr. (1960). Preliminary report on ball lightning, second Ann. Div. of Plasma Phys. Mtg., Am. Phys. Soc., Gatlinburg, Tennessee.
3. Rayle, W. D. (1966). "Ball Lightning Characteristics," NASA TN D-3188.
4. Silberg, P. A. (1965). A review of ball lightning, in "Atmospheric and Space Electricity" (*Conf. Atmospheric Space Elec., 3rd, Montreux, Switzerland*, 1965), p. 436. Elsevier, Amsterdam.
5. Manwaring, J. F. (1966). Personal communication, Radio Frequency Co., 20 Park Street, Medfield, Massachusetts.
6. Rodewald, M. (1954). *Z. Meteorol.* **8**, 27.
7. Wormell, T. W. (1939). *Phil. Trans. Roy. Soc. London* **238A**, 249.
8. Brook, M., Kitagawa, N., and Workman, E. J. (1962). *J. Geophys. Res.* **67**, 649.
9. Young, G. A. (1962). A Lightning Strike of an Underwater Plume, Noltr 61-43, US Naval Ordnance Laboratory.
10. Finkelstein, D., and Rubinstein, J. (1964). *Phys. Rev.* **135A**, 390.
11. Stekol'nikov, I. S. (1943). "Fizika Molnii i Grozozash Nita." Moscow–Leningrad; Shafranov, V. D. (1959). *Zh. Eksperim i Theor. Fiz.* **36**, 478. [(1960) *Soviet Phys. JETP (English Transl.)* **9**, 333].
12. Neugebauer, H. N. (1937). *Phys. Z.* **106**, 474.
13. Hill, E. L. (1960). *J. Geophys. Res.* **65**, 1947.
14. Dauvillier, A. (1957). *Compt. Rend.* **245**, No. 25, 2155-6.
15. Fehr, Y. (1962). Product of ball lighting environment. M.S. thesis, Hebrew University, Jerusalem.
16. Thornton, W. M. (1911). *Phil. Mag.* **21**, 630.
17. Kapitza, P. L. (1955). *Dokl. Acad. Nauk. UzSSR* **101**, 245.
18. Horner, F., and Bradley, P. A. (1964). *J. Atmospheric Terrest. Phys.* **26**, 1155.
19. Powell, J., Zucker, M., Manwaring, J. F., and Finkelstein, D. (1966). Laboratory production of self-sustained atmospheric luminosities, Am. Phys. Soc., Div. of Plasma Phys., Boston, Massachusetts.
20. Uman, M. A., and Helstrom, C. W. (1966). *J. Geophys. Res.* **71**, 1975.
21. Rayleigh, Lord (Strutt), (1911). *Proc. Roy. Soc. (London)* **85A**, 219.
22. Kivel, B. (1961). *J. Aerospace Sci.* **28**, 96.
23. Penner, S. (1959). "Quantitative Molecular Spectroscopy and Gas Emissivities." Chapter 14, Addison-Wesley, Reading, Massachusetts.
24. Myers, B. F., and Bartle, B. (1967). *J. Chem. Phys.* **47**, 1783.
25. Kenty, C., (1964). Collisions involving metastable N_2 molecules in discharges and afterglows in a rare gas plus N_2, in "Atomic Collision Processes" (M. R. C. McDowell, ed.) (*Proc. Intern. Conf. Phys. of Electronic and Atomic Collisions, 3rd*), p. 1133. North-Holland, Publ., Amsterdam.
26. Young, R. A., and Black, G. (1967). *J. Chem. Phys.* **47**, 2311.

27. Hornbeck, G. A., and Hopfield, H. S. (1949). *J. Chem. Phys.* **17**, 982.
28. David, W. T., Leah, A. S., and Pugh, B. (1941). *Phil. Mag.* **31**, 156.
29. Leah, A. S., Rounthwaite, C., and Bradley, D. (1950). *Phil Mag* **41**, 468.
30. Falick, A. M. (1965). *J. Chem. Phys.* **42**, 1837.
31. Ette, A. I. I. (1966). *J. Atmospheric Terrest. Phys.* **28**, 983.
32. Kamara, A. K., and Varshneya, N. C. (1967). *J. Atmospheric Terrest. Phys.* **29**, 1519.
33. Phelps, A. V., and Pack, J. L. (1961). *Phys. Rev. Letters* **6**, 111.
34. Brown, S. C. (1966). "Basic Data of Plasma Physics," 2nd ed. rev., p. 196. MIT Press, Cambridge, Massachusetts.
35. Hasted, J. B. (1966). Recent trends in heavy particle collision research, *Intern. Conf. Phenomena in Ionized Gases, Belgrade, 7th, 1965* Vol. I., p. 9.
36. Prasad, A. N. (1959). *Proc. Phys. Soc.* **74**, 33.
37. Taylor, W. C., Chown, J. B., and Morita, T. (1968). *J. Appl. Phys.* **39**, 191.
38. Prasad, A. N., and Craggs, J. D. (1960). *Proc. Phys. Soc.* **76**, 223.
39. Ryzko, H. (1966). Ionization and electron attachment coefficients in humid air, *Intern. Conf. Phenomena in Ionized Gases, 7th, Belgrade,* 1965, Vol. I., p. 97.
40. Haseltine, W. (1941). *Bull. Am. Phys. Soc.* **15**, 188.
41. Kunkel, W. B., and Gardner, A. L. (1962). *J. Chem. Phys.* **37**, 1785.
42. Gatz, C. R., Young, R. A., and Sharpless, R. L. (1964). *J. Chem. Phys.* **39**, 1234.
43. Landolt, H., and Börnstein, R., (1961). "Spezifische Wärmen von Gasen im Plasmazustand," Zahlenwerte und Funktionen, 6th ed., Vol. II, Pt. 4, p. 717, Springer, Berlin.
44. Rapp, D., and Englander-Golden, P. (1965). *J. Chem. Phys.* **43**, 1464.
45. Bauer, E., and Bartky, C. (1965). Calculation of inelastic electron-atom and electron-molecule collision cross sections by classical methods. Rept. No. U-2043, Aeroneutronic (Div. of Philco Corp.), Newport Beach, California.
46. Von Engel, A. (1956). Ionization in gases by electrons in electric fields, *in* "Handbuch der Physik" (S. Flügge, ed.), Vol. XXI, p. 548. Springer, Berlin.
47. Stein, H. P. (1964). *Phys. Fluids* **7**, 1641.
48. Frommhold, L. (1964). Simultaneous determination of the electron attachment and detachment rates in oxygen, *in* "Atomic Collision Process" (M. R. C. McDowell, ed.) (*Proc. Intern. Conf. Phys. of Electronic and Atomic Collisions*, 3rd), p. 556. North-Holland Publ., Amsterdam.
49. Chalmers, J. A. (1957). "Atmospheric Electricity." p. 221. Macmillan (Pergamon), New York.

ENERGETICS OF THE MIDDLE ATMOSPHERE*†

Conway Leovy

The RAND Corporation, Santa Monica, California

	Page
1. Introduction	191
2. The Ozone Distribution	193
2.1. Observations of the Ozone Distribution	193
2.2. Photochemical Theories of the Ozone Distribution	196
2.3. Ozone and Atmospheric Dynamics	200
3. The Water Vapor Distribution	202
4. Infrared Radiative Transfer	204
4.1. Curtis' Method for Calculation of Heating Rates	205
4.2. Influence of Line Shape and Line Spacing	207
4.3. Vibrational Relaxation	208
4.4. Heating Rate Caused by Infrared Transfer	211
4.5. Infrared Radiative Exchange and Atmospheric Dynamics	212
5. Heat Sources and Sinks above 80 km	214
List of Symbols	216
References	217

1. INTRODUCTION

Let us agree to consider the region which is too high to be dominated by tropospheric circulation systems, and too low to be dominated by molecular diffusion processes as an entity—the "middle" atmosphere. It comprises the mesosphere, most of the stratosphere, and the lower portion of the thermosphere, and normally extends from about 25 to somewhat above 100 km. The thermal structure of this region is determined by heat sources and sinks, and by circulation systems which produce temperature changes by adiabatic compression or expansion, or by advection of the temperature field. This review will be concerned with actual heat sources and sinks in the region as distinguished from those factors which produce temperature changes adiabatically.

Up to at least 80 km, the major heat sources and sinks are caused by radiation, the most important radiative source being the absorption of solar

* This work was supported in part by USAF, Project RAND, and was completed while the author was a visiting faculty member of the Department of Atmospheric Sciences, University of Washington, Seattle, Washington.

† Based in part on a paper presented at the Survey Symposium on Indirect Atmospheric Measurements, Fourteenth General Assembly of IUGG, Lucerne, Switzerland, September 1966.

ultraviolet energy by ozone, which is balanced to a large extent by emission of infrared radiation by carbon dioxide. Early work on these heat balance factors has been summarized by Craig [1, Chapters 5 and 6]. In 1958, Murgatroyd and Goody [2] published a comprehensive study of the distribution of net heating and cooling in this part of the atmosphere. Their calculation extends upward to 100 km and, in addition to ozone absorption and carbon dioxide emission, it includes absorption of solar radiation by molecular oxygen, which is important above 90 km, and infrared emission by ozone which makes a secondary contribution to cooling near the stratopause. More recent calculations [3, 4] support the essential features of their work, and also agree with earlier, more limited calculations by Plass [5, 6]. Figure 1 summarizes the result of Murgatroyd and Goody which was for summer and winter

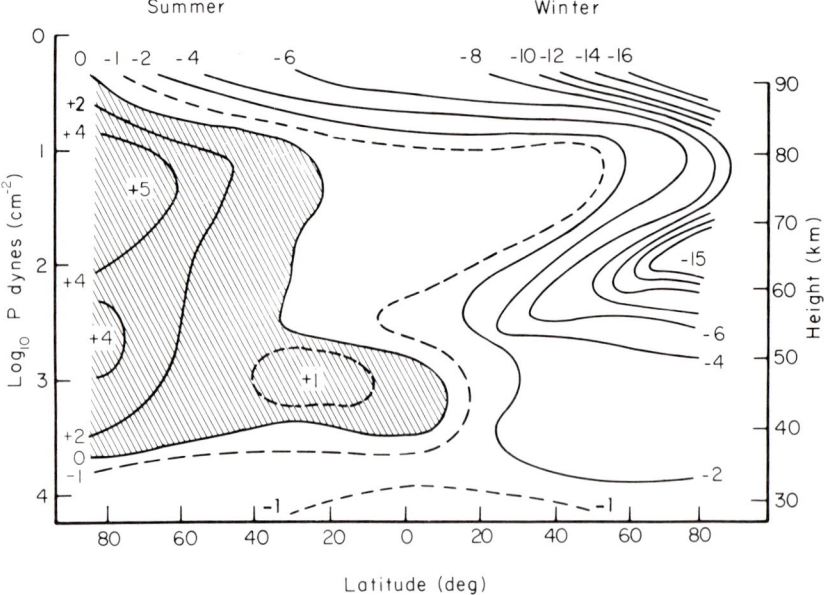

FIG. 1. Heating rates in the middle atmosphere at summer and winter solstices (°K/day), after Murgatroyd and Goody [2].

solstices, the seasons of maximum heating and cooling in most of the region. Significant heat sources and sinks appear at high latitudes, but most of the region appears to be close to radiative equilibrium. The approximately antisymmetric distribution of heat sources and sinks is associated with a nearly antisymmetric distribution of zonal winds—easterly in the summer hemisphere and westerly in the winter hemisphere—which reverses between summer and winter solstices [7–9]. The very intense net cooling during the winter gives rise to a highly baroclinic flow and is characterized by vigorous disturbances [10].

Although the calculation of heat sources and sinks below 80 km is relatively straightforward, and the results of different calculations are consistent with each other and appear to be consistent with the gross features of the wind field as well, some uncertainties in the details remain. These uncertainties are associated primarily with uncertainties in the distribution of ozone; the net heating rate is very sensitive to this factor. Furthermore, water vapor, which was not included in the study of Murgatroyd and Goody, may play a radiative role. The precise level at which the 15 μ band of CO_2 becomes subject to vibrational relaxation is an additional source of uncertainty above 70 km.

Above 80 km, the specification of the distribution of heat sources and sinks is far more difficult because nonradiative processes begin to play a major role. Vertical wave energy propagation associated with gravity, tidal, and Rossby waves appears to dominate the dynamics and may even contribute significantly to heat input through dissipation. These energy propagation processes are not yet well understood, and even order-of-magnitude estimates are difficult. Downward transport of heat by conduction, and transport of chemical energy in the form of atomic oxygen are also likely to be important. Although radiative processes continue to play a role, the upper part of the middle atmosphere, above 80 km or so, is certainly farther from radiative equilibrium than the lower part, and an understanding of its energetics will be correspondingly more difficult to achieve.

2. The Ozone Distribution

2.1. Observations of the Ozone Distribution

Figure 2 summarizes the observational information on the daytime ozone distribution between 20 and 80 km. The range of number density obtained from several different experiments [11–17] is included in the shaded region. With the exception of some data below 35 km obtained by balloon-borne chemiluminescent ozonesondes [17], all of the observations were obtained either from rockets or satellites by measurement of the attenuation of ultraviolet radiation passing through the ozone layer. The radiation monitored is either direct solar radiation, or radiation scattered by the atmosphere. In the latter case, Rayleigh scattering is assumed, and the interpretation of the data is not straightforward. Also shown in Fig. 2 are the results of a single satellite experiment giving data above 70 km [13], and the two bars at 40 and 50 km which indicate the seasonal and latitudinal range in ozone concentration at these levels obtained from the analysis of scattered radiation [18]. The bars indicate the range of observed ozone concentration under the assumption that the local ozone scale height is half that of the major constituents. These data together with the range of results obtained from the chemiluminescent ozonesondes suggest that about half of the variability indicated by the shaded zone in Fig. 2 between 30 and 50 km may be attributable to real ozone variations. On the other hand, a series of satellite observations by Miller and Stewart [16], covering a broad range of latitudes, indicate remarkable lack of variability in the ozone concentration between 40 and 60 km. Small variability in this height range with latitude, season, and synoptic situation is a feature of both photochemical ozone theories to be described in the next section [19, 20]. Thus, it is likely that most of the range indicated by Fig. 2 is due to errors in measurement or interpretation of results.

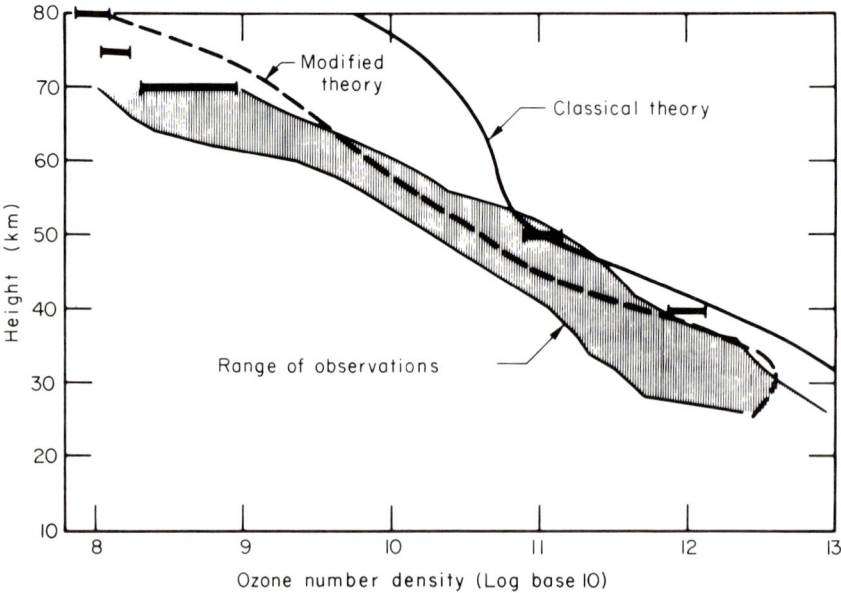

FIG. 2. Observational range in ozone number densities (shaded area) based on References [11–17], and observations by Rawcliffe et al. [13], and by Rawcliffe and Elliott [18] (horizontal bars). The theoretical curves are based on calculation by Hunt [26, 33].

The heating associated with this range in ozone density is shown in Fig. 3. These heating rates correspond to a solar elevation angle of 90° and are based on a single atmospheric density distribution [21]. Some of the ozone density variability shown may be correlated with air density variations and therefore may not be reflected in the variability of the rate of temperature change. Nevertheless, it is clear that the ozone distribution and its real variations must be better defined, in order to obtain an adequate picture of the distribution of heat sources and sinks.

Very recently, interest in the measurement of ozone above 30 km has intensified, and several new measurements by optical methods have been reported [22, 23]. The chemiluminescent method has also recently been extended upward to 60 km using small meteorological rockets [24]. Although results

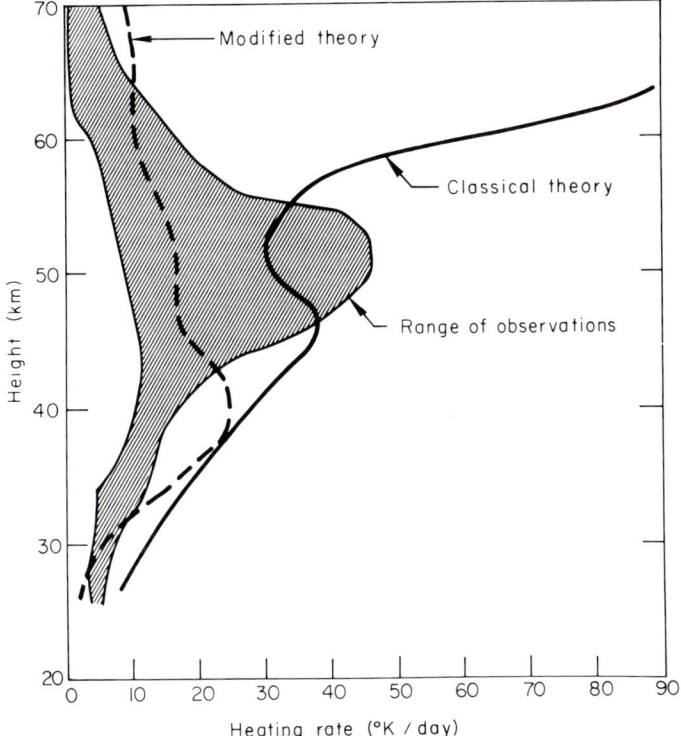

FIG. 3. Heating rates for different ozone number densities. The shaded area and solid and dashed curves correspond to those in Fig. 2.

of measurements by this technique thus far are not consistent with other observations in this height range or with heat balance requirements, combination of the chemiluminescent technique with small meteorological rockets holds promise of becoming a useful system for the investigation of ozone above 30 km.

2.2 Photochemical Theories of the Ozone Distribution

Let us consider the status of the photochemical theory of the ozone distribution. What may be called the "classical" theory of ozone photochemistry invokes only dissociation and kinetic reactions of the oxygen allotropes O, O_2, and O_3 [19, 25, 26]. The reactions and approximate rates are

(2.1) $O + O + M \rightarrow O_2 + M$ $k_1 \approx 2.7 \times 10^{-33}$ cm^6/sec

(2.2) $O + O_2 + M \rightarrow O_3 + M$ $k_2 \approx 8 \times 10^{-35} \exp(890/RT)$ cm^6/sec

(2.3) $O + O_3 \rightarrow 2 O_2$ $k_3 \approx 6 \times 10^{-11} \exp(-5700/RT)$ cm^3/sec

(2.4) $O_3 + h\nu \rightarrow O + O_2$ $J_3 \sim 10^{-2}\text{--}10^{-4}$ sec^{-1}

(2.5) $O_2 + h\nu \rightarrow 2 O$ $J_2 \sim 10^{-8}\text{--}10^{-12}$ sec^{-1}

Reactions (2.5) and (2.2) lead successively to the formation of ozone, and reactions (2.3) and (2.4) destroy ozone; since reaction (2.4) produces another oxygen atom which is then available for ozone formation, its role is to control the ratio of O to O_3. Reaction (2.1) is important above 55 km in removing oxygen atoms which might otherwise be available for ozone formation. The equilibrium concentration of ozone according to this system can be accurately approximated by

(2.6) $$[O_3] = [O_2]\left\{J_2 k_2[M] \Big/ \left(J_3 k_3 + \frac{J_3^2 k_1}{k_2[O_2]}\right)\right\}^{1/2}$$

or

(2.7) $$[O_3] \cong [O_2]\{J_2 k_2[M]/(J_3 k_3)\}^{1/2}$$

The further approximation (2.7) is valid below 55 km where reaction (2.1) is not effective. Note that in this approximation, the equilibrium ozone concentration is highly sensitive to temperature through the temperature dependence of the reaction rate k_3; we shall discuss an important implication of this in Section 2.3. Consideration of the time constants involved in the ozone photochemistry indicates that the ozone density ought to be very close to its equilibrium value between 40 and 70 km and should have a marked tendency to approach equilibrium concentrations as low as 30 and as high as 80 km. In Fig. 2, the equilibrium ozone concentration based on the classical reaction system as computed by Hunt [26] is shown by the solid curve. According to this system, the ozone number density exceeds the observed number density at almost all heights; the total ozone in an atmospheric column exceeds the observed total ozone by a factor of 3 or 4. Evidently, additional loss mechanisms are needed to explain the observed ozone concentration.

Above 60 or 70 km, it is clear that hydrogen atoms and hydroxyl radicals produced by the dissociation of water by ultraviolet radiation can accomplish the required reduction in ozone concentration [27–29]. The most important reactions form a chain:

(2.8) $\quad\quad H + O_3 \to OH + O_2 \quad\quad k_8 = 2.6 \times 10^{-4} \quad cm^3/sec$

(2.9) $\quad\quad OH + O \to H + O_2 \quad\quad k_9 = 5 \times 10^{-11} \quad cm^3/sec$

The quantities of H and OH needed to bring about the required removal of ozone are consistent with the probable quantities produced in the ultraviolet dissociation of water. They are also consistent with the observed intensity and height distribution of the hydroxyl nightglow emission, if it is assumed that the latter is a consequence of vibrationally excited hydroxyl radicals produced in reaction (2.8), and if mixing and diffusion processes are also taken into account [29, 30].

This mechanism cannot explain the additional loss of ozone below 60 or 70 km, because the ultraviolet radiation needed to initiate the process by dissociating H_2O does not penetrate deeply enough into the atmosphere. Hampson [31] has proposed an additional mechanism for water vapor dissociation based on laboratory evidence [32].

In the laboratory, it is found that the quantum yield for photolysis of ozone by ultraviolet radiation in the presence of water vapor is much higher than the corresponding quantum yield for dry ozone. On the other hand, the quantum yield for ozone photolysis by radiation in the visible region (wavelength > 4000 Å) is not enhanced by the presence of H_2O. This evidence, coupled with the observation of the absorption spectrum of OH in the wet photolysis experiment, suggests a reaction chain initiated by excited atomic oxygen in the ^1D state. An atmospheric photochemical system based on such a chain has been studied numerically by Hunt [33]. The treatment of this scheme given here is that of Leovy [20]; a similar approach has been developed independently by Dütsch [34].

Consider reactions (2.1)–(2.5), and (2.9), together with

(2.10) $\quad\quad O_3 + h\nu \quad\quad \to O(^1D) + O_2 \quad\quad J_3^* \sim 10^{-4}\text{–}10^{-2} \quad sec^{-1}$

(2.11) $\quad\quad O(^1D) + H_2O \to 2\,OH \quad\quad k_{11} = 10^{-11} \quad cm^3/sec$

(2.12) $\quad\quad OH + O_3 \quad\quad \to O_2 + HO_2 \quad\quad k_{12} = 5 \times 10^{-13} \quad cm^3/sec$

(2.13) $\quad\quad HO_2 + O_3 \quad\quad \to 2\,O_2 + OH \quad\quad k_{13} = 10^{-14} \quad cm^3/sec$

(2.14) $\quad\quad H + O_2 + M \to HO_2 + M \quad\quad k_{14} = 7.4 \times 10^{-32} \quad cm^6/sec$

(2.15) $\quad\quad HO_2 + O \quad\quad \to OH + O_2 \quad\quad k_{15} = 10^{-11} \quad cm^3/sec$

(2.16) $OH + OH \rightarrow H_2O + O$ $k_{16} = 2.8 \times 10^{-12}$ cm^3/sec
(2.17) $OH + HO_2 \rightarrow H_2O + O_2$ $k_{17} = 10^{-11}$ cm^3/sec
(2.18) $O(^1D) + M \rightarrow O + M$ $k_{18} = 10^{-12}$ cm^3/sec

The reaction rates indicated here correspond to the values assumed by Hunt [33]. The dissociation reaction (2.10) differs from (2.4) in that (2.4) includes dissociation of O_3 to form both ground state and excited atomic oxygen. Reaction (2.10) on the other hand only applies to wavelengths shorter than about 3100 Å and produces atomic oxygen in the ^1D state. Reaction (2.18) quenches $O(^1D)$, and is a much more likely process below the mesopause than the radiative transition

(2.19) $O(^1D) \rightarrow O(^3P) + h\nu$ (wavelength 6300 Å)

With these values for the rate coefficients, the following stationary-state concentrations would be obtained in a time scale of 100 sec or less in the ozone layer.

(2.20) $[O(^1D)] \approx J_3^*[O_3]/k_{18}[M]$

(2.21) $[O]/[O_3] \approx J_3/k_2[M][O_2] \equiv r_1$

(2.22) $[H] \approx k_9[O][OH]/k_{14}[O_2][M]$

(2.23) $[HO_2]/[OH] \approx \dfrac{k_{12} + k_9 r_1}{k_{13} + k_{15} r_1} \equiv r_2$

Substituting these stationary-state relations into the kinetic equations for the rates of change of $([O] + [O_3])$ and $([OH] + [HO_2])$, one obtains

(2.24) $\dfrac{d[O_3]}{dt} = A - C[O_3]^2 - B[OH][O_3]$

and

(2.25) $\dfrac{d[OH]}{dt} = D[O_3] - E[OH]^2$

where

$A = 2J_2[O_2]/(1 + r_1)$ $D = 2k_{11}J_3^*[H_2O]/\{k_{18}[M](1 + r_2)\}$
$B = 2(k_{12} + k_{13} r_2)/(1 + r_1)$ $E = 2(k_{16} + k_{17} r_2)/(1 + r_2)$
$C = 2(k_3 + k_1 r_1)r_1/(1 + r_1)$

In deriving (2.24) and (2.25), it has been assumed that $([OH] + [HO_2] \ll [H_2O])$, so that $[H_2O]$ variations need not be considered. In addition, the production of atomic oxygen by reaction (2.16) has been neglected compared with production by photodissociation, and the time dependence of the ratios

r_1 and r_2 associated with the zenith angle dependence of J_3 has been neglected.

When the ozone concentration is controlled by reactions with the hydrogen compounds, the last term on the right-hand side of Eq. (2.25) is larger than the term in $[O_3]^2$, and the equilibrium concentrations of O_3 and OH are approximately given by

(2.26) $$[O_3]_e \simeq (C/B)^{2/3}(E/D)^{1/3}$$

and

(2.27) $$[OH]_e \simeq \{DC/(EB)\}^{1/3}$$

It is convenient to use these equilibrium concentrations to render Eqs. (2.24) and (2.25) dimensionless. Defining the new variables

$$x = (C/B)^{-2/3}(E/D)^{-1/3}[O_3], \quad y = \{EB/(DC)\}^{1/3}[OH], \quad \tau = (CDB^2/E)^{1/3}t$$

Eqs. (2.24) and (2.25) transform to

(2.28) $$dx/d\tau = 1 - \alpha x^2 - xy$$

(2.29) $$\beta \, dy/d\tau = x - y^2$$

where

(2.30) $$\alpha = AC^{1/3}E^{2/3}/(B^{4/3}D^{2/3})$$

and

(2.31) $$\beta = B/E$$

The parameter α measures the relative importance of the "classical" reactions (2.1) and (2.3) to those involving OH and HO_2 in limiting the ozone concentration. When $\alpha < 1$, the free radical reactions control the ozone photochemistry, and the equilibrium ozone concentration is given by Eq. (2.26). The reaction rates assumed by Hunt lead to small values of α everywhere in the height range 15–60 km; for a uniform water vapor mixing ratio of 3×10^{-6} gm/gm and solar elevation angle of 90°, the maximum value of α is 0.2, occurring near 40 km.

The dashed curve labeled "modified theory" in Fig. 2 gives the equilibrium ozone concentration computed by Hunt on the basis of a somewhat more complicated reaction system than the one given here. Hunt's values were obtained by time integration of the kinetic equations, but, below 60 km, they agree closely with the concentration values given by Eq. (2.26).

The improved agreement between the modified theory and the observed ozone concentrations supports the view that ozone is limited by reactions with hydrogen compounds even in the stratosphere. This reaction system depends on a number of unknown reaction rates, however. The rate constant k_{18} for deactivation of $O(^1D)$ has been estimated from the intensity of the 6300 Å airglow line corresponding to the transition (2.19) [35, 36], and from the intensity of the same line in the aurora [37]. These estimates all support a rate $k_{18} \geq 4 \times 10^{-11}$ cm^3/sec, some 40 times faster than the value assumed by Hunt. On the other hand, the rate constant for dissociation of H_2O, k_{11}, is also likely to be more than an order of magnitude faster than the value assumed by Hunt (see Schofield [38]). Since the ozone concentration depends only on the ratio k_{11}/k_{18}, and not on the individual values, the rates assumed by Hunt for these reactions may lead to the correct results. The greatest uncertainty is associated with reactions (2.12) and (2.13); the rates of these reactions are unknown, and below 40 km the results are quite sensitive to these rates.

There are other possible explanations for the discrepancy between observed ozone density and ozone density calculated on the basis of the classical theory. For example, Brewer and Wilson [39] have measured rates of oxygen dissociation below 30 km which are consistent with lower intensity in the relevant portion of the solar spectrum and with somewhat smaller oxygen absorption coefficients than those which have been used in the photochemical calculations [40, 41]. If their values are correct, the ozone density computed on the basis of the classical theory would be reduced to about half of that shown in Fig. 2 and would be more nearly consistent with the observed values between 30 and 55 km. It has also been suggested that reactions with NO and NO_2 may play a role in limiting the ozone concentration in the stratosphere [33].[1]

2.3. Ozone and Atmospheric Dynamics

The question of whether the ozone density below 60 km is controlled by the hydrogen compounds or by the classical reaction system is of great importance in understanding the interaction between ozone and circulation processes. One important aspect of the photochemical process is the time required for equilibrium to be reached in the ozone concentration. From an analysis of Eqs. (2.29) and (2.30), it follows that when $\alpha \gg 1$, and the classical photochemical system determines the equilibrium ozone concentration, the equilibrium time scale is

(2.32) $$t(O_3) \cong \alpha^{-1/2}\{E/(CDB^2)^{1/3}\} = (AC)^{-1/2}$$

---

[1] The high NO concentration (mixing ratio $\sim 10^{-6}$) recently reported by J. B. Pearce in the upper mesosphere supports this suggestion. This observation is described in Pearce's Ph.D. thesis [41a].

On the other hand, if $\alpha \ll 1$, and the equilibrium ozone concentration corresponds to Eq. (2.26), the equilibrium time scale is

(2.33) $\quad t(O_3) \simeq \mathrm{Max}\{(E/(CDB^2))^{1/3}, (B/(CDE^2))^{1/3}\}$

These two time scales are shown in Fig. 4 for a representative low-latitude zenith angle of 48°. The most interesting region is below 30 km. According to the classical system, the equilibrium time is so long that ozone is virtually unaffected by photochemical processes; it should behave as a passive tracer. The equilibrium time according to the modified system is significantly shorter. Below 30 km, it is short enough so that ozone densities would be influenced by both photochemical and circulation processes.

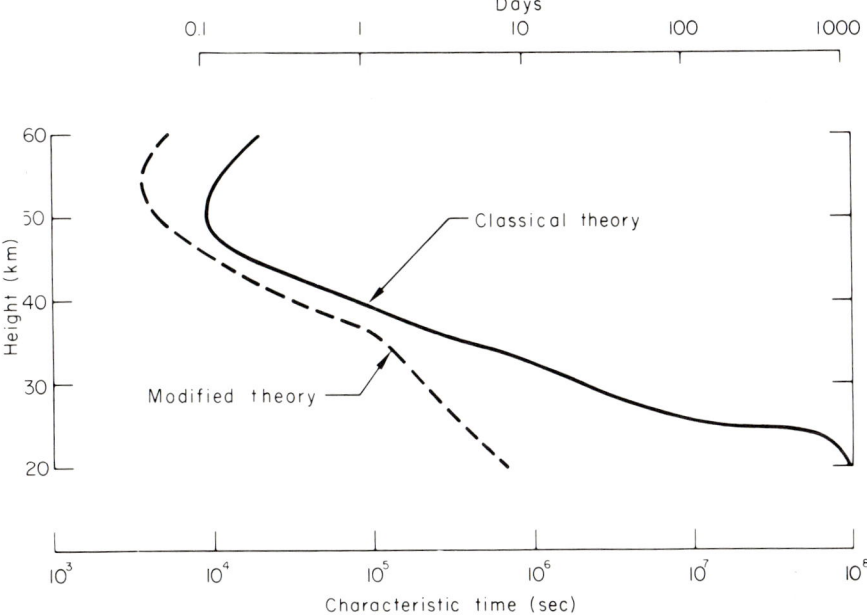

FIG. 4. Photochemical time constants for ozone based on the "classical" and "modified" theories, assuming vertically overhead sun. The reaction rates used to compute these times are those used by Hunt [33].

Another way in which the ozone photochemistry may interact with circulation processes is through the temperature sensitivity of its equilibrium concentration according to the classical system [see Eqs. (2.3) and (2.7)]. Because the equilibrium concentration of Eq. (2.7) decreases with temperature, and because ozone absorbs solar radiation at a rate proportional to its concentration, the net effect of this temperature sensitivity is to damp out any temperature deviations from the local equilibrium temperature. This effect is of maximum importance near the mesopeak where the characteristic thermal damping time is about 5 days. Circulation systems having time scales of the order of 5 days or longer would be strongly influenced [42–47]. Thermal damping associated with ozone density variations does not occur if the ozone concentration is controlled by reactions with hydrogen compounds, provided that none of the key reactions are strongly dependent on temperature. Between 40 and 60 km, reactions (2.9) and (2.15) would comprise the most important ozone-limiting chain and these reactions are known to be too fast to be strongly dependent on temperature [48].

3. The Water Vapor Distribution

From our point of view, the water vapor content of the middle atmosphere may be important for two reasons. In the first place, it interacts photochemically with the oxygen compounds, and its dissociation products exert a control on the ozone concentration in the mesosphere. If the modified reaction system is correct, it may also control the ozone concentration in the stratosphere. If the stratospheric water vapor mixing ratio is more than a few parts per million, the stringent requirements on reaction rates required for the modified reaction system to be effective are relaxed; furthermore, if there is any variability in water vapor concentration, it might be reflected in variations in ozone concentration, although, according to Eq. (2.27), the equilibrium ozone concentration only depends on the cube root of the water vapor concentration.

In the second place, water vapor has a direct influence on radiative heating rates, since it emits and absorbs in the infrared. Stratospheric water vapor also absorbs solar radiation, but the heating effect of this absorption is probably very small above 25 km [49]. Figure 5 illustrates the effect of different assumptions about the water vapor concentration on the rate of temperature change owing to infrared emission by water vapor according to calculations by Rodgers and Walshaw [50] and by Kuhn [3]. Both calculations involve "wet" and "dry" stratospheres (mixing ratios 10^{-4} gm/gm and 10^{-6} [3] or 2×10^{-6} [50] gm/gm above 30 km, respectively). Differences between the two calculations arise from small differences in assumed temperature and moisture distributions and in the infrared band models assumed for the calculations. In addition, Kuhn's calculation did not include the comparatively small effect of the 6.3 μ band in the dry case. Despite these differences, both calculations indicate cooling rates reaching a maximum of approximately 4°K/day near the stratopause for the wet case, and not exceeding 1°K/day for the dry case. In the wet case, this constitutes an important but not

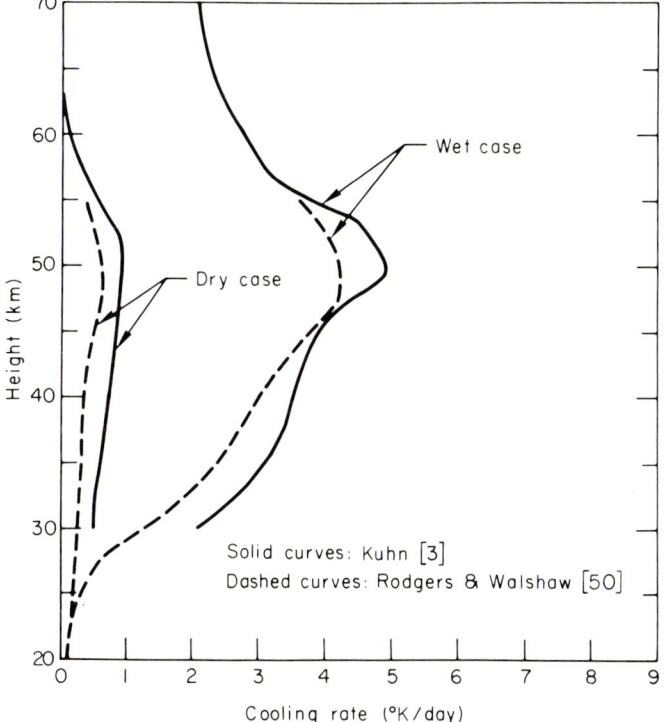

FIG. 5. Cooling rates due to water vapor for "wet" and "dry" cases.

dominant contribution to total radiative cooling. In the dry case, the cooling contribution of water vapor is negligible.

The answer to the question of what the actual water vapor concentration is has been one of the most elusive in stratospheric meteorology. Gutnik [51] summarized the confused state of stratospheric moisture observations in 1961. Since that time, the situation has improved, particularly in the lower stratosphere, but doubt remains as to the water vapor concentration above 25 km. Two methods have been employed for most of the observations: frost point hygrometers [52–56] or spectroscopic measurements of absorption of solar radiation by near infrared water vapor bands [57–60]. The instruments have been flown in aircraft as well as on balloons and the difficulties of measuring the very low stratospheric water vapor concentrations are greatly enhanced by the extreme care which must be taken to prevent contamination by water vapor associated with the instrument system. Both the frost point hygrometers and the spectroscopic method are subject to contamination problems [54, 61, 62].

For observations in which serious contamination effects seem to have been avoided, both hygrometer and spectroscopic observations indicate remarkably uniform water vapor mixing ratios in the range $1\text{--}4 \times 10^{-6}$ gm/gm from just above the tropopause up to 20 or 25 km. Above 25 km, the picture is not so clear. Relatively few measurements have been made at these heights. Spectroscopic measurements are subject to ambiguity of interpretation above 20 km [58, 63]. Some frost-point measurements have indicated continued dryness with mixing ratios of $1\text{--}4 \times 10^{-6}$ gm/gm up to 30 km [64]. In a year-long series of frost point hygrometer ascents at Washington, D.C., and Trinidad, Mastenbrook [55] has found mixing ratios predominantly in the range $1\text{--}4 \times 10^{-6}$ gm/gm above 20 km, but with some seasonal variations, and with greater moisture concentrations at the higher levels on a few ascents. Brown et al. [65] flew a cooled vapor trap on a balloon between 25 and 28 km and found mixing ratios ranging from 9 to 77×10^{-6} gm/gm at these heights. They were able to rule out the possibility of contamination from the balloon by introducing a small amount of deuterium into the balloon as a tracer. The possibility remains, however, of some contamination from other parts of their system, although great care was taken to eliminate all sources of contamination. Recent frost-point measurements by Sissenwine et al. [56] also indicate increasing water vapor concentration above 20 km, with mixing ratios of 20×10^{-6} gm/gm at 25 km. The occasional occurrence of mother-of-pearl clouds in this portion of the stratosphere has been cited as indirect evidence for mixing ratios as high as 10^{-4} gm/gm [66].

It is difficult to account for an increase in the water vapor mixing ratio at the higher stratospheric levels, since the most obvious source for stratospheric water vapor is the troposphere. Only one attempt to measure water vapor at mesospheric levels has been reported. Fedynskii et al. [67] used the humidity dependence of the resistivity of a tungsten-rhenium filament to measure humidity between 70 and 88 km. They report extremely high mixing ratios—up to 7×10^{-3} gm/gm at 78 km. The difficulties in measuring ambient concentrations at such low atmospheric pressures are extreme, however, and the question of the actual water vapor concentration in both the upper stratosphere and the mesosphere remains unresolved.

4. Infrared Radiative Transfer

Since one of the dominant terms in the heat budget of the middle atmosphere is energy exchange by infrared radiative transfer, considerable attention has been devoted to the computational aspects of this complex problem. The spectral features contributing significantly to the exchange process are the 15 μ bands of carbon dioxide and the 9.6 μ band of ozone, and, as noted in Section 3, possibly also the rotational band of water vapor. These bands exhibit extremely complex fine structure, the details of which influence the energy exchange. One of the fundamental problems in heating rate calculations has been to decide which aspects of detailed band structure must be accurately taken into account and which can be accounted for by relatively crude modeling assumptions. In the middle atmosphere, it seems to be necessary to take rather accurate account of the strength and width of spectral lines, and of their shape in the wings. The exact line positions and the line profiles near line centers are not so important. Several aspects of the infrared transfer problem in the middle atmosphere distinguish it from the comparable problem for the lower atmosphere, and these will be discussed in this section.

4.1. Curtis' Method for Calculation of Heating Rates

Flux divergences in the stratosphere and mesosphere are very small, so that methods of computation based on taking differences in fluxes such as are commonly useful in the lower atmosphere [68] require excessively accurate transmission functions. A formulation involving direct calculation of the heating rate, which was introduced by Curtis [69], largely avoids this problem, however.

The monochromatic heating rate per unit volume at frequency ν and height z_0 is

$$(4.1) \qquad h_\nu(z_0) = -\int dI_\nu/ds \, d\Omega$$

where dI_ν/ds is the rate of change with distance of monochromatic radiation intensity in the solid angle element $d\Omega$, and the integration is over all solid angles. If n_r is the number density of radiatively active molecules, and σ_ν is the monochromatic absorption cross section, the transfer equation takes the form

$$(4.2) \qquad dI_\nu/ds = -n_r \sigma_\nu (I_\nu - S_\nu)$$

for a monochromatic source function S_ν, so that the monochromatic heating rate can be written

$$(4.3) \qquad h_\nu(z_0) = -4\pi n_r \sigma_\nu (\bar{I}_\nu - \bar{S}_\nu)$$

where $\bar{I}_\nu$ and $\bar{S}_\nu$ are angular averages of I_ν and S_ν. For thermal radiation, S_ν is isotropic up to heights well above 100 km, so that $\bar{S}_\nu \equiv S_\nu$. For a plane, horizontally stratified, and semi-infinite atmosphere, the solution of Eq. (4.2) leads to the following expression for $\bar{I}_\nu$ [70, Chapter 2]:

(4.4) $$\bar{I}_\nu = -\frac{1}{2}\bigg\{\int_0^{a_\nu} S_\nu(u_\nu)\frac{dE_2(u_\nu)}{du_\nu}du_\nu$$
$$+ \int_0^{b_\nu} S_\nu(u_\nu)\frac{dE_2(u_\nu)}{du_\nu}du_\nu$$
$$- S_\nu(0)E_2(b_\nu)\bigg\}$$

where

(4.5) $$u_\nu(z, z_0) = \left|\int_{z_0}^z n_r \sigma_\nu\, dz\right|$$

and

$$a_\nu(z_0) = u_\nu(z_0, \infty), \qquad b_\nu(z_0) = u_\nu(z_0, 0),$$

and $E_2(u_\nu)$ is the second exponential integral, defined by

(4.6) $$E_n(u_\nu) = \int_1^\infty \exp(-vu_\nu)v^{-n}\, dv.$$

$S_\nu(0)$ is the source function for the lower boundary, ground or cloud-top, at $z=0$. Substituting (4.4) into (4.3), and integrating over a band, or a portion of a band, of width $\Delta\nu$ which is narrow enough so that the variation of S_ν with frequency can be neglected, we obtain the band heating rate h_r,

(4.7) $$h_r = -2\pi n_r(z_0)\bigg\{-\int_0^{z_0}(S_r(z) - S_r(z_0))\bigg(\int_{\Delta\nu}\sigma_\nu(z_0)\frac{dE_2(z, z_0)}{dz}d\nu\bigg)dz$$
$$+ \int_{z_0}^\infty (S_r(z) - S_r(z_0))\bigg(\int_{\Delta\nu}\sigma_\nu(z_0)\frac{dE_2(z, z_0)}{dz}d\nu\bigg)dz$$
$$+ (S_r(z_0) - S_r(0))\bigg(\int_{\Delta\nu}\sigma_\nu(z_0)E_2(b_\nu)\,d\nu\bigg)$$
$$+ S_r(z_0)\bigg(\int_{\Delta\nu}\sigma_\nu(z_0)E_2(a_\nu)\,d\nu\bigg)\bigg\}$$

where S_r is the average value of S_ν for the band. The integrals in this expression can be evaluated by quadrature, in the form

(4.8) $$\int_{z_j}^{z_m}(S_r(z_i) - S_r(z_j))G(z_i, z_j)\,dz$$
$$\cong \sum_{i=j}^m (S_r(z_i) - S_r(z_j))\{w_{ji}G_{ji}(z_i, z_j)\}$$

where

(4.9) $$G(z_i, z_j) = \int_{\Delta\nu}\sigma_\nu(z_j)\frac{dE_2(z_i, z_j)}{dz}d\nu$$

and w_{ji} is an appropriate weighting function depending on the particular quadrature technique. The matrix elements $(w_{ji} G_{ji})$ depend on temperature only through the temperature dependence of line strength, and it is probably sufficient to evaluate them using a standard atmosphere temperature distribution. These elements can then be evaluated once and for all for a given vertical distribution of absorbing gas. This evaluation is particularly straightforward when spectral lines do not overlap, since, in this case, only integrations over individual lines and summations over lines of different strength are needed to carry out the frequency integrations appearing in G_{ji}. These matrix elements have been evaluated by Curtis [69] for nonoverlapping lines having Poisson distributions of line strengths, and for a constant mixing ratio of radiatively active gas. The matrix elements were used by Murgatroyd and Goody [2] to calculate the heating rate due to CO_2 infrared transfer, and they were used by Leovy [19] in computing the radiative equilibrium temperature distribution in the middle atmosphere.

The formulation of Eq. (4.7) explicitly displays the contribution of "cooling to space" to the total heating rate. This effect is represented by the last term in Eq. (4.7). For an isothermal atmosphere above a lower boundary at the same temperature, this term is the only one contributing to the heating rate, and Rodgers and Walshaw [50] have shown that it is an excellent approximation for the actual atmosphere in almost all cases from the lower troposphere up to at least 55 km.

4.2. Influence of Line Shape and Line Spacing

The line shape is described in terms of the line-shape factor $f(\nu)$,

(4.10) $$f(\nu) = \sigma_\nu \Big/ \int \sigma_\nu \, d\nu ,$$

where the integral extends over the entire line. At high pressures, the pressure broadening effect dominates, and the shape corresponds approximately to the Lorentz profile,

(4.11) $$f(\nu) = \frac{1}{\pi} \frac{\alpha_L}{(\nu - \nu_0)^2 + \alpha_L^2} ,$$

where ν_0 is the central frequency of the line, and α_L is the Lorentz line width. At the very low pressures, thermal motion of the molecules is the main cause of the line dispersion, and the Doppler profile is a good representation of the shape:

(4.12) $$f(\nu) = \alpha_D^{-1} \pi^{-1/2} \exp(-(\nu - \nu_0)^2/\alpha_D^2),$$

where α_D is the Doppler line width. For the H_2O, CO_2, and O_3 infrared bands, the Lorentz and Doppler widths become equal within the range of heights 25–35 km, the exact height depending on the particular band [71, p. 72]. Over a range of several scale heights above this level, both pressure and thermal effects are important, and a mixed line profile must be used. This profile is given by

$$(4.13) \qquad f(\nu) = \frac{p}{\pi} \int_{-\infty}^{\infty} \frac{\exp(-v^2)\, dv}{p^2 + (q-v)}$$

where $p = \alpha_L/\alpha_D$, and $q = (\nu - \nu_0)/\alpha_D$. Deviations from these profiles occur because of the complexity of real molecular interaction processes. The effect of such profile deviations on heating rates is probably small [70, p. 166].

Up to heights at which the lines have essentially a pure Doppler profile (60 km or more), the lines become increasingly narrow with increasing height, and Shved [72] has shown that line widths are sufficiently small above 35 km that line overlap can be neglected in heating rate computations for the 15 μ band with a maximum error of only a few per cent. This conclusion holds even for the very closely spaced lines of the Q branch of the 15 μ fundamental. As mentioned in Section 4.1, this leads to a considerable simplification in the calculations, since the relatively complex band models designed to take overlap into account need not be used. The validity of this conclusion is supported by the good agreement in the results of calculations using quite different assumptions about line overlap and line strength distributions [2–6].

4.3. Vibrational Relaxation

At high levels in the mesosphere, departures from local thermodynamic equilibrium alter the rates of radiative energy exchange. Local thermodynamic equilibrium (LTE) is the state in which the molecular energy levels are populated in accordance with the Boltzmann distribution corresponding to the local kinetic temperature. Both the radiation field and collisions affect the populations; collisions tend to produce a distribution of energy levels in accordance with LTE, while radiation tends to produce a distribution which is in equilibrium with the local energy density of the radiation field. When collisions become sufficiently infrequent, the distribution of molecular energy levels takes a finite time to adjust to the local kinetic temperature. This time is the *vibrational relaxation time*, and when it becomes equal to the lifetime of the excited molecular states (which is inversely proportional to the band strength), the distribution of these states is determined by the radiation field as well as by the local temperature. The height at which this occurs is the vibrational relaxation level. For the 15 μ fundamental of carbon dioxide, which is the most important infrared band in the stratosphere and mesosphere, the vibrational relaxation level occurs between 75 and 85 km [73]. Since radiative energy exchange depends on the difference between the distribution of energy among the molecular energy levels and the energy content of the local radiation field, energy exchange, and hence heating and cooling rates will be reduced above the vibrational relaxation level.

Kuhn [3] has shown that a molecular model involving only two vibrational levels, the ground state and one excited level, provides an adequate model of the vibrational relaxation process for the 15 μ fundamental of CO_2. (The rotational energy distribution is determined by collisions to very great heights, so that we need not concern ourselves with this distribution [73].) For such a model, we can write down an approximate equation for the heating rate in terms of the number density of molecules in the excited state, n_2, following Shved [74]:

$$(4.14) \quad h_r = -\int_{\Delta\nu}\int \frac{dI_\nu}{ds} d\Omega\, d\nu$$

$$= n_r \int_{\Delta\nu}\int \sigma_\nu (I_\nu - S_\nu)\, d\nu\, d\Omega$$

$$\cong n_r \int_{\Delta\nu}\int \sigma_\nu \{I_\nu - (n_2/\bar{n}_2) B_\nu\}\, d\nu\, d\Omega,$$

where $\bar{n}_2$ is the population of the excited state in local thermodynamic equilibrium ($\bar{n}_2$ corresponds to the Boltzmann distribution), and B_ν is the blackbody intensity. In local thermodynamic equilibrium $n_2 = \bar{n}_2$, and the source function is simply the blackbody intensity. When radiation determines the distribution,

$$(4.15) \quad (n_2/\bar{n}_2) \cong \int_{\Delta\nu}\int \sigma_\nu I_\nu\, d\Omega\, d\nu \bigg/ \left(4\pi \int_{\Delta\nu}\int B_\nu\, d\nu\, d\Omega\right)$$

and the rate of temperature change vanishes. For intermediate cases, $(n_2/\bar{n}_2)$ is determined by the rates at which the vibrational energy levels are populated by collisions and by radiation. These processes take place on a much faster time scale than that of temperature changes, so they can be assumed to be in equilibrium with each other. Then

$$(4.16) \quad -\int\int (dI_\nu/ds)\, d\Omega\, d\nu + h\nu_0 (dn_2/dt)_c \cong 0$$

where the first term is the rate of change of vibrational energy caused by radiation and the second is the rate of change owing to collisions. Here ν_0 is the frequency at the band center. For the two level model, the rate of change of n_2 by collisions $(dn_2/dt)_c$ is proportional to n_2, so that we can close the system consisting of Eqs. (4.14) and (4.16) with the relation

$$(4.17) \quad (dn_2/dt)_c = -\lambda^{-1}(n_2 - \bar{n}_2)$$

where λ, the vibrational relaxation time, is inversely proportional to air pressure. For the plane, horizontally stratified, and semi-infinite atmosphere, this system of equations can be solved by finite difference and matrix techniques [3], or by finite differences and successive approximations to the source function [73].

Combining Eqs. (4.14), (4.16), and (4.17), and making use of the fact that λ is inversely proportional to pressure, we can exhibit the pressure dependence of h_r for a given field of radiant intensity, I_ν:

$$(4.18) \quad h_r \cong \frac{n_r}{1 + (P_\mathrm{v}/P)} \left\{ \int_{\Delta\nu} \int \sigma_\nu I_\nu \, d\Omega \, d\nu - 4\pi \int \int B_\nu \, d\Omega \, d\nu \right\}$$

where P is the total air pressure, and P_v the air pressure at the vibrational relaxation level. Thus when $P \ll P_\mathrm{v}$, h_r decreases with height in proportion to the air pressure. For the 15 μ fundamental of CO_2 and sea-level pressure, λ lies between 10^{-5} and 10^{-6} sec. The uncertainty in its value is about one order of magnitude and it is probably temperature dependent [3, 75].

Vibrational relaxation is also important for absorption of solar radiation by near infrared bands. This absorption could become intense at high levels where the lines are unsaturated, but McClatchey has shown that vibrational relaxation probably occurs far below those heights at which the stronger lines become unsaturated, and the resulting heating rates due to these bands is probably very small throughout the middle atmosphere [76]. Figure 6, from

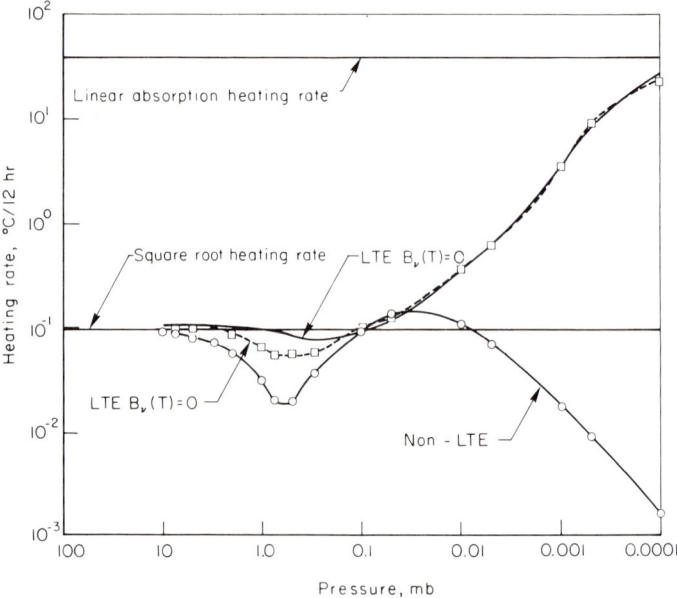

FIG. 6. Heating rates due to the 4.3 μ band of CO_2 for LTE and non-LTE conditions. The solid curves neglect the contribution from atmospheric emission; the dashed curve includes this effect (after McClatchey [76]).

McClatchey's work, illustrates this effect for the 4.3 μ band of CO_2. Heating rates for the absorption of solar radiation are shown with and without the effect of departures from LTE. When vibrational relaxation is neglected, the computed heating rate rises to several tens of degrees per day at low pressure, when, in fact, the actual heating rates are probably negligible at all heights. The dashed curve in Fig. 6 shows the effect of including thermal emission in this calculation. The two horizontal lines are asymptotic heating rates for very weak absorption (linear absorption) and for absorption by the wings of nonoverlapping Lorentz lines (square root absorption). The dip in the curves near 1.0 mb is caused by the predominance of the Doppler line profile at these pressures.[2]

4.4. Heating Rate Caused by Infrared Transfer

Kuhn has taken the special features of the middle atmosphere radiative transfer problem into account and has computed the distribution of cooling and heating for summer and winter seasons due to infrared transfer by carbon dioxide, ozone, and water vapor, assuming a water-vapor mixing ratio of 10^{-6} gm/gm [3]. His results for relaxation time $\lambda = 2 \times 10^{-5}$ sec at sea-level pressure are shown in Fig. 7. They show a broad band of cooling at the stratopause reaching a maximum of 10°K/day in the tropics. Except in winter polar latitudes, his calculations show warming near the mesopause reaching 4°K/day at the summer pole. This heating is probably real, since it is also a feature of other calculations [2, 4]. It is difficult to understand what processes may balance this heating. Neglected diabatic processes in this region only involve heating, and although atmospheric motions may easily produce cooling at some latitudes and seasons, it is not likely that they could produce net cooling in the global mean near the mesopause. Above 85 km, Kuhn's calculations indicate marked cooling especially in high winter latitudes. The magnitude of this high level cooling depends markedly on the assumed vibrational relaxation time, λ. For $\lambda = 2 \times 10^{-5}$ sec at sea-level pressure, the maximum cooling is 12°K/day near 90 km in the winter. For $\lambda = 2 \times 10^{-6}$ sec at sea level, Kuhn has calculated that cooling rates would be as large as 45°K/day at 98 km.

There are some differences between the cooling rates caused by the 15 μ CO_2 band as calculated by Murgatroyd and Goody and as calculated by Kuhn. These are primarily caused by differences in the assumed temperature distributions. The cooling rates are rather sensitive to the assumed temperature distributions, but at most heights it is possible to estimate quite simply,

---

[2] J. T. Houghton of Clarendon Laboratory, Oxford, has recently computed the heating due to absorption of solar radiation by CO_2, and estimates that this amounts to some 2°K (12 hr)$^{-1}$ near the mesopause. A preliminary account of these results was reported at the Survey Symposium on Radiation, IAMAP, IUGG, Lucerne, September 1967, and they are to be reported in full in a paper [76a].

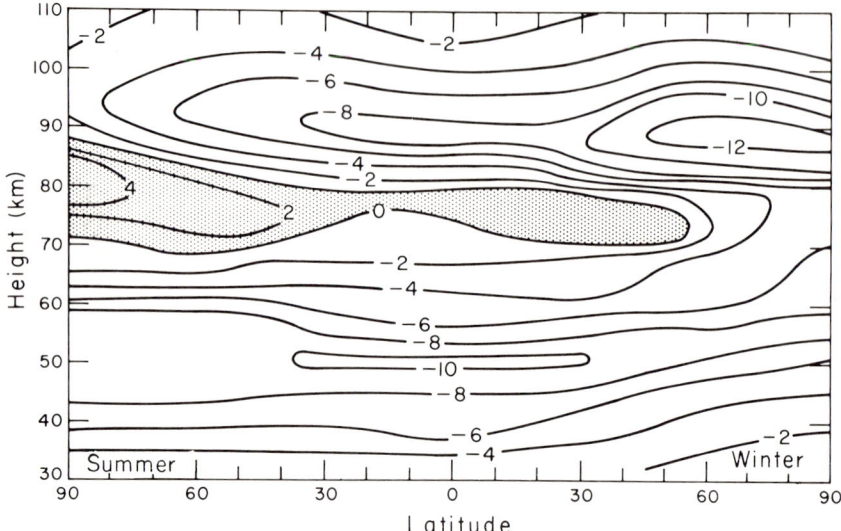

FIG. 7. Infrared cooling and heating rates due to carbon dioxide, ozone, and water vapor for $\lambda = 2 \times 10^{-5}$ sec (after Kuhn [3]).

and with reasonable accuracy, the effect on the cooling rate of small differences in temperature from a given temperature distribution. This point will be discussed further in the next section.

The contribution of water vapor to the total infrared cooling is quite small, if the stratosphere and mesosphere are as dry as Kuhn assumed them to be. On the other hand, if significant concentrations of water penetrate above the mesopause, water vapor could contribute to cooling at the higher levels, since the rotational band is not subject to vibrational relaxation. It is often assumed that water vapor cooling above the mesopause must be insignificant because H_2O is subject to photodissociation at these heights, but the dissociation rate is slow, and transport and turbulent diffusion processes could easily carry water vapor to well above 90 km [30].

4.5. Infrared Radiative Exchange and Atmospheric Dynamics

Murgatroyd and Goody [2] found during the course of their calculations that cooling owing to infrared emission by CO_2 could be represented surprisingly well by a simple linear dependence on the local blackbody intensity, with proportionality coefficients depending on the height. This simple relationship is a consequence of the accuracy of the approximate expression for the heating rate obtained by neglecting all terms except the "cooling to space" term in Eq. (4.7). With this approximation, the heating rate at height z_0 is simply

(4.19) $$h_r(z_0) \cong -2\pi n_r S_r(z_0)\left(\int_{\Delta\nu} \sigma_\nu(z_0) E_2(a_\nu)\, d\nu\right)$$

Rodgers and Walshaw [50] have investigated the accuracy of this approximation in the troposphere and stratosphere for water vapor, ozone, and carbon dioxide, and, with a few notable exceptions, they find that it gives heating rates which are accurate to within 20%. The worst exception is the infrared heating rate for O_3 in the stratosphere below 30 km. This heating rate arises from the predominant role played by exchange of radiation between the lower part of the ozone layer, at comparatively low temperatures, and the ground or cloud-tops at comparatively high temperatures. Most of the heating effect in this case could be accounted for by including the other boundary term from Eq. (4.7),

$$2\pi n_r(S_r(0) - S_r(z_0))\left(\int_{\Delta\nu} \sigma_\nu(z_0) E_2(b_\nu)\, d\nu\right)$$

in Eq. (4.19).

Equation (4.19) expresses the heating rate at z_0 in terms of only the source function at z_0, the volume absorption coefficient at z_0, and the optical depth (a_ν) above z_0. If we neglect the temperature dependence of absorption cross-section and vibrational relaxation effects, we can write (4.19) as

(4.20) $$h_r(z_0) \cong -K_R^*(z_0) B_r(z_0)$$

where

$$K_R^* \equiv 2\pi n_r \left(\int_{\Delta\nu} \sigma_\nu(z_0) E_2(a_\nu)\, d\nu\right)$$

Now consider a temperature distribution in which only small departures, $\delta T(z_0)$, occur from the radiative equilibrium temperature, $T_e(z_0)$, where the radiative equilibrium temperature distribution is taken to be that temperature distribution for which all radiative heating and cooling effects vanish. Then the actual rate of change of temperature due to radiative processes, $\partial(\delta T)/\partial t$, is approximately

(4.21) $$\frac{\partial(\delta T)}{\partial t} \cong -\left(\frac{K_R^*}{n(z_0)c_p}\left(\frac{dB_r}{dT}\right)_{T=T_e(z_0)}\right)\delta T \equiv -K_R(z_0)\,\delta T$$

where

$$K_R(z_0) \equiv \frac{2\pi n_r}{nc_p}\left(\frac{dB_r}{dT}\right)_{T=T_e}\left(\int_{\Delta\nu} \sigma_\nu(z_0) E_2(a_\nu)\, d\nu\right)$$

n is the total air density, and c_p is the constant pressure specific heat per molecule of air. Equation (4.21) expresses the fact that, to the extent that the "cooling to space" approximation is accurate, departures from local radiative equilibrium tend to damp out exponentially with time. For an absorber with fixed concentration, such as CO_2, the damping parameter, $K_R(z_0)$, is a height-dependent quantity independent of time. For CO_2 in the stratosphere, its values can be estimated from the results of Murgatroyd and Goody. From their work for the 15 μ CO_2 bands, $K_R(z_0)$ varies between 10^{-7} and 10^{-6} sec^{-1} in the height range 30–70 km. Its value near the stratopause is $K_R \sim 5 \times 10^{-7}$ sec^{-1}.

This formulation of infrared radiative heating and cooling rates in terms of a simple linear damping term is a very useful one from the point of view of atmospheric dynamics. Any hydrodynamic disturbance associated with temperature perturbations whose time scale is comparable to or longer than K_R^{-1} will be influenced by radiative damping. Furthermore, if photochemical damping associated with temperature dependent reaction rates as described in Section 2.3 is effective, the two processes add linearly to give a net damping parameter equal to the sum of the damping parameters from the individual processes. The dynamical effects of such radiative-photochemical damping processes have been discussed by a number of authors [43–47, 77].

5. Heat Sources and Sinks above 80 km

Infrared emission by carbon dioxide and water vapor may contribute to the heat balance at these heights—the former if the vibrational relaxation time is sufficiently short, the latter if vertical mixing is strong enough to replace water vapor dissociated by solar ultraviolet radiation. It is not known whether significant amounts of water vapor penetrate above 90 km. An additional factor, emission from the upper levels of the triplet ground state of $O(^3P)$, which may be important in the thermosphere, probably is not important below 105 km [78].

The most important radiative heating effect is the absorption of solar radiation in the Schumann–Runge bands and continuum of O_2. The Schumann–Runge heating ranges from less than 1.0°K/day at 80 km to more than 20°K/day at 105 km. It reflects the distribution of incoming solar radiation with a broad flat maximum in the summer hemisphere and decreases to zero in the winter polar region. The total energy absorbed between 80 and 105 km is on the order of 14 ergs/cm^2-sec.

The distinguishing feature of this upper region is the contribution to the heat balance of nonradiative effects other than simple heat advection by organized circulations. Johnson and Wilkins [79] have estimated that the downward transport of energy across 100 km is on the order of 5 ergs/cm^2-sec owing to thermal conduction and turbulent heat transfer. An additional downward energy flux of about 3 ergs/cm^2-sec is associated with the average vertical transport of atomic oxygen which represents an effective supply of chemical energy [80]. If all of the energy crossing 100 km were made available for heating in the 80–100 km region, it would correspond to an average heating rate of almost 10°K/day. Actually, very little of this energy is converted to internal energy within the region. Most of the atomic oxygen recombines below 80 km, where it makes a negligible contribution to the net heat balance [80]. In addition, a large fraction of the total energy input to the 80–105 km region is radiated away in nonthermal emissions. The most important of these is the nightglow emission from the Meinel bands of OH. The energy loss owing to this emission amounts to some 2 ergs/cm^2-sec from the region 80–105 km, corresponding to a cooling rate of about 2°K/day near 90 km [29, 81].

Although heating due to atomic oxygen emission may be comparatively small in the mean, Kellogg [82], and Young and Epstein [83] have shown that it can be enhanced locally by downward flow. Heating rates exceeding 10°K/day are quite likely near 90 km for even modest vertical velocities.

The greatest uncertainty at these heights is the role of dissipative heating. Hines [84], using vertical wind profiles, has estimated that dissipative heating associated with gravity waves generated at lower levels amounts to 10°K/day at 95 km, and is still larger at higher levels. It is not clear that the wind profiles analyzed by Hines were free from contamination by tidal winds. Lindzen [85] has estimated that the upward flux of energy associated with the diurnal tide is as large as 7 ergs/cm^2-sec near the equator, but falls to less than 1 erg/cm^2-sec outside of the tropics. Much of this energy is likely to be dissipated between 80 and 105 km, and it must correspond to substantial heating in the tropics. Finally, there is an upward energy flux associated with the synoptic scale systems, or Rossby waves, generated in the troposphere. The magnitude and spatial distribution of this energy flux is very sensitive to the zonal wind and temperature structure. Furthermore, the theory of this mode of vertical energy propagation is not yet sufficiently well-developed to make quantitative estimates of this energy flux possible [86, 87]. In middle latitudes, it probably has its maximum in the spring and fall. This theoretical seasonal distribution of vertical energy flux may be connected with predominant semi-annual variations in the magnetic index K_p, as suggested by Newell [88], and in the winds observed in this altitude range by the meteor trail technique [89]. It should be noted that if absorption of solar ultraviolet radiation provides the main drive for motions in this region, annual variations would be more prominent than semiannual ones.

The heat balance components in the 80–105 km region are illustrated schematically in Fig. 8. The heat sources and sinks shown depend strongly on motions of both large and small scale. These motions are not well understood so that the numbers indicated are very rough estimates at best. They suggest a net energy input whose order of magnitude is 28 ergs/cm^2-sec,

216 CONWAY LEOVY

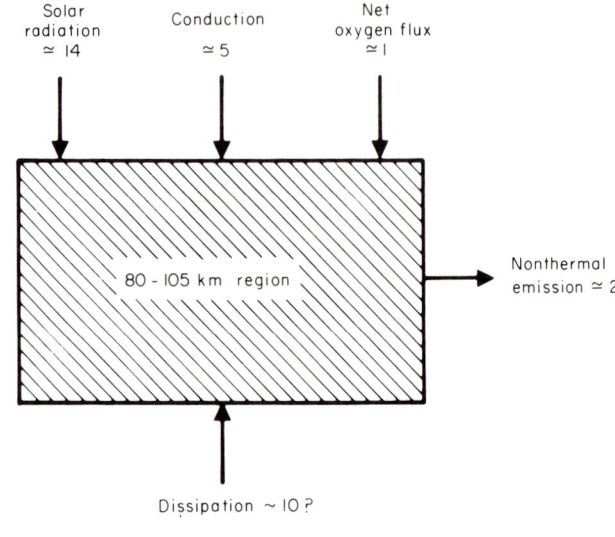

FIG. 8. Schematic representation of heat budget constituents between 80 and 105 km.

which is probably balanced mainly by CO_2 cooling. This order of magnitude is consistent with Kuhn's estimate of CO_2 cooling above 80 km.

It is not now possible to estimate the space and time distribution of net heating in the region, but these estimates should emphasize the fact that the distribution may be very different from the distribution of incoming solar radiation.

LIST OF SYMBOLS

A, B, C, D, E	Photochemical parameters defined in Section 2.2
B_ν, B_r	Blackbody monochromatic radiant intensity at frequency ν and average value of B_ν for band r
c_p	Specific heat of air at constant pressure per molecule
E_n	nth exponential integral [Eq. (4.6)]
G, G_{ji}	Weighting function in integrals for infrared heating rate, defined by Eqs. (4.8) and (4.9)
$I_\nu, \overline{I}_\nu$	Monochromatic radiation intensity at frequency ν and angular average of I_ν
dI_ν/ds	Spatial derivative of I_ν in an arbitrary direction
J_1, J_2, J_3	Photodissociation rates for water vapor, molecular oxygen, and ozone
K_R, K_R^*	Infrared radiative damping parameters, defined in Section 4.5
P	Total air pressure
P_v	Air pressure at the vibrational relaxation level
R	Universal gas constant

$S_\nu, \bar{S}_\nu$	Infrared radiative monochromatic source function, and angular average of S_ν
$T, \delta T$	Absolute temperature, and departure of temperature from a reference value
a_ν, b_ν	Monochromatic optical path parameters defined in Section 4.1
e	Subscript used to denote an equilibrium value
f	Line shape factor defined in Eq. (4.1)
h_ν, h_r	Volume heating rate due to radiation at monochromatic frequency ν, and volume heating rate for absorption band r
k_i	Rate coefficient for the ith reaction
n	Total number density
n_r	Number density of radiatively active gas
$n_2, \tilde{n}_2$	Number density of molecules in an excited vibrational state, and value of n_2 under local thermodynamic equilibrium conditions
$(dn_2/dt)_c$	Rate of change of n_2 due to collisions
p	Line-shape parameter defined in Section 4.2
q	Line-shape parameter defined in Section 4.2.
r	Subscript indicating representative value for an absorption band or part of an absorption band
r_1, r_2	Photochemical stationary-state concentration ratios, defined by Eqs. (2.21) and (2.23)
t	time
$t(O_3)$	Photochemical equilibrium time-scale for ozone
u_ν	Monochromatic optical path parameter, defined in Section 4.1.
w_{ji}	Quadrature weighting function, Section 4.1
z, z_0	Height above ground or cloud top; z_0 is the height at which heating rate is computed
α, β	Photochemical parameters, Eqs. (2.30) and (2.31)
α_D	Doppler line width
α_L	Lorentz line width
ν	Frequency
$\Omega, d\Omega$	Solid angle, and element of solid angle

References

1. Craig, R. A. (1965). "The Upper Atmosphere, Meteorology and Physics." Academic Press, New York.
2. Murgatroyd, R. J., and Goody, R. M. (1958). Sources and sinks of radiative energy from 30 to 90 km. *Quart. J. Roy. Meteorol. Soc.* **84**, 224–234.
3. Kuhn, W. R. (1966). Infrared radiative transfer in the upper stratosphere and mesosphere. Ph.D. Thesis. Dept. of Astrogeophys., Univ. of Colorado, Boulder, 159 pp.
4. Shved, G. M. (1965). On the heat effect of the 15 micron band of carbon dioxide in the upper atmosphere. "*Problems of Atmospheric Physics*" (Collection 3), pp. 96–104. Leningrad Univ.
5. Plass, G. N. (1956). The influence of the 9.6 micron ozone band on the atmospheric infrared cooling rate. *Quart. J. Roy. Meteorol. Soc.* **82**, 30–44.
6. Plass, G. N. (1956). The influence of the 15 micron carbon dioxide band on the atmospheric cooling rate. *Quart. J. Roy. Meteorol. Soc.* **82**, 310–324.
7. Murgatroyd, R. J. (1957). Winds and temperature between 20 and 100 km. *Quart. J. Roy. Meteorol. Soc.* **83**, 417–458.

8. Batten, E. S. (1961). Wind systems in the mesosphere and lower ionosphere. *J. Meteorol.* **18**, 283–291.
9. Murgatroyd, R. J. (1965). Winds in the mesosphere and lower thermosphere. *Proc. Roy. Soc. (London)* **A288**, 575–589.
10. Teweles, S. (1961). Time section and hodograph analysis of Churchill rocket and radiosonde winds and temperatures. *Monthly Weather Rev.* **89**, 125–136.
11. Johnson, F. S., Purcell, J. D., Tousey, R., and Watanabe, K. (1952). Direct measurements of the vertical distribution of atmospheric ozone to 70 km altitude. *J. Geophys. Res.* **57**, 157–176.
12. van Allen, J. A., and Hopfield, J. J. (1952). *Mem. Soc. Roy. Sci. Liège* **12**, 179.
13. Rawcliffe, R. D., Meloy, G. E., Friedman, R. M., and Rogers, E. H. (1963). Vertical distribution of ozone from a polar orbiting satellite. *J. Geophys. Res.* **68**, 6425–6429.
14. L'vova, A. A., Mikirov, Y. A., and Poloskov, S. M. (1964). Rocket investigations of the vertical ozone distribution above the level of maximum concentration during the total solar eclipse of February 15, 1961. *Geomag. and Aeron.* **4**, 839–843.
15. Yakovleva, A. V., Kudryatseva, L. A., et al. (1964). Spektrometricheskoe Issledovahnye Ozonova Sloi do Visoti 60 km. *Isskustvennye Sputniki Zemli* No. 14, 57–68.
16. Miller, D. E., and Stewart, D. H. (1965). Observations of atmospheric ozone. *Proc. Roy. Soc. (London)* **A288**, 540–544.
17. Hering, W. S., and Borden, Jr., T. R. (eds.) (1964). Ozonesonde Observations over North America, Vols. I and II, Air Force Cambridge Research Laboratories, Environmental Research Papers, No. 38.
18. Rawcliffe, R. D., and Elliott, D. D. (1966). Latitude distribution of ozone at high altitudes deduced from a satellite measurement of earth's radiance at 2840 Å. *J. Geophys. Res.* **71**, 5077–5089.
19. Leovy, C. (1964). Radiative equilibrium of the mesosphere. *J. Atmos. Sci.* **21**, 238–248.
20. Leovy, C. (1969). Atmospheric ozone: an analytic model for photochemistry in the presence of water vapor. *J. Geophys. Res.* **74**, 417–426.
21. USAF, Air Force Cambridge Research Laboratory (1962). "Handbook of Geophysical Data." L. G. Hanscom Field, Bedford, Massachusetts.
22. Nagata, T., Thomatsu, T., and Tsuruta, H. (1967). Observations of mesospheric ozone density in Japan. Paper presented at IQSY-COSPAR Symp., London, July, 1967.
23. Kreuger, A. J. (1967). Rocket measurements of equatorial and southern hemisphere ozone. Paper presented at the Specialist Symp. on Ozone, IAMAP, IUGG Meeting, Lucerne, September 1967.
24. Randhwa, J. S. (1966). Ozone measurements with rocket-borne ozonesondes. *J. Geophys. Res.* **71**, 4057–4059.
25. Craig, R. A. (1950). "The Observations and Photochemistry of Atmospheric Ozone" (Meteorol. Monographs, Vol. I, No. 2). Am. Meteorol. Soc., Boston, Massachusetts.
26. Hunt, B. G. (1966). The need for a modified photochemical theory of the ozonosphere. *J. Atmos. Sci.* **23**, 88–95.
27. Bates, D. R., and Nicolet, M. (1950). The photochemistry of atmospheric water vapor. *J. Geophys. Res.* **55**, 301–327.
28. Ballif, J. R., and Venkatswaran, S. V. (1963). On the temporal variations of the OH nightglow. *J. Atmos. Sci.* **20**, 1–4.
29. Wallace, L. (1962). The OH nightglow emission. *J. Atmos. Sci.* **19**, 1–16.
30. Konashenok, V. N. (1967). Models of an oxygen-hydrogen mesosphere. *Izv. Akad. Nauk, Fiz. Atmosfery i Okeana* **3**, 227–235.

31. Hampson, J. (1964). Photochemical behavior of the ozone layer. Tech. Note 1627/64, CARDE, Valcartier, Quebec.
32. McGrath, W. D., and Norrish, R. G. W. (1960). Studies of the reactions of excited atoms and molecules produced in the flash photolysis of ozone. *Proc. Roy. Soc. (London)* **A254**, 317–326.
33. Hunt, B. G. (1966). Photochemistry of ozone in a moist atmosphere. *J. Geophys. Res.* **71**, 1385–1398.
34. Dütsch, H. U. (1967). The photochemical theory of ozone. Paper presented at the Survey Symposium on Ozone. IAMAP, IUGG, Lucerne, September 1967.
35. Dalgarno, A., and Walker, J. C. G. (1964). The red line of atomic oxygen in the day airglow. *J. Atmos. Sci.* **21**, 463–474.
36. Hunten, D. M., and McElroy, M. B. (1966). Quenching of metastable states of atomic and molecular nitrogen. *Rev. Geophys.* **4**, 303–328.
37. Rees, M. H., Walker, J. C. G., and Dalgarno, A. (1967). Auroral excitation of the forbidden lines of atomic oxygen. *Planetary Space Sci.* **15**, 1097–1110.
38. Schofield, K. (1967). An evaluation of kinetic rate data for reactions of neutrals of atmospheric interest. *Planetary Space Sci.* **15**, 643–670.
39. Brewer, A. W., and Wilson, A. W. (1965). Measurements of solar ultraviolet radiation in the stratosphere. *Quart. J. Roy. Meteorol. Soc.* **91**, 452–461.
40. Detwiler, C. R., Garrett, D. L., Purcell, J. D., and Tousey, R. (1961) The solar spectrum in the extreme ultraviolet. *Ann. Geophy.* **17**, 9–18.
41. Ditchburn, R. W., and Young, P. A. (1962). The absorption of molecular oxygen between 1850 and 2055 Å. *J. Atmospheric Terrest. Phys.* **24**, 127–139.
41a. Pearce, J. B. (1968). Rocket measurement of nitric oxide in the earth's atmosphere between 60 and 96 kilometers. Ph.D. thesis. Department of Astro-Geophysics, Univ. of Colorado, Boulder, Colorado.
42. Craig, R. A., and Ohring, G. (1958). The temperature dependence of ozone radiational heating rates in the vicinity of the mesopeak. *J. Meteorol.* **15**, 59–62.
43. Leovy, C. (1964). Simple models of thermally driven mesospheric circulation. *J. Atmos. Sci.* **21**, 327–341.
44. Lindzen, R., and Goody, R. (1965). Radiative and photochemical processes in mesospheric dynamics: I, models for radiative and photochemical processes. *J. Atmos. Sci.* **22**, 341–348.
45. Lindzen, R. S. (1966). Radiative and photochemical processes in mesospheric dynamics: II, vertical propagation of long period disturbances at the equator. *J. Atmos. Sci.* **23**, 334–343.
46. Lindzen, R. (1966). Radiative and photochemical processes in mesospheric dynamics. III, stability of a zonal vortex at midlatitudes to axially symmetric disturbances: *J. Atmos. Sci.* **23**, 344–349.
47. Lindzen, R. (1966). Radiative and photochemical processes in mesospheric dynamics: IV, stability of a zonal vortex at midlatitudes to baroclinic waves. *J. Atmos. Sci.* **23**, 350–359.
48. Kaufman, F. (1964). Aeronomic reactions involving hydrogen, a review of recent laboratory studies. *Ann. Geophys.* **20**, 157–176.
49. Houghton, J. T. (1963). The absorption of solar infrared radiation by the lower stratosphere. *Quart. J. Roy. Meteorol. Soc.* **89**, 319–331.
50. Rodgers, C. D., and Walshaw, C. D. (1966). The computation of infrared cooling rate in planetary atmospheres. *Quart. J. Roy. Meteorol. Soc.* **92**, 67–92.
51. Gutnik, M. (1961). How dry is the sky? *J. Geophys. Res.* **66**, 2867–2872.
52. Brewer, A. W., Cwilong, B. M., Dobson, G. M. B. (1948). Measurement of absolute humidity in extremely dry air. *Proc. Phys. Soc. (London)* **60**, 52–70.

53. Mastenbrook, H. J., and Dinger, J. E. (1961). Distribution of water vapor in the stratosphere. *J. Geophys. Res.* **66**, 1437–1444.
54. Mastenbrook, H. J. (1963). Frost point hygrometer measurements in the stratosphere and the problem of moisture contamination. *In* "Humidity and Moisture" (E. J. Amdur, ed.), Vol. 2, pp. 480–485. Rheinhold, New York.
55. Mastenbrook, H. J. (1968). Water vapor in the stratosphere and high troposphere. *J. Atmos. Sci.* **25**, 294–311.
56. Sissenwine, N., Grantham, D. D., and Salmela, H. A. (1968). Mid-latitude humidity to 32 km. *J. Atmos. Sci.* **25**, 1129–1140.
57. Houghton, J. T., and Seeley, J. S. (1960). Spectroscopic observations of the water vapor content of the stratosphere. *Quart. J. Roy. Meteorol. Soc.* **86**, 358–370.
58. Houghton, J. T. (1963). Absorption in the stratosphere by some water vapor lines in the v_2 band. *Quart. J. Roy. Meteorol. Soc.* **86**, 332–338.
59. Kiseleva, M. S., and Neporant, B. S. (1965). Measurement of the moisture of gas mixtures by infrared absorption spectra. *Opt. Spectr. USSR (English Transl.)* **17**, 513–518.
60. Murcray, D. G., Murcray, F. H., and Williams, W. J. (1966). Further data concerning the distribution of water vapor in the stratosphere. *Quart. J. Roy. Meteorol. Soc.* **92**, 160–161.
61. Houghton, J. T. (1966). The effect of contamination on spectroscopic determinations of stratospheric water vapor. *Quart. J. Roy. Meteorol. Soc.* **92**, 281–283.
62. Kondrat'yev, K. Ya., Badinov, I. Y., Gaevskaya, G. N., Nikolsky, G. A., and Shved, G. M. (1966). Radiative factors of the heat regime and dynamics of the upper atmospheric layers. *In* "Problems of Atmospheric Circulation" (R. V. Garcia and T. F. Malone, eds.) (Intern. Space Sci. Symp., 6*th*, Mar de La Plata, Argentina), pp. 44–75. Macmillan, New York.
63. Williamson, E. J., and Houghton, J. T. (1965). Radiometric measurements of emission from stratospheric water vapor. *Quart. J. Roy. Meteorol. Soc.* **91**, 330–338.
64. Goldsmith, P. (1964). Measurements of humidity up to 30 km using a new hygrometer. *Proc. Ozone Symp. Albuquerque,* **1964** (H. U. Dütsch, ed). WMO, Geneva.
65. Brown, F., Goldsmith, P., Green, H. F., Holt, A., and Parham, A. G., (1961). Measurements of the water vapor, tritium and carbon-14 content of the middle stratosphere over southern England. *Tellus* **13**, 407–416.
66. Newell, R. E. (1963). Transfer through the tropopause and within the stratosphere. *Quart. J. Roy. Meteorol. Soc.* **89**, 167–204.
67. Fedynskii, A. V., Perov, S. P., and Chizhov, A. F. (1967). An attempt to measure directly the concentration of water vapor and atomic oxygen in the mesosphere. *Izv. Akad. Nauk, Fiz. Atmosfery i Okeana* **3**, 557–561. (AGU Transl. **3** (5), 315–317).
68. Elsasser, W. M., and Culbertson, M. F. (1960). Atmospheric radiation tables, AFCRL Tech. Rpt. 60-236, School of Science and Engineering, Dept. of Physics, Univ. of Calif., La Jolla.
69. Curtis, A. R. (1956). The computation of radiative heating rates in the atmosphere. *Proc. Roy. Soc. (London)* **A236**, 156–159.
70. Goody, R. M. (1964). "Atmospheric Radiation," Vol. I, Theoretical Basis, Oxford Univ. Press (Clarendon), London and New York.
71. Kondrat'yev, K. Ya. (1965). "Radiative Heat Exchange in the Atmosphere" (English edition), Pergamon Press, Oxford, 1965.
72. Shved, G. M. (1964). Approximation of the single line during the transfer of radiant energy in the upper atmosphere, *Bull. Leningrad Univ. Ser. Phys. and Chem.* **1**, No. 4, 79–81 (English Transl.: The RAND Corporation, p. 3595, May 1967).

73. Curtis, A. R., and Goody, R. M. (1956). Thermal radiation in the upper atmosphere, *Proc. Roy. Soc. (London)* **A236**, 193–205.
74. Shved, G. M. (1965). Method of computing the deviation from Kirchhoff's law in the mesosphere for the transfer of radiation in the 15-micron band of carbon dioxide. *Bull. Leningrad Univ. Ser. Phys. and Chem.* **1**, No. 4, 67–79. (English translation: The RAND Corporation, p. 3597, May 1967).
75. Marriott, R. (1964). Molecular collision cross sections and vibrational relaxation in carbon dioxide. *Proc. Phys. Soc. (London)* **84**, 877–888.
76. McClatchey, R. A. (1966). The effect of vibrational relaxation on atmospheric heating in the 4.3 micron CO_2 band. Ph.D. Thesis, Dept. of Meteorol. Univ. of Calif., Los Angeles.
76a. Houghton, J. T. (1969). Absorption and emission by carbon dioxide in the mesophere. *Quart. J. Roy. Meteorol. Soc.* **95** (to appear).
77. Dickinson, R. E. (1968). On the excitation and propagation of zonal winds in an atmosphere with Newtonian cooling. *J. Atmos. Sci.* **25**, 269–279.
78. Mahoney, J. R. (1966). A study of energy sources for the thermosphere. Rept. No. 17, Planetary Circulations Project, Dept. of Meteorol. MIT, Cambridge, Massachusetts.
79. Johnson, F. S., and Wilkins, E. M. (1965). Thermal upper limit on eddy diffusion in the mesosphere and lower thermosphere. *J. Geophys. Res.* **70**, 1281–1285.
80. Colegrove, F. D., Hanson, W. B., and Johnson, F. S. (1965). Eddy diffusion and oxygen transport in the lower thermosphere. *J. Geophys. Res.* **70**, 4931–4941.
81. Houghton, J. T. (1965). Infrared emission from the stratosphere and mesosphere. *Proc. Roy. Soc. (London)* **A288**, 545–555.
82. Kellogg, W. W. (1961). Chemical heating above the polar mesopause in winter. *J. Meteorol.* **18**, 373–381.
83. Young, C., and Epstein, E. S. (1962). Atomic oxygen in the polar winter mesosphere. *J. Atmos. Sci.* **19**, 435–443.
84. Hines, C. O. (1965). Dynamical heating of the upper atmosphere. *J. Geophys. Res.* **70**, 177–183.
85. Lindzen, R. S. (1967). Thermally driven diurnal tide in the atmosphere. *Quart. J. Roy. Meteorol. Soc.* **93**, 18–42.
86. Charney, J. G., and Drazin, P. G. (1961). Propagation of planetary-scale disturbances from the lower into the upper atmosphere. *J. Geophys. Res.* **66**, 83–110.
87. Dickinson, R. E. (1968). Planetary Rossby waves propagating vertically through weak westerly wind wave guides. *J. Atmos. Sci.* **25**, 984–1002.
88. Newell, R. E. (1966). Thermospheric energetics and a possible explanation of some observations of geomagnetic disturbances and radio aurorae. *Nature* **211**, 700–703.
89. Müller, H. G. (1966). Atmospheric tides in the meteor zone. *Planetary Space Sci.* **14**, 1253–1272.

THE NATURE AND PROPERTIES OF GASEOUS IONS ENCOUNTERED IN ATMOSPHERIC STUDIES*

Leonard B. Loeb

Department of Physics, University of California, Berkeley, California

Introduction 223
1. Forces Influencing Ionic Behavior 224
2. Mobility Theory and Equations 225
3. Cluster Ions.. 227
4. Reactions between Ions and Molecules 229
   4.1. Positive Ion Generation 229
   4.2. Negative Ion Generation 231
   4.3. Charge Exchange Reactions............................. 231
   4.4. Chemical Reactions 232
5. Ion Mobility and Ion Identification 233
6. Accuracy of Mobility Determinations as Diagnostic Tools ... 234
7. Summary of Influences Affecting the Presence of Ion Species . 235
8. Normal Ions in the Atmosphere 236
9. Reaction Rates of Ions 243
10. Langevin and Intermediate Ions............................. 244
References 252

Introduction

The nature of small ions encountered in the atmosphere, differentiating between the terms, center of force, solid elastic interaction, and composite interaction, is presented. There follows a discussion of the small atom, and complex, cluster, intermediate, and Langevin ions. A consideration of the initially ionized molecules or atoms and their charge exchange, aggregation, and evolution to final forms likely to be encountered follows. Laboratory studies covering a larger range of variables including field strength to pressure ratios, temperatures, densities, gas types, and conditions of generation are important for prediction. Atmospheric ions observed will depend largely on the composition of the atmosphere, including some minority species. In time, positive charges reside on species of lowest ionization potential. Negative ions reside on species of greatest stability. Both form different complex ions and hydrate with H_2O down to 10^{-4} Torr. They are stable in field strength to pressure ratios used in most atmospheric measurements, $E/p_0 \leq 5$. The faster ions may show discrete mobilities. As aggregates of H_2O, or chemically

* This work was supported by the U.S. Office of Naval Research under Contract N00014-67-A-0114-0014.

nucleated H_2O, and other aerosols pick up ions, continuous or band spectra are the rule. Upper atmosphere ions may exhibit discrete mobilities. One cannot identify atmospheric ions by their mobilities. Mass spectroscopy is imperative.

1. Forces Influencing Ionic Behavior[1]

One must begin this discussion by introducing the forces existing between charged atomic and molecular ions and neutral molecules of the ambient gas. As a result of these forces, the atomic or molecular ion will, in its motions, interact more effectively with the surrounding molecules. On occasion, this interaction can lead to some form of aggregation with molecules to larger complexes.

The forces are chiefly electrostatic polarization forces between the charge on the ion and either a polar molecule or one which becomes an induced dipole or quadrupole in the ionic force field. The induced dipole forces vary inversely as the fifth power of the distance between ion and molecule centers and as the bulk dielectric constant of the gas diminished by unity. The potential energy at a distance r varies inversely as the fourth power of r. There are also induced higher order or quadrupole forces acting which vary inversely as the seventh power of the distance and the potential for which varies inversely as the sixth power. Finally, for very large and polarizable ions and molecules one must include the so-called van der Waals' cohesive forces, which are similar in action to the quadrupole forces but act independently and can be added independently of them. For Xe and Cs^+ ions, the latter forces are even larger than the ion-induced quadrupole forces by a factor of five, but fall off so rapidly with r that they are inferior to the dipole forces. In addition to these attractive forces, there are the universal so-called repulsive or quantum mechanical exchange forces which properly vary inversely exponentially as the distance between centers. They are usually for convenience expressed as inverse repulsive functions of the separation, varying from about r^{-9} to r^{-14}. Probably the most frequent one is the inverse twelfth power law. They come from the Pauli exclusion principle when the outer electronic shells approach too closely. A variant of the repulsive force law is the so-called solid elastic repulsive law, which is zero until the centers of the ion and molecule approach to within a distance σ. At this point the repulsive force becomes suddenly indefinitely great. In general, the difference between the inverse twelfth power law, which is most frequent, and the solid elastic law is not too different. In a certain number of gases, such as with inert gas *atoms* in their own gas and O_2^+

---

[1] The summary of material and applicable references will be found in greater detail in Loeb [1]. This material, presented in summary form in this article, is essential to the discussions to follow.

ions in O_2, resonant transfer of the electrons on close approach from neutral atom to positive ion leads to a neutral atom with the ion momentum and an atom ion at rest. Similar reactions are now recognized in the case of proton exchange between an H_3^+ ion and a resting H_2 molecule at low energies and in other gases. These inelastic exchange force collisions in principle produce the same effect as solid elastic impacts. One may thus treat charge exchange collisions as equivalent to solid elastic impacts. At very high values of the ratio of field strength E to pressure p in all gases where dipole interactions are weak, the collisions are sensibly solid elastic.

2. Mobility Theory and Equations[2]

In 1905, Langevin derived a general theory of mobility for ions having a solid elastic collision radius σ given by $\sigma = (\sigma_M + \sigma_m)/2$, where σ_M is the solid elastic collision radius of the molecules and σ_m that of the ions, irrespective of the nature of the ion (cluster or monomolecular) and without regard to whether exchange forces are active. He also included the inverse fifth power law polarization forces, but did not include the quadrupole or van der Waals' forces. This led to a value for the ionic mobility at any gas density ρ given by

$$(2.1) \qquad k = \frac{3}{16 Y} \frac{0.462[(M+m)/m]^{1/2}}{\rho/\rho_0[(D-1)_0 M_0]^{1/2}}$$

Here M and m are masses of molecules and of ions, ρ and ρ_0 are the densities of the ambient gas and of the gas at 0°C and 760 Torr, $(D-1)_0$ is the dielectric constant of the gas less unity at 0°C and 760 Torr, and M_0 is the molecular weight of the gas. The quantity $3/16 Y$ is a complicated function of a quantity $1/\lambda$ calculated by Langevin from the orbits of the interacting ion and molecule. Here

$$(2.2) \qquad 1/\lambda = \left[\frac{8\pi N_0 \sigma^4 MC^2}{3(D-1)_0 e^2}\right]^{1/2} = \left[\frac{8\pi p_0 \sigma^4}{(D-1)_0 e^2}\right]^{1/2} = \left[\frac{KE}{PE}\right]^{1/2}$$

the ratio of the kinetic energy of the ion and the potential energy of ion and molecule at the distance between centers of σ. The value of $3/16Y$ rises from about 0.51 at $1/\lambda = 0$ to a peak of about 0.59 at $1/\lambda = 0.6$, falling rapidly thereafter to 0.185 at $1/\lambda = 4$.

It can be seen that if σ and m/M are known, the mobility can be computed as long as the fields are such that the average energy of the ions is about that of the gas molecules. At higher fields, the energy of the ion is not known, though Wannier (see Loeb [1]). has shown how it can be estimated if the mobility of the ion is known. The low energy criterion applies to most conditions of study of atmospheric ions.

[2] See Loeb [1].

There are two extreme forms of this equation. The one applies when the influence of the polarization forces is very small compared to that of the solid elastic impacts. This is the case for charge exchange force collisions and for very little polarizable molecules or for those having high energies and for very large ions. This is known as the solid elastic equation and has the form

$$(2.3) \qquad k = 0.815 \frac{e}{M} \frac{\lambda}{C} \left(\frac{m+M}{m} \right)^{1/2}$$

Here C is the root mean square molecular velocity at the ambient temperature T, and λ is the mean free path as given by $\lambda = 1/\sqrt{2}\,\pi\sigma^2 N$. The factor $\sqrt{2}$ comes from the relative velocity correction, which is valid only if both ion and molecule have thermal velocities and if $m = M$. As m exceeds M, it reduces to values nearer unity; and if M is greater than m, as for H^+ ions in a heavy gas, one should use the thermal velocity C of that ion. Here the whole influence, besides the mass factors and relative velocity correction, depends on the cross section for collision $\pi\sigma^2$, and this must be known from atomic dimensions for solid elastic impact and from the number of atoms in the ionic configuration.

In the other extreme, it is assumed that the forces are sufficient so that trajectories end with impact motion along the line of centers. In this event, the value of σ plays no role, and the attractive forces predominate. This leads to the point centers of the force equation. For Langevin's first order dielectric polarization force, the equation simplifies to read

$$(2.4) \qquad k = \frac{0.235[(m+M)/m]^{1/2}}{\rho/\rho_0[(D-1)_0 M_0]^{1/2}} \quad \text{in cm}^2/\text{V sec}$$

Note that here k is independent of the charge e and depends only on D, M_0, M and m. Many ions obey this theory very well for low E/p. For a constant D, i.e., for ions of different mass in a gas like N_2 or Ar, the value of k varies as $[(m+M)/m]^{1/2}$, and this is called the mass dispersion law.

Although in principle the Langevin theory disregards quadrupole forces and the esoteric repulsive laws as well as quantum mechanical theory which would alter the mass dispersion law, experimental data for many reasons are of insufficient accuracy to benefit much by these refinements. Since, under most conditions, atmospheric ionic problems concern fields where E/p is such that ions are essentially in thermal equilibrium, this discussion will largely concern itself with conditions which conform to the Langevin type of equation. Quantum mechanics also is too deficient to yield the correct cross sections, especially where polarization forces alter momentum exchange without direct impact.

In passing, it may be added that the question of ionic mobilities has been treated at greater length by Wannier (see Loeb [1]) extending to high values of E/p and to more generalized conditions, which are of no concern here.

From this work it appears that the drift velocity depends on two different generalized cases: the one where the free path is constant, independent of velocity; the other in which the mean free time between collisions is constant, that is, the faster the ion moves, the greater its free path. It is clear that constant free path represents solid elastic impacts, while constant free time is typical of polarization forces. Since at high E/p the polarization forces become negligible, one has only solid elastic impacts with more or less constant free path. Under these conditions, the drift velocity varies as $(E/p)^{1/2}$ such that $v = 1.147\,[(Ee/m)\lambda]^{1/2}$. Hence the mobility k varies as $(p/E)^{1/2}$.

It must be recognized that with the center of force point theory, one must know m/M. It should also be added that the effect of quadrupole forces in larger molecules and ions may alter the effective value of D, a matter which could be best developed by quantum mechanics if we had accurate knowledge of the force fields. For moderately polarizable gases where *known alkali* ions have been studied in N_2, the mass dispersion law curve appears to predict values of the mobility through the mass factor alone quite well. However, the constant term from theory must be normalized to one well-established experimental mobility. Such a relation would indeed prove useful in identifying ionic species through their masses. However, it is really helpful only for the alkali ions in very pure gases, measured after short time intervals.

The solid elastic relation leaves two unknown factors; the mass ratio m/M, and the value of σ. If m/M is known, then evaluation of k gives a value of σ, but this is inaccurate unless the relative velocity correction $\sqrt{2}$ is replaced by a value based on m/M. However, *the value of σ tells us nothing about the ion*. It cannot serve to identify it, though the values of m may do so. Identification from the value of σ would require a knowledge of the diameter of ionic constituent atoms, the gas molecules, and the shape of the ion configuration at thermal energies, none of which are well known. In a smaller measure, the same condition applies to the use of the full Langevin equation also, because there the attractive forces play a greater role and the influence of σ is less. Thus, basically, theory of this character is of little help in ion identification. The same applies to the more sophisticated theories, including quantum mechanics, owing to inadequate knowledge of force fields.

3. Cluster Ions[3]

At this point one must consider the value and significance of σ. Very early in mobility study, it was noted that experimentally k was of the order of about one-fifth that expected from the value of λ for monomolecular ions. It was thus at once assumed that λ was decreased by an increase in σ owing to

---

[3] See Loeb [1].

the formation of cluster ions or ionic aggregates. These will form if the polarization potential or binding energy at σ between ionic centers is greater than the average kinetic heat energy $\frac{3}{2}kT$ at a temperature of $T°$K. Aggregation thus appeared to apply for $O_2{}^+$ and other common ions in their parent gases. Unfortunately, the theory is more involved than a mere dependence on the ratio of the energies. This follows, since the entropy of the molecules of an ion in the dispersed state has a greater value than it has in the condensed state, that is, the statistical weights of clustered and unclustered states must be included. Calculations by Margenau and Bloom (see Loeb [1]) have shown that under most conditions, cluster formation at room temperatures is unlikely when statistical weights are included in the involved semiclassical calculation. Experiment has borne this out.

Recently, Hohla and Mohnen [2] made an extensive study of the Li^+ ions in Xe and H_2O, using quantum theory methods based on the earlier fundamental theory of Hirschfelder. Their calculations can be carried out for the Li^+Xe combination. Here it happens that the addition of the vibrational and rotational degrees of freedom leads to factors exceeding the entropy difference of the dispersed and clustered states. Similar considerations apply to the Li^+H_2O combination where, in addition, the hydrogen bonding must be considered in higher polymers. Theory is inadequate for an extension to more than one Xe or H_2O atom or molecule in the complex. In any event, the favorable conditions of the calculation do not extend beyond two or three clustered atoms or molecules. Experiments of Munson et al. [3] indicate that as many as three Xe or H_2O molecules may be clustered on Li^+.

It now appears that caution must be used in generalizing on the basis of dipole moments or polarizability alone, since this unquestionably involves in addition structure and valence bonds when certain molecular species are considered. For example, the H_2O molecule, with its strong hydrogen bonds, represents a structural complexity in behavior that is not germane to the general question of cluster formation. Such action may also apply to the action of alcohols and possibly amines, as well as ether, etc. It appears that for purely polarizable *atomic* configurations, clustering at around 300°K occurs only for small atomic ions like H^+ and Li^+ with relatively larger polarizable atoms such as Kr and Xe. Here possibly three Xe atoms may form a labile complex with Li^+ and H^+. At 300°K, however, K^+ does not cluster in Ar. For pure molecular gases such as O_2 and N_2, as well as for inert gases, according to Yang and Conway [4], $O_4{}^+$ appears below 150°K and is relatively stable at 76°K. These workers show that polarization forces alone cannot account for a binding energy of 0.41 eV. Likewise, $O_4{}^-$ appears in the presence of $O_2{}^-$ below about 150°K. This could be called a limited cluster formation for O_2 at low temperatures. At room temperatures and at low E/p values, $N_2{}^+$ polymerizes to $N_4{}^+$ with a binding energy of some 0.5 eV [5]. A transient ion of N_6^+ of short life has been detected by Fehsenfeld and Ferguson [6]. At 80°K, these workers report that in He various N_n complexes up to N_9 have been observed, with the N_2^+ preponderating since N_2^+ ions are created by the He_2^+, He^+, and He^m configurations. However, N_2 does not form a stable N_2^- ion, and no evidence exists that N_4^- is stable. Here one notes again strongly indicated structural affinities as contrasted with sign of charge independent cluster formation, as in O_2 at low temperatures.

It should be noted that the O_n and N_n ions with n exceeding 4 are very few, and that both of these gases at 80°K are near their condensation points in a gas of low atomic mass. Thus there are present small numbers of unstable neutral aggregates, which can temporarily stabilize by acquisition of charge, thus accounting for the observed complexes.

Some remarks must be made concerning classical clustering when it occurs. Classical clustering is *not charge specific*, since polarity does not alter pure dielectric attraction. It could influence quantum mechanically considered forces [2]. To form the cluster ion requires triple impact to remove the heat of cluster formation. This energy is generally small compared to that of most chemical binding. Clustering at any temperature continues to add molecules until the rate of attachment of molecules equals the rate of detachment by impact. Thus at *any given temperature the cluster ion has an average number $\bar{n}$ of molecules which is exceedingly temperature- or energy-sensitive*. Such clusters thus are amenable to analysis by chemical equilibrium methods. A slight increase in E/p or temperature reduces the average sojourn of the last molecule attached and decreases $\bar{n}$. As $\bar{n}$ decreases, $\bar{\sigma}$ decreases; λ increases as $1/\bar{\sigma}^2$ and k increase. Cluster ions, therefore, show a continuous and rather rapid variation of mobility with E/p; and such clusters constitute labile ions, differing from the $(p/E)^{1/2}$ law when E/p is high.

It is thus evident that the true *classical cluster* ion concept applies only to a limited number of ions, that few molecules or atoms are involved, and that this phenomenon is limited to low temperatures for the more common ions.

4. Reactions between Ions and Molecules[4]

4.1. Positive Ion Generation

Ions are generated in most cases by ionizing impacts of electrons in gases. They may on occasion be created by photoelectric ionization of atoms and molecules by photons of sufficiently high energy. Ionization by fast positive

---

[4] See Loeb [1].

ions in gases requires too much energy and is unlikely in most gaseous breakdown processes. Ions may also be created by interactions of high energetic excited states or metastable states of one species with a species of lower ionizing potential. Where atomic or molecular ions are created by interaction with metastable states, the phenomenon is called the Penning effect. Where *molecular* ions are created by collision of an excited atom with an unexcited one, the reaction is the Hornbeck–Molnar reaction. Under some conditions, both of these mechanisms are highly efficient. These reactions all give rise to positive ions and electrons. The ions under most conditions encountered are singly charged atoms or molecules. Energetic electrons may give doubly or multiply charged ions. These do not persist for any length of time in gases at any appreciable pressures, since they readily share their excess charges by stealing electrons from neutral atoms. The Ar^{++} ion has recently been observed in the early afterglow of Ar plasma at lower pressures, lasting for perhaps 10^{-3} sec [7]. Most gases to be encountered exhibit only singly charged ions. Table I, listing relative ionizing potentials of atmospheric ions, appears below.

In monatomic gases these will either be atomic ions or, for certain species, molecular ions created by the Hornbeck–Molnar process.

In molecular gases, the ions generated can be of multiple species. Thus, for example, in N_2 gas, stable N^+, N_2^+, N_3^+, and N_4^+ are encountered [8] at room temperatures. The N^+ comes from dissociation and ionization of N_2 by energetic electrons. The N_2^+ is created by direct electron impact or photoionization of N_2. The N_3^+ is caused by creation of an energetic, excited N_2^{+*} ion in electron impact which in a subsequent impact with an N_2 molecule leads to N_3^+. The N_4^+ occurs at lower energies and higher pressures through combination of N_2^+ and N_2, apparently in three-body impacts. It has a binding energy of 0.5 eV. N_3^+ is also formed by interaction of N^+ and N_2.

The concern here is largely with the creation of ions in a gas at relatively high pressures, compared to those where the mean free path is of the dimensions of the apparatus; the problem is one of electron swarms moving in fields in gases with some sort of an energy distribution. There, the parameter field strength E divided by pressure p or by the number density n is the factor determining the average electron energy and the distribution law. The electrons have a range of energies distributed about a mean for each E/p value. Thus as E/p varies, the conditions for generating one ion or the other change. Using relatively low pressures of N_2 and mixing it with Ar at higher pressures through the Penning effect creates almost exclusively N_2^+ at $<10^{-3}$ Torr N_2. Above 1 Torr N_2, the ion is largely N_4^+ [9]. If higher pressures of N_2 are used and the gas is N_2, all four ions appear in varying relative proportions as E/p increases, the N^+ and N_3^+ predominating at higher E/p.

4.2. Negative Ion Generation[5]

Negative ions are created in one of two principal ways. They may in some gases be created by electron attachment to molecules in triple impacts. In much greater measure, they are created by dissociative attachment of energetic electrons in impact with molecular gases. The halogens, for example, will attach thermal electrons with large cross sections, yielding atomic ions such as Cl^-, Br^-, etc. In gases like HCl, the creation of Cl^- requires impacts of around 0.8 eV. However, as E/p is raised, both H^- and Cl^- appear, and at very high E/p one has the creation of H^+ and Cl^-. In water vapor, at an Ep of 10, the OH^- ion appears, and at an E/p of 30, both OH^- and H^- ions appear. Some negative ions are created by sputtering of adsorbed layers of ions on metals by high energy positive ion impact. It is thus clear that here the nature of the ions generated depends on E/p, p, and other surrounding circumstances.

4.3. Charge Exchange Reactions[6]

Once ions are created in a gas, one cannot be sure that their identity is sustained. An important mechanism is charge exchange. If an ion has an ionization potential higher than that of the surrounding gas molecules or that of a minor impurity constituent thereof, it will steal the electron, creating an ion of the molecules of lower ionizing potential. Since ions make on the order of 10^9 collisions per second at 760 Torr and room temperature, most ionized gases will, in the presence of as little as 10^{-3} Torr of Hg vapor, be Hg^+ ions within some 10^{-3} to 10^{-2} sec, depending on the probability of charge exchange in a kinetic collision. Thus, with the use of McLeod gauges without cold trap, many ions observed in earlier studies were Hg^+. A similar action will occur for many negative ions: The electron is transferred from an ion of lower binding energy to a more stable ion in impacts. Perhaps, results of Moruzzi and Phelps [10] will illustrate this situation. Photoelectrons enter O_2 at about 1 to 4 Torr, where they interact with O_2 molecules. Below an E/p of 1, and at pressures of 1 Torr, electrons attach to O_2 molecules in triple impact to yield O_2^- ions. Above an E/p of 1, the O_2^- ions do not form very readily, as the electrons have too much energy; but some of the more energetic ones in the swarm get the needed 3.1 V to form O^- ions by dissociative attachment. Concurrently with this, some of the O^- ions in impact with O_2 form O_3^- ions. Thus many O^- and some O_3^- ions appear up to an E/p of 5, at which point O_3^- is no longer stable and only O^- ions persist. At an E/p of 10, many of the O^- ions gain an energy of 1.6 eV and, in impact, *form O_2^- or else shed their*

---

[5] See Loeb [1].
[6] See Loeb [1].

electron at an energy of around 2 eV. Thus O^- gains enough energy in the field to begin to shed its electrons so that, between E/p of 10 and 20, O^- ions begin to disappear, leading *either to free electrons or to* O_2^- *ions*. The O_2^-, once formed, is stable up to an E/p of about 100. Above this, it begins to shed its electron. However, the free electrons created yield O^- and the O^- again *regenerates* O_2^-, whence O_2^- is the only negative ion persisting in detectable amounts up to an E/p of near 200, together with free electrons [11]. Strangely, at low E/p, as gas temperature increases in O_2, the O_2^- ion *begins to break up* at around 550°K [12] while the O^- ion persists up to about 1200°K. Between these temperatures, mobility studies would measure values of mobilities hybrid between those of electrons and O^- ions [13]. The binding energy of O_2^- is about 0.4 eV; that of O^- is 1.45 eV. However, in a field, because of mobility and mass differences, O^- acquires more energy from the field than does O_2^-, and thus sheds electrons more readily. At low E/p, however, O_2^- forms slowly from attachment, but begins to shed its electron at 500°K while O^- requires much higher thermal impact energies. At temperatures between 76°K and 150°K, O_4^- ions appear [4].

The only ions escaping charge exchange, and by means of which our basic theories have been tested, are the positive alkali ions [1]. Once generated, they remain intact in most *pure*, simple gases, inert gases, N_2, H_2, and other weakly polar gases where they do not cluster because their ionization potentials are lower than those of the gases cited.

4.4. Chemical Reactions[7]

Beyond charge exchange and the rare cluster formation, ions may form charge specific complexes. Thus, for example, positive ions such as H^+ or Na^+ with NH_3 and amines of the form RNH_2 combine to form ammonium-type ions. On the other hand, negative ions have a strong affinity for H_2O and the alcohols of the form ROH to form negative complex ions. In fact, electrons in O_2 between an E/p of 0.1 and 1, with 2.16 Torr of O_2 and 0.44 Torr of H_2O vapor, yield complexes of the form $(H_2O)_n O_2^-$ [10], with n from 1 to 4. The value $n = 2$ is the most abundant, $n = 3$ is next, and $n = 1$ follows, with $n = 4$ the least abundant. The $n = 2$ complex is ten times as abundant as O_2^- and five times as abundant as $n = 4$. The $(H_2O)O_2^-$ ion persists down to 10^{-4} Torr of H_2O. Likewise the OH^- ion between E/p of 5 and about 20 forms complexes $(H_2O)_n OH^-$ with n ranging from 5 for the least abundant and least stable complex, with 4 the next, 3 following, and $n = 1$ the most stable, but somewhat less prominent at low E/p [10]. Strangely, the H^- ion

---

[7] See Loeb [1].

does not form $H^-(H_2O)_n$ complexes, while $H^+(H_2O)_n$ complexes do form. This indicates that the situation is a chemical one of hydrogen bonding—not a cluster formation.

The H_2O^+ should exchange charge with O_2, yielding O_2^+. Lampe et al. [14] have shown that H_2O^+ reacts very rapidly with H_2O to yield the H_3O^+ oxonium ion and OH. In a personal communication dated November 3, 1968, E. E. Ferguson announced that, as had been predicted by Mohnen [2], O_2^+ in a three-body impact rapidly forms $O_2^+(H_2O)$, which in turn also rapidly reacts with H_2O to yield H_3O^+, OH, and O_2. Thus in moist air the dominant O_2^+ ion eventually goes to H_3O^+.

Many other peculiar electrochemical affinities that appear as plasmas are investigated with the mass spectrometer. Pressures of gases up to 20 Torr can now be studied with the quadrupole mass spectrometer. Thus small amounts of H_2 as low as 1% in Ar yield the peculiar ion ArH^+ [15], and N_2H^+ has recently been observed. Similarly, with Ne in He gas, the ion $NeHe^+$ is observed.

5. Ion Mobility and Ion Identification

It is clear from the data given that in any but the simplest gases of extreme purity and within relatively short times, the identity of the ions, the mobility of which has been determined, is in doubt. It is only during the last five years that mass spectrograph study has been made in connection with mobility determinations, and only in the last two years that pressures up to 40 Torr have been investigated in this fashion. The results have been astonishing in the variety and changes in relative ion types with varying parameters. Among the gases, Ne and He are those in which ion mobilities are most clearly and widely dispersed in values. Thus one would imagine that in a simple gas like pure He one would find only two ions, He^+ and He_2^+ of unique mobili-, ties and widely separated values readily identifiable. The He^+ undergoes frequent charge exchange collisions and has a mobility of about 11 cm^2/V sec. The He_2^+ does not, and should have about double the value. Strangely, there are *two* He_2^+ ions [16], one created in the early afterglow by the Hornbeck–Molnar process with a value of about 19, and a second He_2^+ ion created by triple impacts in the late afterglow, of mobility 16.2. The only possible difference is that the one is in a vibrationally excited state and does not undergo charge exchange reactions. The He_2^+ in the vibrationally unexcited state appears to undergo an occasional transfer of He^+ to a resting He atom, making an He_2^+ relatively at rest and a fast atom. It is thus slower than theory indicates. The excited one has the correct value. At 76°K an He_3^+ ion appears. However, if there are *traces* of Ne present in He, there is a complex sequence of charge changes in which He_2^+ interacts with Ne to give Ne^+. The Ne^+ in triple impact with He atoms yields $HeNe^+$. Eventually the $HeNe^+$ interacts with an Ne atom to give Ne_2^+. [17] At different periods of the afterglow, these various ions appear. Thus the values of the mobilities leave ion identification in doubt.

Conditions are equally confused in most other gases, and today the mass spectrograph is the only certain source of ion identification. Under certain very carefully specified circumstances as to pressure, E/p, nature of the ionizing source, and ascertained purity of the gas, one can on the basis of values of mobility assume that an ion identified under those conditions in the past is the ion in question.

6. Accuracy of Mobility Determinations as Diagnostic Tools

One must next raise the issue of the accuracy of mobility determinations in relation to the unique discriminatory value of a given mobility. While mass spectroscopy yields sufficiently clearly identifiable values, this may not be so for the mobilities. In gases of low atomic or molecular weight, such as H_2, Ne, and He, mobilities differ over a considerable range of values. This ceases to be the case in most other gases. For example, for the negative ions in O_2, identified as O^-, O_2^-, O_3^-, and O_4^-, the values obtained by diverse but reputable methods extrapolated to low E/p ranged respectively as follows: 3.4, 3.2 $\pm$ 0.3, 3.02, 3.0, 2.8, and 2.4 for O^-; with 2.48, 2.56, 2.5 $\pm$ 0.1, 2.5, 2.39, and 2.0 for O_3^-; and 1.95, 2.25 $\pm$ 0.15, 2.46, 2.13, 2.28, 2.19, and 2.1 for O_2^-; and finally 2.00 for O_4^-. Similar discrepancies appear for ions in N_2. The difficulty here lies in the fact that with mass identified ions, the mobilities undergo strange alterations other than the expected transition from the k remains constant regime to one in which k varies as $(p/E)^{1/2}$ at ion energies above thermal. These changes have to do with the formation of complex molecules like N_4^+, the measurement of ions undergoing transition in the measuring field, and the appearance of diverse charge exchange reactions as a function of energy. The best approximate values for the negative ions in O_2 would be 3.0, 2.51, 2.39, and $\sim$ 2.4 for the O^-, O_3^-, O_2^-, and O_4^- ions, respectively, in which all four have been identified by mass spectrographs of one form or another [11]. Even under the best conditions, it is hard to obtain values of k that are valid to better than 5% with any one method at a fixed E/p. With diverse methods, 10% is the best one can do. Where measurements yield mobility spectra, certainly no separation better than 10% is possible.

Attempts at theoretical evaluation of mobilities on the basis of identified ions under well controlled conditions in gases are not any better. Thus for He_2^+ in He, wave mechanical theory gives 24, the Langevin equation 21.6, and Wannier–Varney 18.2. The two values observed are 19.0 and 16.2. For

He$^+$ the quantum mechanical theory with exchange forces gives 12.0, whereas Wannier's solid elastic theory gives 13.4. The observed value is about 11.2. For Ne$_2^+$ the Wannier–Varney relation gives 6.21, with an observed value of 6.30; and for Ne$^+$ the calculated values were 4.85 and 4.05, but the observed value was 4.20. Kr$_2^+$ gave a calculated value of 1.18, with that observed about 1.20; and Kr$^+$ gave a calculated value of 1.01 and an observed value of 0.93. In virtually all other gases, the calculated values are in poor agreement with observed values, chiefly because of the changing nature of charge exchanges with E/p which does not permit extrapolation to low E/p value.

7. Summary of Influences Affecting the Presence of Ion Species

One may briefly summarize the preceding as follows. What ions are found in a given gas species depends on several factors.

(a) The gas purity is a factor of primary importance. With 10^8 to 10^9 impacts per second, traces of less than one in 10^7 of impurity reacting at high rates may, within 10^{-1} to 10^{-2} sec, yield a completely foreign set of ions. Mass spectrometers have revealed that at a few Torr, the H$_2$O present in *really dry* N$_2$ or air yields oxonium ions from the traces of H$_2$O emitted by the baked out glass and electrodes [18]. A finding of Luhr [19] in 1933 has recently been confirmed by Varney [18], that apparatus in which N$_2$ has been used, when supposedly clean and filled with O$_2$, will under some ionizing impacts yield surprising amounts of NO$_2^+$.

(b) It has recently been established by Varney [18], in a drift tube with aging chamber, mass spectrographic detection, and ionization by electrons of known energy, that in O$_2$ with traces of N$_2$, *or in dry air*, only O$_2^+$ ions appear at 12.1 eV. At an energy of 13.7 eV, NO$_2$ appears, probably created in a delayed reaction which yields electrons to O$_2^+$. Thus NO$_2^+$ ions appear. Above 19.8 eV, NO$^+$ ions appear. These probably come from three metastable nitrogen atom types: one in a D state, another in a P state, and a third in a $2D$ state. Ionization of N$_2$ occurs at 15.6 V. At 19.8 V all types of ions appear in enhanced quantities (at least five times). Above 22 eV, N$_3^+$ could appear, but the O$_2$ prevents this. Some O$_2^+$ is formed by direct impact, but some is also created in delayed action by some excited N or N$_2$ states. The formation of NO$_2^+$, however, is a reaction appearing some time after the ionizing pulse; hence, it seems to be the result of chemical reaction between excited N states and O$_2$. Similarly, N$_3^+$ does not appear in pure N$_2$ until electrons deliver about 21.5 eV or more at impact. Thus one must realize that the mode or energy of ion generation can influence the ions observed under otherwise similar conditions.

(c) Ions alter in time by charge exchange or by chemical combination.

Hence the age of the ions at observation will strongly influence the ions present. This depends on possible reactions and relative reaction rates.

(d) The ionic changes, being of a chemical nature, are altered by the partial pressures, by the total pressure, and by the temperature, following the well-known laws of chemical equilibrium.

(e) Where ion changes occur in the measuring field, hybrid mobilities between the two species will be observed, which will shift as equilibria force one ion phase or the other to be present over longer intervals.

(f) Finally, the imposed measuring fields or, better, the E/p ratio, will influence the composition for ions of low binding energy. If E/p is such that the ions remain in near thermal equilibrium with the gas, its influence is not important.

(g) Since complex ions of a chemical nature are very common, and since such ions are charge- and species-specific, they are not to be confused with, or *misnamed*, "clusters." In any case, they involve at most a few of the gas molecules in question, rarely exceeding five. They show no evidence of growth to larger aggregates. Such ions had best be called complex molecular ions. True clustering, in the classical sense, is limited to small atomic ions with larger polarizable atoms, or to a few species at relatively low temperatures.

8. Normal Ions in the Atmosphere

One may now turn to the question of ions to be found in the atmosphere. In this case, one need only consider the most prevalent gases. These are N_2, O_2, H_2O, Ar, CO_2, CO, NO, and NO_2. Minor constituents are perhaps SO_2, NH_3, CH_4, and hydrocarbons, H_2S, and their decomposition products.

In order to discuss intelligently the ionic situation, one must take stock of the ionization and appearance potentials of the major constituents, and some special impurities. These, in electron volts, are listed in Table I.

It should also be indicated that in gases containing O_2, dissociative attachment of electrons leads to such ions as O^-, OH^-, and H^-, in many cases at relatively low electron energies ranging between 3.5 and 10 eV. Free electrons

TABLE I. Relative ionizing potential of atmospheric ions.

Molecule	Potential in eV	Molecule	Potential in eV	Atom	Potential in eV
Ar	15.68	H_2O	12.6	N	14.54
N_2	15.51	O_2	12.07	O	13.61
CO_2	13.7	NO_2	9.78	H	13.6
CO	14.1	NO	9.25	Hg	10.4
SO_2	13.1				

attach at low energies to O_2, and SO_2 in triple impacts, and to the dimer $(NO)_2$ at low energies to yield O_2^-, SO_2^-, and NO^-.

Relatively little is known about the binding energy of negative ions and the relative displaceability, except for halogen ions. However, the following values relative to atmospheric ions are more or less well established. They are: O^-, 1.47 eV; O_2^-, 0.43 eV; OH^-, 1.83 eV; O_3^-, 1.9 eV; NO^-, less than 0.43 eV; NO_2^-, around 3.8 eV; CO_3^-, stable but unknown; CO_4^-, 1.22 eV; and NO_3^-, stable but unknown.

Positive ions encountered in these gases are as follows:

(a) H_2. In H_2 there are H^+, H_2^+, H_3^+, and H_5^+ [20]. The H^+ appears only at higher E/p where recombination at the walls is not likely. Its mobility in H_2 by theory is 18; its measured extrapolated mobility is around 16.6. The H_2^+ appears only at low pressures or at higher E/p, as it reacts strongly with H_2 to give H_3^+ and H. Its k is not well defined. At higher pressures, H_3^+ predominates. At low E/p, H_3^+ undergoes proton exchange with $k \approx 11$. At intermediate E/p its k increases to higher values, as proton exchange is less frequent. It undergoes electron exchange with H_2, leading to transitional H_2^+ which at once reforms H_3^+ in impact with H_2 as E/p increases further, leading to a decline in k. Its zero field mobility could extrapolate to the theoretical value 22, were it not for the interactions. H_4^+ has never been observed as a stable ion. H_5^+ appears by triple impact of H_3^+ in H_2 at low energies and has a k of about 9. H_2 has two negative ions, H^- of binding energy 0.74 eV, and H_2^-. The H^- ion is encountered in the ionization of H_2O and of H_2 under proper conditions. Neither H^- nor H_2^- is frequently encountered in mobility studies. It is noted that these ions are all discrete chemical combinations with charge specificity and do not represent clusters of the classical sort.

(b) N_2. This has four relatively stable ions at room temperatures; N_2^+ of a k in N_2 gas of about 1.87 cm^2/V sec, N^+ of $k \approx 2.97$, N_3^+ of k ranging from 2.26 at low E/p to 2.6 at high E/p, owing to charge exchange at low E/p [8]. The N_4^+ ion is created in triple impacts from N_2^+ and N_2 and has a mobility of 2.34. It exists at room temperatures and low E/p and undergoes transitions to N_2^+ as E/p increases, as its binding energy is 0.5 eV. The N_3^+ comes from impacts of excited N_2^+* ions in impact with N_2. It is also created by interaction of N^+ with N_2. N^+ is not frequently encountered, as it takes high energies to ionize and dissociate the N_2 molecule. A transient ion N_6^+ of very short lifetime has been reported and may be involved in the breakup of N_4^+. N^2 forms negative ions of a lifetime less than 10^{-14} sec, so that these are not observed. The N_2 ions will not persist long in air, since they will transfer their charge to O_2 in a short time. If H is present in any combination in N_2, the ion N_2H^+ appears, and in very pure N_2 with traces of H_2O, the ion

N_2H^+ predominates over N_2^+. If traces of H_2O are present in N_2, neither N_4^+ nor N_3^+ appears at low pressures. N_2H^+ with H_2O gives H_3O^+ and N_2.

(c) O_2. There is only one positive ion, O_2^+, observed in O_2. The O^+ ion forms, but at once exchanges charge to give O_2^+. The mobility of O_2^+ at low E/p is in doubt, and appears to lie at 2.25 ± 0.2 cm^2/V sec but could be as high as 2.4 [21]. At very low temperatures $<150°K$, and low E/p, the O_4^+ ion may appear. As noted, O_2 has four negative ions, with mobilities of 3.1, 2.5, 2.39, and ~ 2.4 for the ions O^-, O_3^-, O_2^-, and O_4^-, respectively. The circumstances of their appearance and transformations have already been indicated. In O_2 electrical breakdown will create O_3 molecules; these will react exothermically with O_2^- to give O_3^-. The O_2^+ ion should react with NO_2 to give NO_2^+, and with NO to give NO^+ rather readily. In air, these ions are rare at low excitation energies, because little NO_2 and NO are present, unless the exciting electrons have an energy exceeding 19.8 eV. NO has a mobility of 3.6 ± 1 cm^2/V sec. O_2^+ steals an electron from NO_2 to yield NO_2^+. NO_2 appears to form if traces of N_2 are present in O_2, and if electron impacts exceed 13.7 eV [18]. NO_2^+ will steal an electron from NO to yield NO^+. NO_2^+ has a mobility in air of about 3.0; O_2^+ in air has about the same mobility as NO_2^+.

The reactions of oxygen ions with H_2O, which are really important, will be considered under H_2O. If H_2 or H_2O is present, it may react with O_2^+ to give $(H_2O)H^+$. This ion could be the first stage of clustering about the H^+ ion, though according to Pahl [15] during formation it dissociates off H_2^+. Varney [18] recently reported that the common positive ion in really dry air is O_2^+, certainly below 13.7 eV. No other ions appear in dry air at lower excitation energies.

Actually, the behavior in air is exceedingly complex, as the combined drift tube-mass spectrographic study with variable ionizing electron energies by Varney [18] appears to indicate. Pressures here are low, up to a few Torr. It appears that in dried air there is enough moisture from the glass to yield some H_2O. Thus, as indicated, depending on the conditions, O_2^+ alone appears below 13.7 eV, with later appearance of traces of H_3^+O if water vapor can come from the glass. At very low pressures, N_2H^+ appears below 10^{-2} Torr. As pressures increase, the N_2H^+ disappears and O_2^+ appears. If H_2O is present, the N_2H^+ gives way to H_3O^+. This follows a reaction postulated by Shahin [22] in which $N_2H^+ + H_2O$ gives $H_3O^+ + N_2$. As electron energies increase to 13.7 eV, NO_2^+ appears. At 19.8 eV, NO^+ appears. At 19.8 eV all species strongly increase. The NO_2^+ appears to break up at high values of $E/p > 100$, leading to NO^+. Using air taken from an air-conditioned laboratory, in which in the process of dehumidification, SO_2, H_2S, and pollutants were removed, the air was afterward humidified to 40% at 23°C with pure H_2O. With 1% H_2O present, the only ion was oxonium at low pressures such as 1 micron. At a pressure of more than 0.8 Torr, the $H(H_2O)_2^+$ ion appeared, and above 1.5 Torr, the ion $H(H_2O)_3^+$ was noted as well. No masses greater than the latter were observed up to mass numbers 250. In air with 1% H_2O at 760 Torr, at low E/p, one could expect ions of the form $H^+(H_2O)_5$ to be present.

It is clear that in ordinary clean air with normal water vapor content, the ions at higher pressures and at the humidities shown will be largely of the group of oxonium ions. If the gas is exceptionally dry, then depending on energies of electrons from the ionizing agency, NO_2^+ and NO^+ will be observed; otherwise, it will be O_2^+. With H_2O, $NO_2(H_2O)_n^+$ and $NO(H_2O)_n^+$ appear.

(d) Ar. The Ar^+ ion, as well as Ar_2^+, will not survive in air for any length of time. They will transfer their charges to N_2 and O_2 molecules.

(e) H_2O. This gas or, better, vapor is one of the most active agents in regard to ions found in the atmosphere. On electron impact at various energies it yields H^+, H^-, O^+, OH^+, O^-, O_2^-, H_2O^+, and OH^-. Its high dipole moment and its hydrogen bond linkages, etc., make it prone to undergo multiple reactions with ions. One difficulty is that it cannot be studied well by itself because of its low vapor pressure. Usually it is studied either in an inert gas, or in O_2, or some other gas like CO_2. Eiber [11] believed that with 1% H_2O vapor in O_2, the charge goes from O_2^+ to H_2O^+. This is incorrect, since H_2O^+ in impact with O_2 yields O_2^+ and H_2O. Some H_2O is directly ionized by electron impact to H_2O^+. The O_2^+ reacts rapidly in a three-body reaction with H_2O to form $O_2^+(H_2O)$. This, in impact with H_2O, leads to H_3O^+, OH, and O_2. Eiber was then observing spectra containing various proportions of the complexes $H^+(H_2O)_n$ and O_2^+ as E/p and H_2O vapor content changed. With the mass spectrograph, Varney [18] observed the H_3O^+ ion to have a mobility extrapolating toward 5 to 6 at zero field. The curves of Eiber and Kandel [23] for supposed H_3O^+ show a similar trend. In air ionized by a positive point corona, Bricard et al. [36] observed a mobility of 6 if much H_2O vapor was present.

The question of the positive ions in H_2O has been settled by Knewstubb and Tickner [24], who analyzed ions in H_2O vapor at 0.3 to 0.6 Torr in a dc glow discharge of around 0.4 mA as a function of distance from the cathode with a mass spectrometer. The wall potentials varied along the wall, descending from high values at the cathode to low values at some 6 cm distant. The fields in the gas near the cathode were also high. Thus creation of simple ions by electron impact occurred in the cathode fall region. In the more remote lower fields, these ions grew more complex by ternary collisions. As analyzed at the wall, their growth was limited by impacts which removed H_2O molecules in the wall fields. Near the high field cathode, a few H_2O^+ ions were noted ($\approx 5\%$), but these declined at greater distances. Some O_2^+ ions were also noted in this region. They came from dissociation of H_2O vapor; beyond 1 cm there was less than 2%. The predominant ion near the cathode was H_3O^+ or, better, $H^+(H_2O)$. Other ions present were $H^+(H_2O)_3$ and $H^+(H_2O)_4$. The ion $H^+(H_2O)_4$ was the predominant ion at 4 cm, but ions with larger values of n also appeared. Traces of ions out to $H^+(H_2O)_6$ were observed. This leads to the conclusion that the $H^+(H_2O)$, which is known to be created very rapidly through the reaction

$$H_2O^+ + H_2O \rightarrow H_3O^+ + OH$$

established by Lampe et al. [14], is formed near the cathode by H_2O ionized by electron impact. Following the creation of $H^+(H_2O)$, a succession of ternary impact reactions of the form

$$H_3O^+ + 2H_2O \rightleftarrows H_5O_2^+ + H_2O$$

leads to larger complex aggregates which appear to reach a limit of stability in the glow discharge at the fields and temperatures existing with a value of n equal to 6. Depending on the value of the electrical fields at the wall through which ions created have to pass to reach the spectrometer, the ions recorded are limited to an upper value of n. Evidence from the curves indicated that the binding energy of the transition from H_3O^+ to $H_5O_2^+$ lies around 1 eV since at this energy the $H_5O_2^+$ ion does not exist. Thus $H_5O_2^+$ is not seen near the cathode. The energy for regression of $H_{11}O_5^+$ to $H_9O_4^+$ appears to be around 0.2 eV.

Thus the nature of the ion species created in H_2O depends on the H_2O vapor pressure, the temperature, the field strength, or more than one of these. These determine the conditions of chemical equilibrium between rate of formation by triple impacts and rate of dissociation by thermal impacts, thus delineating the shape of the spectrum of n values present.

It has been indicated by model studies of Siksna [25] that the more stable complexes are $H^+(H_2O)_3$ and $H^+(H_2O)_4$ which have triangular symmetry.

While these ion complexes might be considered classical polarization force clusters about the small proton, the situation is probably more complex as a consequence of the hydrogen bonding, which requires quantum mechanical analysis. In fact, one might more correctly call this type of reaction *aggregation by complex ion formation*.

Since the size of the aggregate depends on decreasing binding energy as n in the aggregate $H^+(H_2O)_n$ increases, and since collisions with any molecules of adequate energy can dissociate H_2O molecules, it is unlikely that in mixtures such as air the aggregates with 0.4 Torr of H_2O will reach the size of $n = 4$ in the numbers that have been observed in pure H_2O gas. Narcisi and

Bailey [26] using a rocket (containing a mass spectrometer) shot into the atmosphere, noted H_3O^+ predominating with some $H_5O_2^+$ below 82.5 km at about 220°K. Above this altitude, photodissociation of H_2O appeared to occur. Numerous NO^+ ions and trace amounts of N_2^+ were observed. Some H_2O^+ was present at 70 km but decreased with altitude. Above 82.5 km, O_2^+ and NO^+ were dominant. Below 82.5 km, masses greater than 45 were observed, which could have been $H^+(H_2O)_n$ aggregates with larger n.

Very recent studies by the ESSA group under Ferguson have shown that the Narcisi and Bailey observations below 82.5 km can be explained as follows. As already mentioned, the H_3O^+ ion comes from the reactions (reported by Ferguson) of O_2^+ and H_2O leading to H_3O^+, OH, and O_2. The $H^+(H_2O)_2$ is derived by the reaction of NO^+ with H_2O, forming $NO^+(H_2O)$, then $NO^+(H_2O)_2$, and finally $NO^+(H_2O)_3$, which dissociates into $H^+(H_2O)_2$ and HNO_2.

Kebarle and Hogg [27], using a mass spectrograph, with inert gas vehicles for 0.5 Torr H_2O, as well as D_2O, ionizing by α particles, observed instead of H_2O^+, the ions $H^+(H_2O)_n$, in agreement with Knewstubb and Tickner [24]. Mass spectroscopic studies at about 1 Torr, reported by Pahl [15], indicate that H_2O^+ and H_2O molecules react to give $H^+(H_2O)$ or the H_3O^+ ion.

As noted previously, if electrons attach to form O_2^-, or if O_2^- ions are present, the ion $(H_2O)O_2^-$ will appear even at 10^{-4} Torr of H_2O. At higher pressures, the $(H_2O)_n O_2^-$ ions appear [10]. If water vapor alone is ionized at low pressures, dissociative attachment with an $E/p > 10$, ions of the form $(H_2O)_n OH^-$ appear, as earlier noted.

Thus it is clear that the presence of H_2O vapor in air will seriously modify the ion mobility spectra in unknown ways. The complexity of the reactions possible are such that, until a mass spectroscopic analysis is made of the ion mobility spectra in air at different E/p and different values of p with varying concentrations of H_2O, on both positive and negative ions, we only speculate.

(f). CO. This gas has been studied fairly completely with mass spectroscope in relation to mobility by Saporoschenko [28]. The CO^+ ion reacts with CO to give the $(CO)_2^+$ ion at low E/p. Its mobility extrapolates to 1.93. The theoretical value is 2.23. As E/p increases, it reacts to yield either CO^+ or CO_2^+. Dissociation of the $(CO)_2^+$ to CO^+ and CO and the formation of CO_2^+ occur simultaneously, and relative amounts cannot be estimated very accurately. Both have the same zero field mobility of 1.16. At higher E/p, the reaction

$$CO^+ + CO_2 \rightleftarrows CO_2^+ + CO$$

is reversible, the energy of this reverse reaction being 0.3 eV.

As regards negative ions, the situation in CO is bizarre [29]. At a relatively low energy, dissociative electron attachment to yield O^- occurs for CO. However, CO reacts very readily with O^- to *reliberate the electron* and form CO_2. Thus, although inelastic impacts occur and a few O^- ions are observed, these rarely last long enough to be detected in quantity. CO^- ions do not form.

(g) CO_2. CO_2 has been most extensively studied by Saporoschenko and Wisner [30], using the four-gauze shutter method and mass spectrograph from E/p of 10 to 700 V/cm Torr and pressures of 0.5 to 1.0 Torr. In the ionization of CO_2 gas, O_2, O_2^+, C^+, O^+, CO^+, C_2O^+, $C_2O_4^+$, and CO_4^+ are produced and detected. However, the chief ions are only CO_2^+ and O_2^+. At low pressures, O_2^+ decreases and CO_2^+ increases relatively as E/p increases. At higher pressures, the O_2^+ ion predominates even at an E/p of 200. At an E/p of 150, O_2^+ increases relative to CO_2^+ as pressure increases. The curves clearly indicate the charge exchange of CO_2^+ in interaction with the O_2 created in the discharge. The abundance of the ions CO_2^+ and trace ions CO^+ and C^+ at an E/p of 60 at 0.45 Torr and 305°K declines exponentially with distance in the drift space, while O_2^+ increases to reach a value of more than 90% beyond 8 mm travel. Reactions postulated are charge transfer from C^+ to CO_2 to give CO^+ and CO; reaction of CO^+ with CO_2 to give CO_2^+ and CO; and a complex series of reactions between CO_2^+ and CO_2 to yield C_2O_2 and O_2^+. The zero field mobility of O_2^+ in CO_2 lies between 1.2 and 1.3; and the mobility is less than 1.2 for CO_2^+ in CO_2.

Data on mobilities of negative ions in CO_2 are not satisfactory. According to Schlumbohm [20], with poor resolution and no mass spectrograph, negative ions in CO_2 have the same mobility values as positive ions in CO_2. He could not distinguish CO_2^+ and O_2^+ in CO_2. At 3.85 eV, CO_2 dissociatively attaches electrons to yield O^-. This occurs at an E/p greater than 7. The O^- ions then react with $2CO_2$ to give a stable ion CO_3^- and CO_2, as indicated by Moruzzi and Phelps [10]. If O_2 is added to CO_2, these workers observe four ions. Between E/p of 1 and 10, O_2^- and CO_4^-, presumably a combination of O_2^- and CO_2, are present. Above an E/p of 10, the ions are only CO_3^- and O^-.

(h) Negative ions in the upper atmosphere. A study by Fehsenfeld [31], with the flow method, gives more light on reactions of negative ions appropriate to the upper atmosphere. In the region where O_3 is present, O^- and O_2^- react with O_3 to yield O_3^-. The O_3^- reacts with CO_2 and yields CO_3^- and O_2. If O is present, the CO_3^- interacts with O to create O_2^- and CO_2. The CO_3^- reacts with NO to yield NO_2^- and CO_2. The O_3^- reacts with NO to yield NO_3^- ions, which are very stable. Thus, instead of the heretofore assumed O^- and O_2^- in the D region of the atmosphere, the ions will probably be NO_2^-, NO_3^-, and CO_3^-, with less O^- and O_2^-. The ratios are determined by the O_3/O ratio. Negative ion density relative to electron density can thus be expected to increase if the ratio $O_3/O > 1$.

(i) NO. Since this gas has the lowest ionization potential of the group, NO^+

might be expected in time to be the dominant ion. Since NO is only a minor constituent of air at lower altitudes, little NO^+ is seen. Its presence depends on electron energy. In air, NO is formed at 19.8 eV. It will lose electrons to all positive ions, in pure dry air. Where it is present, this molecule plays a very important role [37, 38]. In NO gas the two ions, NO^+ and $NO^+(NO)$, are both present and have an ambipolar diffusion coefficient equivalent to a mobility of 0.43 cm^2/V sec, which is very low. As indicated, the NO^+ undergoes a cycle of reactions with traces of H_2O, leading to the formation of $H^+(H_2O)_2$ and HNO_2. The negative ions in NO are NO_2^-. The O^- forms in dissociative attachment of electrons to NO. The O^- reacts with NO, presumably in a fast three-body reaction, to yield NO_2^-. At very low E/p, less than 2, NO^- forms from electrons in a three-body process; but it cannot form at $E/p > 6$. The NO_2^- ion is dominant. However, NO_2^- forms complexes with H_2O in a fast three-body reaction, yielding $NO_2^-(H_2O)$. This in turn reacts with NO to yield $NO_2^-(HNO_2)$ and OH with a mass number 93.

(j) NO_2. This gas, with its low ionization energy, is again likely to yield the constituent NO_2^+, because it can also take electrons from O_2^+. It will lose its charge to NO if present. The NO_2 forms above an electron impact energy of 13.7 eV by a chemical process unknown at present [18]. However, as NO appears under conditions of the atmosphere with NO_2, the NO^+ ion will dominate. Thus it is usually present only in small amounts. The NO_2^+ breaks up at high E/p to yield various ions [18]. At electron energies as low as 1.35 to 1.4 eV, NO_2 interacts readily to give O^- ions and NO. The O^- reacts rapidly with NO_2 to give NO_2^- and O.

One may conclude from what has gone before that the reactions of atmospheric ions are far from simple. Until mobility spectra are connected to mass spectroscopic spectra as associated with varying conditions of E/p, p, and relative gaseous concentrations in air itself, all speculations about what occurs or is present in the upper atmosphere and at higher pressures are essentially futile. The situation is even more complex for the ambient polluted air with its SO_2, H_2S, CH_4, etc., in our environment.

9. REACTION RATES OF IONS

This article cannot conclude the small ion discussion without calling attention to the many studies of reaction rates for charge exchange, etc., of the small ions. The most fruitful of these have been derived from a flow method developed by a group headed by E. E. Ferguson[8] and including, notably, among others, Fehsenfeld and Schmeltekopf. They have delineated the reaction rates of most common ion species in different environments, includ-

[8] Ion reaction papers of the ESSA group under Ferguson [32a–j].

ing particularly both positive and negative ions to be found in the atmosphere. The work was done with ion identification by mass spectrograph, largely at low pressures. Study of their tables of values for various reactions is essential for determining what ions are present in a given situation.

10. Langevin and Intermediate Ions[9]

Aside from the so-called normal ions in air, it appears that with adequate moisture there are several classes of larger ions. Some of these may be associated with charged aerosols, such as dust particles, which may or may not have films of moisture. Many products of oxidation and other sources of atmospheric pollution coming from industrial activities or natural geological and meteorological processes, including HCl, HNO_3, H_2SO_3, H_2SO_4, and salts such as $NaCl$, which dissolve in water and lower the vapor tension, will serve as nuclei. All of these in time may acquire electrical charges from ions present in the atmosphere. Depending on their size, they will have a Gaussian charge distribution. For particles of $NH_4Cl + H_2O$ of 10^{-4} cm diameter appearing as spherical solution droplets, the charge of a few may reach as high as 10 electrons. As radii become smaller, the thermal energy of motion is unable to bring as many charges up to the surface. For particles as small as 10^{-5} cm in radius, charges in excess of 3 would be unlikely.

In the early days, Langevin observed two classes of ions. The first class, properly called Langevin ions, were those that appeared when water was condensed on them by adiabatic expansion. On warming, the visible drops disappeared, but the charged nuclei persisted as invisible particles which would condense water vapor in cloud chambers on later supersaturation. They were stable in normally moist air of relative humidity of 40 to 60% at 20°C and might just be visible in the ultramicroscope. They had a fairly broad mobility spectrum which centered around 5×10^{-3} cm^2/V sec and probably less. They were assumed to be spherical water droplets stabilized by an electrical charge, as indicated by J. J. Thomson (see Loeb [1]). They are not sensitive to sign of charge. From their observed mobilities, they should have radii of the order of 10^{-6} cm and may contain from 10^5 to 10^6 H$_2$O molecules with well developed spherical surfaces controlled by surface tension. They are usually assumed singly charged, but may have as many as two or three electrons of charge.

A group of faster ions with a considerable spread of mobilities is noted in moist, dust-free air, with values centering around 6×10^{-2} cm^2/V sec. These ions have radii, calculated from the solid elastic mobility equation, of 3×10^{-7} cm. They contain on the order of some hundreds to a thousand molecules. For these ions, the Stokes–Cunningham law gives radii of 7×10^{-7} cm,

[9] See Loeb [1].

which is again too large. They do not have enough molecules to build a surface tension layer. They are in some sort of ordered state as crystallites of H_2O with hydrogen bond linkages, such that negative charges at their centers, on supersaturation, orient oncoming H_2O molecules to favor binding on the surface and condensation. Similar crystallites in alcohols favor positive charges for condensation. Once adiabatic condensation has taken place, and the ions have changed to charged visible drops, they reevaporate to invisible particles on warming and remain stable as Langevin ions. They are formed through the acquisition of small ions by the existing crystallite condensation nuclei always present about impurities. It is probable that in all cases the crystallite condensation nuclei have formed about hygroscopic molecules in the atmosphere. A Langevin ion with one molecule of H_2SO_4 would constitute a 0.01 molar solution which might be in equilibrium at 60% supersaturation at 20°C. Likewise, one molecule of NaCl on an intermediate ion would yield a concentration of the order of tenths of a mole per liter.

Vassails (see Loeb [1]) observed large ions with mobilities ranging from 1.2×10^{-3} to 0.52×10^{-3} cm^2/V sec with apparent definite groups of ions. These he attributed to essentially a single sized droplet which had multiple charges. The radii were assumed to be about 8.8×10^{-6} cm, and the particles were visible in the ultramicroscope. He could produce them in filtered, moist, dust-free air, by exposure to ultraviolet light through quartz optics for some minutes. Besides these, Langevin ions of 5.8×10^{-6} cm radius were produced. The gases which in the presence of moisture led to these ions were SO_2, CS_2, H_2S, CCl_4, C_2HCl_2, and NO_2. All of these, by ionization, dissociation, or oxidation by the ozone from ultraviolet light, can give hygroscopic substances such as SO_2, SO_3, H_2SO_3, H_2SO_4, HNO_3, HCl, etc. According to Vassails, they may give colloidal particles of sulfur from CS_2 and H_2S. It was these nuclei that yielded what Vassails called Langevin ions. In addition, the entrainment of sea salts from minute bursting bubbles on the vast ocean surfaces gives rise to salt nuclei NaCl, $MgCl_2$, $CaCl_2$, Na_2SO_4, etc., which have been collected as far inland as the deserts of New Mexico and Arizona.

More puzzling is the sequence of ions observed by Chapman (see Loeb [1]) in the bubbling of distilled water and observed in a stream of clean, dry air of 35% relative humidity at 20°C. Negative mobilities of 1.8, 0.95, 0.45, 0.37, and 0.2 cm^2/V sec were observed as distinctly resolved peaks. No larger ions were observed. Positive ion peaks were more blurred, with values centering around 1.0, 0.4, and 0.2 cm^2/V sec; the first peak started at 1.8 cm^2/V sec, but it was composed largely of the slower components. These ions were undoubtedly the various hydrated oxonium ions, $H^+(H_2O)_n$. The O_2^+ and O_2^- ions had a slightly higher mobility. The slower negative ions were possibly $(H_2O)_nOH^-$ and $(H_2O)_nO_2^-$, and the positive ions must have been $(H_2O)_nH^+$ ions torn off by disruption of the surface double layers.

Those results may be related to the studies from Bricard's laboratory [33a–f] of mobilities of ions created by decay of thoron and radon in filtered room air with a relative humidity of 40%, using the Erickson method but with detection of ions by radioactivity at the collecting plates. The ions were largely positive. Mobility peaks occurred at 2.1, 1.5, 0.5, and 0.3 cm^2/V sec. The fast thoron- and radon-produced ions are the positively charged ThB$^+$ and RaA$^+$ ions, the mobility of which is predicted by the Langevin theory as 2.15 cm^2/V sec. The fallibility of present theory and the inaccuracy of all mobility measurements render such an accurate agreement fortuitous, although the identification is probably correct. The slower ions correspond unquestionably to hydrated complexes, with 1.5 possibly being ThB$^+$(H$_2$O). The others, down to 0.5, can be complexes with (H$_2$O)$_n$, with n as high as 3 or 4. Diffusion studies of these ions indicated that with age, in air, the ions degraded to mobilities of 0.2 cm^2/V sec *in 30 minutes*. In shorter times, only a few fast ions were observed, but most ions had values of 0.8 and 0.2 cm^2/V sec. The time scales of complex ion formation for these ions clearly indicate that they *do not occur by the relatively rapid progressive condensation of H$_2$O vapor molecules on the ions*. The formation can occur only if the RaA and the ThB ions diffuse to encounter large aggregates of H$_2$O, probably condensed around hygroscopic molecules created in the air by the radioactive ionization. Cluster formation by condensation would occur on a time scale of between 10^{-3} and 10^{-2} sec. Studies made on diffusion coefficients of the neutral radioactive particles, which arise through encounters with negative ions produced by radioactive ionization, indicated that these have radii of about 6×10^{-8} cm; and at this radius, the solid elastic impacts far outweigh the dipole forces. Thus, the mass dispersion law may not be used for such ions. On the solid elastic impact theory, with inactive dipole attractions, a mobility of 0.2 cm^2/V sec implies a radius of 1.7×10^{-7} cm. Such an ion, if spherical, could accommodate on the order of 200 or less H$_2$O molecules.

Renoux [33f] using the techniques of this group, observed with unfiltered air the following radioactive ionic distributions: small positive ions, 2.6%; large positive ions (Langevin), 4.7%; large ions (presumably solids), 2.6%; radioactive neutral nuclei, 33.7%; other particles with radii greater than 2×10^{-6} cm, 58.7%. Thus in 10 to 15 min in unfiltered air, almost all ThB$^-$ and RaA$^-$ ions had been captured by water or solid nuclei. Many of the positive ThB$^-$ aggregates were neutralized by ambient negative ions.

Billard *et al.* [33e] recently made an extensive study on the formation of nuclei, both charged and uncharged, in a large vessel filled with *filtered* room air, such that diffusion loss was slow. It was shown that particles of a radius of 2×10^{-7} to 2×10^{-6} cm were generated. After admission of filtered nucleus-free air, in the *dark*, these increased in number to a variable maximum and declined slowly by loss by aggregation to larger particles. Exposed to sunlight through plexiglass or glass, the number again increased to a peak after the initial loss by aggregation. Similar results were obtained by the addition of radon and thoron. The size of the initially growing particles as a function of time could be ascertained by diffusion measurement in an associated tube system. These changed from the initial values of small ions to those of particles of a radius of 1.7×10^{-7} cm in the time scale of some further minutes. Addition of thoron produced nucleation of the same sort. The initial particles had a radius of 1.4×10^{-7} cm. On introduction of thoron, in 5 min the average radius was 2.2×10^{-7} cm. After an additional 10 min, the radius had increased to 10^{-6} cm. Plots were made of the reciprocal of the number of nuclei as a function of time after introduction of thoron. The plots are sensibly linear, representing a *coefficient of coagulation* value of 10^{-8} cm^3/sec. This corresponds to the coagulation time of particles of 2×10^{-7} to 2×10^{-6} cm radius. A Zeleny mobility tube indicated that the ions which were eventually formed by capture of small ions by those particles had a radius of 10^{-6} cm; that is, they were Langevin ions.

It is to be noted that the nucleation occurred for air taken from outside the room, in the region of Paris. In order to show the nucleation in the dark, which varied widely in different samples studied, the air had to be collected *in full sunlight* [34]. If air was collected at night, no nuclei were observed to form in the dark. It was shown that an initial sojourn in the dark, which could extend from seconds to minutes, rendered the air incapable of later forming nuclei in the dark. The nuclei invariably formed under the influence of sunlight. It appeared that an unfiltered catalyzer which was activated by sunlight was needed to develop nuclei in the dark. The initially filtered air, even that collected in the dark, on illumination produced a number of primary nuclei. These developed and coagulated to produce a maximum, then declined some time after illumination, as no more primary nuclei were formed and the earlier nuclei coalesced and reduced in number with time.

Formation of nuclei on illumination had also been observed by Verzar and Evans [35] in 1959. These workers had shown that previously filtered SO$_2$ gas leads to nucleus formation in the dark after illumination by sunlight through glass at the same speed observed by Bricard's group. Verzar and Evans had also shown that the natural concentration of SO$_2$ in the air, which can be around 2.7×10^{13} molecules per cm^3, is more than enough to account for the nucleation. Rough quantitative measurements by Bricard's group yielded a coefficient of coagulation of the order of 10^{-8} cm^3/sec for those nuclei. This corresponds to the fixation of particles of 10^{-7} cm radius on nuclei of 10^{-6} cm radius, or to the coefficient of coagulation between particles of 10^{-7} cm and particles of 5×10^{-7} cm radius. Thus the intermediate ion nuclei are observed to coagulate to lead to Langevin-ion sized nuclei.

A study of ionization of these nuclei by radiolysis from the disintegration of thoron or ThA revealed the following. The small, fast ion density is initially high and declines rapidly after the source is removed. These ions were not detected in the Zeleny tube, but larger ions of both signs of charge were studied with the Zeleny tube. Ions did not appear in that tube until 50 min after the thoron was removed. These ions, from the value of their coefficient of diffusion, were shown by that time to be Langevin ions. In any event, not more than 10% of these large particles appeared as ions. The small intermediate ions appeared not to be charged, as the counter did not detect them. This was doubtless because, during their transient lifetime of a few minutes, the ambient ion densities of positive and negative small ions and the neutralization by coalescence of ions of opposite charge gave so few charged intermediate nuclei that they were not counted. On the other hand, the few remaining charged ions after 50 min were picked up by 10% of the large nuclei which had formed.

It appears that with the SO_2 content of the air used, the nucleation proceeds about an activated SO_2 molecule through the condensation of water molecules. This nucleus then may pick up an atmospheric or a radioactive ion. However, it is as likely from the observations that the activated SO_2 molecule could have acquired a radioactive or other ion and then nucleated by condensation of water vapor. The latter is the more likely, with the intense thoron ionization. Ordinarily, in air, nucleation proceeds about activated SO_2. The ions of themselves show no tendency to nucleate and grow, *unless associated with a nucleating impurity.*

Recent further experiments by Bricard's [36] group indicate that the nucleation in their studies probably involves sulfur. One fact is certain; and that is that filtering air over activated charcoal, which absorbs SO_2, almost completely eliminates the formation of photolytic particles. Charcoal filtration is less effective if thoron is used. They believe that filtration does not completely remove SO_2, and that thoron is more effective by its radiolysis than is sunlight. On the other hand, the intense *thoron* ionization may be creating products like NO_2, or other nitrogen oxides, that lead to a photolytic production of nucleation and that cannot be created by photolysis, so that carbon filtration does not affect the result.

Recent unconfirmed results [39] indicate that the formation of aerosol particles in filtered air in the dark is strongly increased if SO_2 is added. Further addition of SO_2 then increased the nuclei, but a third addition failed to show much increase. This points toward the exhausting of some essential catalytic agent introduced with the air. It had been indicated that filtration over activated charcoal strongly reduced nucleus formation. This was known to remove SO_2. Addition of SO_2 to air after filtration over activated charcoal did not greatly augment the small amount of nucleation observed. This indicates that the charcoal had indeed removed some other agency for activation in the dark, and/or the photolytic activity. Addition of NO_2 to the air after filtration over activated charcoal, together with the addition of SO_2, markedly increased the nucleation. This pointed to NO_2 as one of the photolytically essential ingredients for nucleation with SO_2. On addition of NH_3, after a first addition of NO_2, the creation of nuclei ceased; the number of particles decreased rapidly, while particle size increased by coagulation. The NH_3, reacting to form NH_4NO_3 with H_2O present, destroyed the photolytic properties of NO_2.

In transition to the study of the large ions, e.g., the intermediate and Langevin, it pays to regard recent results presented at the Tokyo Conference in the Spring of 1968 by Bricard et al. [36]. They perfected a Tyndall–Powell four-gauze shutter method for the study of ions in room air generated by a positive point corona discharge. The unique feature of this study was that besides its high resolving power, it could be extended to investigate ions in the lower mobility range. It was also possible to vary the age of the ions studied from around 10^{-3} to 10^{-2} sec. The results are quite interesting, and are presented in Table II. The ions observed in the shorter

TABLE II. Aging data on mobilities in air.[a]

Aging time (sec)	Mobility (cm^2/V sec)					
10^{-3}	2.1	1.9	1.5	1.35		
5×10^{-3}	2.1	1.9			0.7	
10^{-2}		1.9				0.4

[a] From Bricard et al. [36].

times are akin to those observed in room air freshly ionized by thoron or ThD. The mobility of 2.1 decreases at the expense of that at 1.9, while the ions at 1.5 and 1.35 give way to an ion of mobility 0.7. Within 10^{-2} sec, only the ion at 1.9 and those at 0.4 appear. Varney [18] inclines to the belief that the common mobility of ions in air at around 1.9 cm^2/V sec may be the stable $H^+(H_2O)_4$ ion.

At any quick aging time, if one proceeds in a closed space filled with air, and progressively adds more H_2O vapor or SO_2, the groups at 2.1 and 1.9 disappear successively rapidly, and the remainder of the spectrum moves toward lower mobilities approaching 0.4. If excessive H_2O vapor is added, an additional ion group at mobility 6 is distinguishable. This is the H_3^+O, or oxonium ion, observed by Varney at about this mobility when adequate water vapor was added to permit direct or indirect ionization of the H_2O sufficient to give a record in the mobility apparatus. Changes in air pressure from about 760 to 540 Torr at a given age led to mobility values inversely proportional to pressure.

Regarding Varney's [18] results, since high fields exist near the point in the corona discharge one can expect NO_2^+ and NO^+, as well as O_2^+ and smaller amounts of H_3^+O. Conditions of the measurement appear to have been such that there was accumulation of the NO_2 molecules and perhaps NO molecules created by neutral chemical reactions. In the presence of H_2O and sunlight, SO_2 forms very hygroscopic large neutral nuclei which start with a few hundred molecules and grow in time by condensation and aggregation. Bricard [36] has recently shown that NO_2 is needed to increase the amount and rate of such nucleation. Thus it is clear from the work of Varney, of Shahin [22], and of those in the ESSA group [33], that the dominant ions in the early stages will be O_2^+, some NO_2^+, some NO^+, and the series of hydrates of types $O_2^+(H_2O)_n$, $NO^+(H_2O)_n$, $NO_2^+(H_2O)_n$, as well as the assertedly hygroscopic HNO_2 via the Ferguson cycle.

Any detailed interpretation of the data is futile, since mass identification associated with mobilities of ions has not been achieved, since there are even then doubts because of inaccuracies in measurement as well as poor discrimination between species of nearly the same mobility, and since the calculation of mobilities other than in order of magnitude is impossible. Definitely observed is the unique mobility 6, mass identified as the oxonium ion. The mobilities of 2.1 and 1.9 could be associated with hydrates of the type $H^+(H_2O)_n$ with n at least 2 and probably 3 and 4, or similar hydrates, such as $O_2^+(H_2O)_n$. It is doubtful if the ions NO_2^+ and NO^+, and their hydrates, are sufficiently numerous to account for the mobility peaks. However, it must be recognized that the values cited represent the *peaks* of the observed time-dependent current curves. These do not even reach zero current between peaks, which indicates transitions in the drift space and the presence of minority ions. The peaks at 1.5 and 1.35 again represent larger aggregates of a more or less unique composition which are probable in collisions of an ion with an H_2O molecule in about 10^{-7} sec. The further hydration of the 1.5 and 1.35 mobility ions in some 5×10^{-3} sec is not impossible. On the basis of a solid elastic sphere collision, the mobilities of 0.4 and 0.7 indicate agglomerates, of closely packed *spherical form*, containing tens to hundreds of H_2O molecules. Even a mobility of 1.35 with a spherical radius of about two molecular radii might represent ten molecules. While such a concept would appear to indicate a continuous "clustering" or rate of growth and condensation about an ion, this *cannot be so* in the face of the evidence. Note that these ions *appear as distinct groups associated with unique mobilities*. There is no reason for such groupings in a continuous agglomeration process. The author suggests a much simpler explanation. An *effective* radius of three molecular diameters as indicated by mobility data *does not require* the particularly naive concept of a solid, densely packed, spherical cluster. Long ago the author pointed out that in virtue of the hydrogen bond linkages, the small aggregates of H_2O in near saturated H_2O vapor would not have a spherical shape, but would resemble small well-oriented and organized crystallites with (logically) considerable spatial extension. As he has followed the recently assumed structural linkages of the ionic hydrates reported, he is convinced that these chemically bonded structures have elongated chainlike forms. To the author's knowledge, no one has studied, on the basis of a solid elastic theory, the reduction in free paths of such structures relative to those of tightly packed spheres. Thus, it would not be surprising to discover that these mobility groups, including the 0.7 and 0.4 values, represent very specific chain structures with relatively few H_2O molecules. The influence of SO_2 and of H_2O vapor on accelerating the spectral changes and in producing the 0.4 mobility group suggests possible complexes of six to eight H_2O molecules in chainlike hydrogen bond linkages with such molecules. In addition, no one has considered the retarding action of the binding and release of "end-of-the-chain" H_2O molecules, in impact, which act much as charge-exchange reactions do in reducing mobilities.

Meanwhile, in such air, the generation of hygroscopic neutrals in the discharges, which are growing rapidly by condensation of H_2O vapor, will produce notable concentrations of such nuclear aggregates as observed by Bricard and his associates. These will later play a great role in the generation of the Langevin ions. However, it is difficult to imagine such protonuclei as being present in sufficient number to register as mobility classes in the studies of Bricard *et al.*, and of others. If they are present to the extent of some few thousand or more per cm^3, they are adequate to account for nucleation observations. More could be present, but only a small fraction will be charged.

The influence of ionic electrical charges on the formation of nuclei is best illustrated by the work of Barreto [40]. In this study a supersonic jet of slightly moist air is emitted from a convergent-divergent metal nozzle with an exit Mach number of about 1.5 and a throat diameter of about 7 mm. A concentric needle is supported from the convergent subsonic side of the nozzle. This needle is electrically insulated from the nozzle and its tip is located very near the throat about 0.5 cm away from the exit plane. It can be raised to a high negative or positive potential with respect to the grounded metal nozzle. The throttling of the air at the nozzle produces an enormous supersaturation of the water vapor at the nozzle. If the potential is *not* applied to the needle, thus not yielding a corona, no condensation occurs, and one has merely a supersonic air jet. If the potential is raised on the needle to the point at which a corona discharge takes place, homopolar ions, positive or negative, are generated. There is then at once a condensation of water about each ion. This leads to the formation of a jet of invisible Langevin ions, of mass about 10^6 water molecules, which are entrained in the air blast. The condensation of charge on the droplets is so complete that all the negative or positive corona current is projected through the nozzle and along the aerodynamic flow pattern from the jet without any of the mobile ions or electrons reaching the nozzle. The remarkable properties of this aerosol cloud have been described adequately in Barreto's several publications [40].

Of interest here is the physics of the formation of the aerosol cloud. A study of the aerodynamics and thermodynamics of the jet in the region of the needle indicates the following: While the supersaturation attendant on the cooling of the moist air on expansion is many-fold greater than would normally cause general cloudy condensation of the vapor, it does *not* occur. The reason is principally that, if in the absence of corona charges the general condensation at the densities involved occurred, the heat of condensation would be liberated so rapidly and so widely in the gas as to warm the surrounding air and prevent condensation until supersaturation increased still further. The corona discharge creating ionic populations of the order of perhaps 10^{12} or more ions per cm^3, compared to the some 10^{17} H$_2$O protonuclei per cm^3, will be captured by the embryonic drops. These protonuclei exist in statistical equilibrium but cannot all act as condensation nuclei without too much general heat liberation. The relatively sparse population of 10^{12} droplets per cm^3 that have captured ions will have the electrostatic potential added to their van der Waals' potential. Thus the lives of these embryonic nuclei will be prolonged, and condensation about them increased, relative to the surrounding uncharged nuclei, and there will be adequate opportunity to give off the heat of condensation without too much heating of the ambient gas. Thus progressive condensation, on such protonuclei as capture ions, proceeds to the development of the large Langevin ions on each charged nucleus. These are entrained in the air blast and carried downstream past the grounded nozzle. Thus one sees a very striking example of the stabilizing effect of the ionic charge in the condensation and growth of a protonucleus to the Langevin ion aggregate, illustrating clearly the role of charge in such processes. In this example each small ion fulfills its function in Langevin ion production. However, note the homopolar nature of the charges and the high density of local protonuclear production. Note also that it required a situation in which the protonuclei were created in profusion at the region where ions were produced, and that the small ions would not of themselves produce the condensation without the high supersaturation and presence of protonuclei.

REFERENCES

1. Loeb, L. B. (1955). "Basic Processes of Gaseous Electronics." Univ. of California Press, Berkeley, California.

2. Hohla, C. and Mohnen, V., (1967). Berechnungsmöglichkeiten für die Bildung von Cluster-Ionen; Bundesministerium für wissenschaftliche Forschung, Forschungsbericht K 67, Kernforschung, Sektion Physik der Universität München, January, 1967, pp. 1–45; Mohnen, V. (1967). The nature of tropospheric ions. I.U.G.G. Atmos. and Space Electricity Conf., 14th Genl. Assembly, Lucerne, *Intern. Conf. Universal Aspects Atmos. Electricity, 4th, Tokyo, 1968*, Session 7B, Paper 8.
3. Munson, R. J., Tyndall, A. M., and Hoselitz, K. (1939). *Proc. Roy. Soc.* **A172**, 28–54.
4. Yang, J., and Conway, D. C. (1964). *J Chem. Phys.* **40**, 1729; (1965). *J. Chem. Phys.* **43**, 2900; Chanin, L. M., Phelps, A. V., and Biondi, M. A. (1962). *Phys. Rev.* **128**, 219; Pack, J. L., and Phelps, A. V. (1966). *J. Chem. Phys.* **44**, 1870; Pack, J. L., and Phelps, A. V. (1966). *J. Chem. Phys.* **45**, 4316.
5. Varney, R. N. (1968). Rates for the Reversible Reaction $N_4^+ \rightleftarrows N_2^+ + N_2$. Electronic Sciences Laboratory, Lockheed Palo Alto Research Laboratory, Report 6-79-68-5, April 1968 [also in *Phys. Rev.*, **174**, 165 (1968)]; Bohme, D. K., Dunkin, D. B., Fehsenfeld, F. C., and Ferguson, E. E. (1968). Observation of saturation in three body ion-neutral association reactions. To appear in *J. Chem. Phys.* (1968–1969).
6. Fehsenfeld, F. C. and Ferguson, E. E. (1968). Reported observation of N_6^+ and. later, up to N_9^+ at 80°K (personal communication to R. N. Varney, April 4, 1968), *Ann. Gas. Electron. Conf., 21st, Boulder, Colorado, October 1968*, Paper H3.
7. Madson, J. M. and Oskam, H. J. (1967). Mobility of Ar ions in Ar. *Physics. Letters* **25A**, 407.
8. Saporoschenko, M. (1965). Mobility of mass-analyzed N^+, N_2^+, N_3^+, and N_4^+ ions in nitrogen gas. *Phys. Rev.* **139**, A352; Moseley, J. T., Snuggs, R. M., Martin, D. W., McDaniel, E. W., and Miller, T. M. (1968), and McDaniel, E. W., Martin, D. W., Snuggs, R. M., and Moseley, J. T. (1968). *Ann. Gas. Electron. Conf., 21st Boulder, Colorado, October 1968*, Papers J-6 and J-7.
9. Kasner, W. H. and Biondi, M. A. (1965). Electron-ion recombination in nitrogen. *Phys. Rev.* **137**, A317; Asundi, R. K., Schulz, G. J., and Chantry, P. J. (1967). Studies of N_4^+ and N_3^+ ion formation in nitrogen using high-pressure mass spectrometry, *J. Chem. Phys.* **47**, 1584.
10. Moruzzi, J. L. and Phelps, A. V. (1966). *J. Chem. Phys.* **45**, 4617.
11. Eiber, H. *Z. Angew. Phys.* (1963). **15**, 103, 461; Frommhold, L. (1964). *Fortschr. Physik* **12**, 597; McKnight, L. G., McAfee, K. B., and Sipler, D. P. (1967). Low-field drift velocities and reactions of nitrogen ions in nitrogen. *Phys. Rev.* **164**, 62; For values see Huxley, L. G. H., Crompton, R. W., and Bagot, C. H. (1959). *Australian J. Phys.* **12**, 303; Rees, J. A. (1965). *Australian J. Phys.* **18**, 41; Shafer, C., and Beaty, E. C. (1968). *Ann. Gas. Electron. Conf. 21st, Boulder, Colorado, October 1968*, Paper J-8.
12. Voshall, R. E., Pack, J. L., and Phelps, A. V. (1965). *J. Chem. Phys.* **43**, 1990.
13. Shale, C. C. (1965). The physical phenomena underlying the negative and positive coronas in air at high temperatures and pressures. *IEEE Intern. Conv. Rec.* **7**, 77–87.
14. Lampe, F. W., Field, F. H., and Franklin, J. L. (1957). *J. Am. Chem. Soc.* **79**, 6132.
15. Pahl, M. (1962) *Ergeb. Exakt. Naturw.* **34**, 209 ff.
16. Beaty, E. C. and Patterson, P. L. (1965). *Phys. Rev.* **137**, A346; Madson, J. M., Oskam, H. J., and Chanin, L. M. (1965). *Phys. Rev. Letters* **15**, 1018; Beaty, E. C., Browne, J. C. and Dalgarno, A. (1966). *Phys. Rev. Letters* **16**, 723.
17. Oskam, H. J. (1957). Microwave investigations of disintegrating discharge plasmas. Thesis, Univ. of Utrecht, Holland, p. 101.

18. Varney, R. N. (1968). Personal communications; (May–August, 1968); Sinnott, G., Golden, D. E. and Varney, R. N. (1968). Positive-ion mobilities in dry air, *Phys. Rev.* **170**, 272; Varney, R. N. (1968). *Ann. Gas. Electron. Conf., 21st, Boulder, Colorado, October 1968*, Paper A-1.
19. Luhr, O. (1933). *Phys. Rev.* **44**, 459.
20. Saporoschenko, M. (1965). Mobility of mass-analyzed H^+, H_3^+ and H_5^+ ions in hydrogen gas. *Phys. Rev.* **139**, A349; Saporoschenko, M. (1965). *J. Chem. Phys.* **42**, 2760; Schlumbohm, H. (1965). *Z. Physik* **182**, 317; Albritton, D. L., Miller, T. M., Martin, D. W., and McDaniel, E. W. (1968). *Phys. Rev.* **171**, 94; Miller, T. M., Moseley, J. T., Martin, D. W., and McDaniel, E. W. (1968). *Phys. Rev.* **173**, 115.
21. Samson, J. A. R. and Weissler, G. L. (1965). Mobilities of oxygen and nitrogen ions. *Phys. Rev.* **137**, A381.
22. Shahin, M. M. (1966). *J. Chem. Phys.* **45**, 2600; Shahin, M. M. (1967). Use of corona discharges for the study of ion-molecule reactions. *J. Chem. Phys.* **47**, 4392.
23. Eiber, H. and Kandel, W. (1968). Mobilities and cross sections of ions in mixtures of oxygen and water vapor. *Z. Angew. Phys.* **25**, 18.
24. Knewstubb, P. F. and Tickner, A. W. (1963). *J. Chem. Phys.* **38**, 464.
25. Siksna, R. (1967). A symmetrical aspiration condenser for simultaneous counting of ions of both polarities. Mimeographed for Atmospheric and Space Electricity Conference of the I.U.G.G. XIVth General Assembly, Switzerland, October 5, 1967; Siksna, R. (1967). The role of water substance in the structure and production of ions in ambient atmospheric air. Paper for Inst. for Högspanningsforskning, Uppsala, Sweden, 1967.
26. Narcisi, R. S. and Bailey, A. D. (1965). *J. Geophys. Res.* **70**, 3687.
27. Kebarle, P. and Hogg, A. M. (1965). *J. Chem. Phys.* **42**, 798.
28. Saporoschenko, M. (1968). Mobility of CO^+, CO_2^+, and $C_2O_2^+$ ions in carbon monoxide gas, *J. Chem. Phys.* **49**, 768.
29. Pack, J. L. and Phelps, A. V. (1966). Electron attachment and detachment, II. Mixtures of O_2 and CO_2, and of O_2 and H_2O. *J. Chem. Phys.* **45**, 4316; Moruzzi, J. L., Ekin, Jr., J. W., and Phelps, A. V. (1968). *J. Chem. Phys.* **48**, 3070.
30. Saporoschenko, M., and Wisner, W. W. (1968). Mobilities of O_2^+ and CO_2^+ ions in CO_2 gas. *Bull. Am. Phys. Soc.* **13**, 206 (also to appear in *J. Chem. Phys.* 1968).
31. Fehsenfeld, F. C., Schmeltekopf, A. L., Schiff, H. I., and Ferguson, E. E. (1967) Laboratory measurements of negative ion reactions of atmospheric interest. *Planetary Space Sci.* **15**, 373; Ferguson, E. E., Fehsenfeld, F. C., Goldan, P. D., and Schmeltekopf, A. L. (1965). Positive ion-neutral reactions in the ionosphere. *J. Geophys. Res.* **70**, 4323.
32a. Ferguson, E. E., Fehsenfeld, F. C., Goldan, P. D., and Schmeltekopf, A. L. (1965). *J. Geophys. Res.* **70**, 4323.
32b. Fehsenfeld, F. C., Schmeltekopf, A. L., Goldan, P. D., Schiff, H. I., and Ferguson, E. E. (1966). Thermal-energy ion-neutral reaction rates, I. Some reactions of helium ions. *J. Chem. Phys.* **44**, 4087.
32c. Goldan, P. D., Schmeltekopf, A. L., Fehsenfeld, F. C., Schiff, H. I., and Ferguson, E. E. (1966). Thermal-energy ion-neutral reaction rates, II. Some reactions of ionospheric interest. *J. Chem. Phys.* **44**, 4095.
32d. Fehsenfeld, F. C., Ferguson, E. E., and Schmeltekopf, A. L. (1966). Thermal-energy ion-neutral reaction rates, III. The measured rate constant for the reaction $O^+(^4S) + CO_2(^1\Sigma) \to O_2^+(^2\Pi) + CO(^1\Sigma)$. *J. Chem. Phys.* **44**, 3022.

32e. Fehsenfeld, F. C., Ferguson, E. E., and Schmeltekopf, A. L. (1966). Thermal-energy ion-neutral reaction rates, VI. Some Ar$^+$ charge-transfer reactions. *J. Chem. Phys.* **45**, 404.

32f. Fehsenfeld, F. C., Schmeltekopf, A. L., and Ferguson, E. E. (1967). Thermal-energy ion-neutral reaction rates, VII. Some hydrogen-atom abstraction reactions. *J. Chem. Phys.* **46**, 2802.

32g. Fehsenfeld, F. C., Ferguson, E. E., and Schmeltekopf, A. L. (1966). Thermal-energy associative-detachment reactions of negative ions. *J. Chem. Phys.* **45**, 1844.

32h. Fehsenfeld, F. C., Schmeltekopf, A. L., Schiff, H. I., and Ferguson, E. E. (1967). *Planetary Space Sci.* **15**, 373.

32i. Schmeltekopf, A. L., Fehsenfeld, F. C., and Ferguson, E. E. (1967). Laboratory measurement of the rate constant for $H^- + H \to H_2 + e$. *Astrophys. J.* **148**, L155.

32j. Schmeltekopf, A. L., Fehsenfeld, F. C., Gilman, G. I., and Ferguson, E. E. (1968). Temperature effects. To be published.

33a. Bricard, J., Girod, P., and Pradel, J. (1965). Mobility spectrum of small radioactive ions in air. *Compt. Rend. Acad. Sci.* **260**, 6587.

33b. Bricard, J., Billard, F., Blanc, D., Cabane, M., and Fontan, J. (1967). Structure detaillée du spectre de mobilite des petits ions radioactifs dans l'air. *Compt. Rend. Acad. Sci.* **263**, 761.

33c. Madelaine, G. (1965). Behavior of descendants of radon and thoron in a dust-free atmosphere. Report of work done under MM. Bricard and Pradel, August 1965.

33d. Bricard, J., Girod, P., and Pradel, J. (1965). Etat de charge des aerosols ultra fins en milieu faiblement ionizé, application aux gros ions atmospheriques. *J. Phys. Appl.* **26**, 141A.

33e. Billard, F., Bricard, J., Cabane, M., and Madelaine, G. (1967). Formulation et evolution des noyaux de condensation qui apparaissent dans l'air préalablement debarrassé d'aerosols. *Compt. Rend. Acad. Sci.**

33f. Renoux, A. (1965). Doctoral Thesis, Faculté des Sciences de l'Université de Paris.

34. Bricard, J. Billard, F., and Madelaine, G. (1968). Formation and evolution of nuclei of condensation that appear in air initially free of aerosols. *J. Geophys. Res.* **73**, 4487.

35. Verzar, F. and Evans, H. D. (1959). Formation of condensation nuclei in air devoid of aerosol particles. *Geofis. Pura Appl.* **43**, 259.

36. Bricard, J. (1968). Personal communication to author, July 1968; Bricard, J., Cabane, M., and Madelaine, G. (1968). *Intern. Conf. Universal Aspects Atmos. Electricity, 4th, Tokyo, May 1968*, Session 7B, Paper 11.

37. Ferguson, E. E. (1968). Personal communication, October 1968; Lineberger, W. C., and Puckett, L. J. (1968). *Ann. Gas. Electron. Conf., 21st, Boulder, Colorado, October 1968*, Papers H-12 and H-13.

38. Stockdale, J. A., Compton, R. N., and Reinhardt, P. W. (1968). *Ann. Gas. Electron. Conf., 21st, Boulder, Colorado, October 1968*, Paper H-8.

39. Bricard, J. (1968). Personal communication by letter, October 1968.

40. Barreto, E., and Mulcahy, M. J. (1965). Production and neutralization of a charged aerosol by corona fields. *J. Geophys. Res.* **70**, 1303; Barreto, E., and Martinot, K. (1967). Nonthermal ionization caused by aerodynamic discontinuities in charged aerosol jets. *Phys. Fluids* **10**, 2155; Barreto, E. (1967). A charged aerosol cloud. Curtiss-Wright Corporation Res. Rept., Wood-Ridge, New Jersey; Barreto, E. (1968). *Intern. Conf. Universal Aspects Atmospheric Electricity, 4th, Tokyo, Japan, May 1968*, Session 7A, Paper 3.

* For English translation, see reference [34].

AUTHOR INDEX

Numbers in parentheses are reference numbers and indicate that an author's work is referred to, although his name is not cited in the text. Numbers in italics show the page on which the complete reference is listed.

A

Abragam, A., 23, *39*
Affleck, J., 13(2a), *39*
Agocs, W. B., 32(3), *39*, 73(1), *76*, 85(1), *86*
Albritton, D. L., 237(20), 242(20), *254*
Alder, J., 125(6), *138*
Alekseev, V. V., 80(2), 86(2), *86*
Alldredge, L. R., 14, *40*
Allen, C. O., 129(18), *138*
Asundi, R. K., 230(9), *253*

B

Badger, R. M., *165*
Badgley, P. C., 108, *110*
Badinov, I. Y., 203(62), *220*
Bagot, C. H., 232(11), 234(11), 239(11), *253*
Bailey, A. D., 241, *254*
Ballif, J. R., 197(28), *218*
Balsley, J. R., 4(6), 7, *39*
Barker, R. A., 42, *79*
Barreto, E., 251, 252, *255*
Barrett, E. W., 122, 124, 129(1), *138*
Barringer, A. R., 41, *76*, *77*
Bartky, C., 177(45), *189*
Bartle, B., 164(24), *188*
Bates, D. R., 197(27), *218*
Bates, R. G., 81(28), *87*
Batten, E. S., 192(8), *218*
Bauer, E., 177(45), *189*
Beaty, E. C., 232(11), 233(16), 234(11), 239(11), *253*
Bell, W. H., 106(29), *111*
Bender, P. L., 17, *39*
Ben-Dov, O., 122, 124, 129(1), *138*
Berbezier, J., 83(3), 84, *86*
Bettinger, R. T., 129(18), *138*
Billard, F., 246, 247(34), 250(33b, 33e), *255*
Biondi, M. A., 228(4), 230(9), 232(4), *253*
Birks, J. B., 80(4), *86*

Bisby, H., 80(5), 82(34), 85(34), *86*, *88*
Black, G., 165, *188*
Blanc, D., 246(33b), 250(33b), *255*
Blangy, B., 83(3), 84(3), *86*
Bleil, D. F., 66(3), *76*
Bloom, A. L., 25(7), *39*
Bock, R. O., 100(41), *102*
Börnstein, R., 177(43), *189*
Bohme, D. K., 228(5), *253*
Boitnott, B. D., 98(2, 3), *100*
Bond, J. W., *165*
Boniwell, J. B., 41, 42(4), *77*
Borden, T. R., Jr., 193(17), 194(17), *218*
Bosschart, R. A., 73, *77*
Bowie, S. H. U., 79(6), 81(6), *87*
Boyd, D., 41, *77*
Boyle, T. L., 81(7, 8), 86, *87*
Bradley, D., 166(29), *189*
Bradley, P. A., 151(18), *188*
Brant, A. A., 41, *77*
Brewer, A. W., 200, 203(52), *219*
Brewer, W., 104(10), *110*
Bricard, J. 239, 246, 247(34), 248, 249, 250, *255*
Brière, R., 24(29), *40*
Brook, M., 149, 172, *188*
Brooks, R. R., 41, *77*
Brown, F., 204, *220*
Brown, S. C., 172(34), 179(34), *189*
Browne, B. C., 90, *100*
Browne, J. C., 233(16), *253*

C

Cabane, M., 239(36), 246(33b, 33e), 248(36), 249(36), 250 (33b, 33e, 36), *255*
Cambray, R. S., 81(32, 33), *88*
Cameron, H. L., 108, *110*
Cantrell, J. L., 106(8), *110*
Carmichael, H., 79(17), 80(17), *87*
Carswell, J., 106(29), *111*

Chalmers, J. A., 186(49), *189*
Chanin, L. M., 228(4), 232(4), 233(16), *253*
Chantry, P. J., 230(9), *253*
Charney, J. G., 215(86), *221*
Cheriton, C. G., 41, *77*
Chermack, E. A., 113(25), *139*
Childs, L., 108(6), *110*
Childs, W. H. J., *165*
Chinnery, M. A., 98, *100*
Chizhov, A. F., 204(67), *220*
Chown, J. B., 174(37), 176(37), 177(37), 183(37), *189*
Clemesha, B. R., 122(14), 125(14), 129(14), *138*
Colbert, C., 109(26), *110*
Colegrove, F. D., 215(80), *221*
Collett, L. S., 41, 42(9, 10, 11), *77*
Collis, R. T. H., 125(6), 126(2), 129, *138*
Colwell, R. N., 104, *110*
Combrisson, J., 23(1, 2), *39*
Compton, R. N., 243(38), *255*
Connock, S. H. G., 79(17), 80(17), *87*
Conway, D. C., 228, 232(4), *253*
Conway, W. H., 107(11), *110*
Cook, J. C., 85(9), *87*
Cosgrove, M. E., 81(27), *87*
Cowper, G., 79(17), 80(17), *87*
Cragg, B. G., 29(8), *39*
Craggs, J. D., 176(38), *189*
Craig, R. A., 192, 196(25), 202(42), *217, 218, 219*
Crane, P. H., 109(26), *110*
Cranshaw, T. E., 100(6), *100*
Crompton, R. W., 232(11), 234(11), 239(11), *253*
Culbertson, M. F., 205(68), *220*
Curtis, A. R., 205, 207, 208(73), 209(73), 210(73), *220, 221*
Curtiss, L. F., 80, *87*
Cutler, D., 17(36), *40*
Cwilong, B. M., 203(52), *219*

D

Dalgarno, A., 200(35, 37), *219*, 233(16), *253*
D'Arcy, D. F., 98(17), *101*
Dauvillier, A., 151, *188*
David, W. T., 166(28), *189*
Davis, F. J., 79(11), 80(11), *87*

Dehmelt, H. G., 25, *39*
Deirmendjian, D., 119(7), *138*
Detwiler, C. R., 200(40), *219*
Dicke, R. H., 107, *110*
Dickinson, R. E., 214(77), 215(87), *221*
Dinger, J. E., 203(53), *220*
Ditchburn, R. W., 200(41), *219*
Dmitriev, V. I., 42, *78*
Dobson, G. M. B., 203(52), *219*
Dolan, W. M., 41(7), *77*
Domzalski, W., 33(10), *39*
Drazin, P. G., 215(86), *221*
Driscoll, R. L., 17, *39*
Dütsch, H. U., 197, *219*
Dunkin, D. B., 228(5), *253*
Dupeyre, R. M., 24(29), *40*

E

Eiber, H., 232(11), 234(11), 239, *253, 254*
Ekin, J. W., Jr., 241(29), *254*
Elliot, C. L., 41(7), *77*
Elliott, D. D., 194(18), *218*
Elsasser, W. M., 205(68), *220*
Elterman, L., 119(8), 121(8, 8a), *138*
Englander-Golden, P., 177(44), *189*
English, J. E., 33(56), *41*
Eötvös, R., 89, *100*
Epstein, E. S., 215, *221*
Ette, A. I. I., 172, *189*
Evans, H. D., 247, *255*
Evans, S., 109, *110*
Evans, W. E., 114(22), *138*
Evenden, G. I., 14(12), 37(12), *39*

F

Falick, A. M., 166(30), *189*
Feder, A. M., 109(16, 17), *110*
Fedynskii, A. V., 204, *220*
Fehr, Y., 151, *188*
Fehsenfeld, F. C., 228(5), 229, 242, 243, *253, 254, 255*
Fenn, R. W., 123, *138*
Ferguson, E. E., 228(5), 229, 242(31), 243 *253, 254, 255*
Fernald, F. G., 125(6), *138*
Field, F. H., 233(14), 240(14), *253*

Finkelstein, D., 150(10), 152, 168, *188*
Fiocco, G., 114, 129, *138*
Fischer, W. A., 104(18), *110*
Fleming, H. W., 41, 77
Fontan, J., 246(33b), 250(33b), *255*
Forward, R. L., 99(8, 9), *101*
Franklin, E., 79(25), 85, *87*
Franklin, J. L., 233(14), 240(14), *253*
Fraser, D. C., 42(54), *79*
Friedman, H., 80, *87*
Friedman, R. M., 193(13), 194(13), *218*
Frischknecht, F. C., 14(12), 37(12), *39*, 41, 54, 59, 60(13), 77
Fromm, W. E., 7, 14, 37, *39*
Frommhold, L., 179, *189*, 232(11), 234(11), 239(11), *253*

G

Gaevskaya, G. N., 203(62), *220*
Gardner, A. L., 177(41), *189*
Garrett, D. L., 200(40), *219*
Gattinger, R., *165*
Gatz, C. R., 177(42), *189*
Gaur, V. K., 41, 77
Geiger, H., 79, *87*
Geleynse, M., 41, 77
Gilbert, R. L., 97(10), *101*
Gilman, G. I., 243(32j), *255*
Giret, R. I., 29(14, 15), *39*, 85, 86(15), *87*
Girod, P., 246(33a, 33d), 250(33a, 33d), *255*
Glicken, M., 37, *39*, 88(29), 89, 92(29), 93(29), *101*
Godby, E. A., 14(17), *39*, 79(17), 80, *87*
Goldan, P. D., 242(31), *243*(32a, 32b, 32c), *254*
Golden, D. E., 235(18), 238(18), 239(18), 243(18), 249(18), 250(18), *254*
Goldsmith, P., 204(64, 65), *220*
Goldstein, N. E., 34(18), *39*
Goody, R. M., 192, 202(44), 205(70), 207, 208(2, 70, 73), 209(73), 210(73), 211(2), 212, 214(44), *217*, *219*, *220*, *221*
Goyer, G. G., 113, *138*
Graf, A., 94, *101*
Graham, L. C., 109(26), *110*
Gramma Kov, A. G., 80(2), 86(2), *86*

Grams, G., 129, *138*
Grant, F. S., 41, 69(16), 77
Grantham, D. D., 203(56), 204(56), *220*
Green, H. F., 204(65), *220*
Gregory, A. F., 84, 86(18), *87*
Guitton, J., 83(3), 84(3), *86*
Gutnik, M., 203, *219*

H

Hackman, R. J., 108, *110*
Hall, E. H., 31 *39*
Hamilton, P. M., 129, *138*
Hampson, J., 197, *219*
Hanson, W. B., 215(80), *221*
Harrison, J. C., 89(28), 90(19), 91, 92(28), 94(14), *101*
Haseltine, W., 176, *189*
Hasted, J. B., 174, *189*
Hatton, A., 84(21), *87*
Hawkins, C. S., 95(42), *102*
Hedstrom, E. H., 41, 77
Heiskanen, W. A., 89(15), 98(15), *101*
Helstrom, C. W., 152(20), *188*
Herbert, R., 27(20), *40*
Hering, W. S., 193(17), 194(17), *218*
Herzberg, G., *165*
Hill, E. L., 151, *188*
Hines, C. O., 215, *221*
Hogg, A. M., 241, *254*
Hohla, C., 228, 229(2), 233(2), *253*
Hollander, J. M., 82(22), *87*
Holt, A., 204(65), *220*
Holter, M. R., 106(20), *110*
Honey, R. C., 114(22), *138*
Hood, P., 39(21), *40*
Hood, P. J., 51, 77
Hopfield, H. S., 165(27), *189*
Hopfield, J. J., 193(12), 194(12), *218*
Hornbeck, G. A., 165, *189*
Horner, F., 151, *188*
Horwood, J. L., 84, *87*
Hoselitz, K., 228(3), *253*
Houghton, J. T., 202(49), 203(57, 58, 61), 204(58, 63), 211, 215(81), *219*, *220*, *221*
Howell, B. F., 109(21), *110*
Howell, H. B., 117, 123, *139*
Hoylman, H. W., 34(22), *39*

Hunt, B. G., 194, 196, 197, 198, 200(33), 201, *218*, *219*
Hunten, D. M., *165*, 200(36), *219*
Hutchins, R. W., 98(16, 17), 99, *101*
Huxley, L. G. H., 232(11), 234(11), 239(11), *253*

I

Irons, H. R., 13, *40*
Isaacson, L., *165*

J

Jensen, H., 7, 27(24), 33, *40*
Johnson, F. S., 193(11), 194(11), 214, 215(80), *218*, *221*
Joklik, G. F., 41, *77*

K

Kamara, A. K., 172, *189*
Kanamori, H., 97(43, 44), *102*
Kandel, W., 239, *254*
Kapitza, P. L., 151, *188*
Kasner, W. H., 230(9), *253*
Kastler, A., 24, *40*
Kaufman, F., 202(48), *219*
Kebarle, P., 241, *254*
Keller, G. V., 41, *77*, 109(22), *110*
Kellogg, W. W., 215, *221*
Kent, G. S., 122(14), 125(14), 129, *138*
Kenty, C., 165, *188*
Khomenyuk, Yu. V., 41, *77*
Kiseleva, M. S., 203(59), *220*
Kitagawa, N., 149(8), 172(8), *188*
Kivel, B., 163(22), *188*
Knewstubb, P. F., 239, 241, *254*
Konashenok, V. N., 197(30), 212(30), *218*
Kondrat'yev, K. Ya., 203(62), 208(71), *220*
Kreuger, A. J., 195(23), *218*
Kruse, P. W., 105(23), *110*
Kudryatseva, L. A., 193(15), 194(15), *218*

Kuhn, W. R., 192(3), 202, 208(3), 209, 210(3), 211(3), 212, *217*
Kunkel, W. B., 177, *189*

L

LaCoste, L. J. B., 88(29, 39), 89(28, 29, 39), 90(19), 91, 92, 93(29), *101*, *102*
Lallemant, C., 83(3), 84(3), *86*
Lampe, F. W., 233, 240, *253*
Landis, G., 104(10), *110*
Landolt, H., 177(43), *189*
Langan, L., 16(26), 27(20), *40*
Langley, P., 104(10), *110*
Lattman, L. H., 106, *110*
Laurila, S., 33(27), *40*
Lawrence, J. D., 116(15), 126, *138*
Leah, A. S., 166, *189*
Leliak, P., 32(28), *40*
Leonov, R. A., 147, *188*
Leovy, C., 194(19, 20), 196(19), 197, 202(43), 207, 214(43), *218*, *219*
Lemaire, H., 24(29), *40*
Levine, D., 108(25), 109(26), *110*
Licastro, P. H., 109(21, 22), *110*
Ligda, M. G. H., 114, 129, *138*
Lindzen, R. S., 202(44–47), 214(44–47), 215, *219*, *221*
Lineberger, W. C., 243(37), *255*
Loeb, L. B., 224, 225, 226, 227, 228, 229, 231, 232, 244, 245, *252*
Logachev, A. A., 6(30), *40*
Long, R. K., *138*
Lozinskaya, A. M., 97(22), *101*
Luhr, O., 235, *254*
Lundberg, H. T., 7, *40*, 98, *101*
L'vova, A. A., 193(14), 194(14), *218*

M

McAfee, K. B., 232(11), 234(11), 239(11), *253*
McClatchey, R. A., 210, *221*
McCormick, M. P., 116, 126(15), *138*
McCormick, P. D., 129, *138*
McDaniel, E. W., 230(8), 237(8, 20), 242(20), *253*, *254*
McElroy, M. B., *165*, 200(36), *219*

McGlauchlin, L. D., 105(23), *110*
McGrath, W. D., 197(32), *219*
Mack, S. Z., 16(42), *40*
MacKay, D. G., 42, *77*
McKnight, B. K., 42(53), 51(53), *79*
McKnight, L. G., 232(11), 234(11), 239(11), *253*
McNally, J. R., Jr., 144, *188*
McQuistan, R. B., 105(23), *110*
Madelaine, G., 239(36), 246(33c, 33e), 247(34), 248(36), 249(36), 250(33c, 33e, 36), *255*
Madson, J. M., 230(7), 233(16), *253*
Mahoney, J. R., 214(78), *221*
Malnar, L., 29(14), *39*
Manwaring, J. F., 147, 152(19), 153, 156, *188*
Marriott, R., 210(75), *221*
Martin, D. W., 230(8), 237(8, 20), 242(20), *253, 254*
Martinot, K., 251(40), 252(40), *255*
Maskell, S. C., 81(32), *88*
Mastenbrook, H. J., 203(53, 54, 55), 204, *220*
Mecke, R., *165*
Melfi, S. H., 166(15), 126(15), *138*
Meloy, G. E., 193(13), 194(13), *218*
Menon, V. K., 107(27), *110*
Mero, J. L., 83(23), *87*
Meuschke, J. L., 14(12), 37(12), *39*
Middleton, W. E. K., 114, 119(19), *138*
Mikirov, Y. A., 193(14), 194(14), *218*
Miller, B., 109(28), *111*
Miller, D. E., 193(16), 194, *218*
Miller, J. M., 79(6), 81(6), *87*
Miller, T. M., 230(8), 237(8, 20), 242(20), *253, 254*
Mizyuk, L. Ya., 42, *77*
Mössbauer, R. L., 100, *101*
Mohnen, V., 228, 229(2), 233, *253*
Morgan, J., 104(10), *110*
Morita, T., 174(37), 176(37), 177(37), 183(37), *189*
Moruzzi, J. L., 231, 232(10), 241(10, 29), 242, *253, 254*
Moseley, J. T., 230(8), 237(8, 20), 242(20), *253, 254*
Moxham, R. M., 80(24), *87*, 106, *111*
Müller, H. G., 215, *221*
Müller, W., 79, *87*
Mulcahy, M. J., 251(40), 252(40), *255*

Munick, R. J., 121, *138*
Munson, R. J., 228, *253*
Murcray, D. G., 203(60), *220*
Murcray, F. H., 203(60), *220*
Murgatroyd, R. J., 192, 207, 208(2), 211(2), 212, *217, 218*
Myers, B. F., 164(24), *188*

N

Nagata, T., 195(22), *218*
Narcisi, R. S., 241, *254*
Nathan, A. M., 113(25), *139*
Neporant, B. S., 203(59), *220*
Nettleton, L. L., 88(29), 89(28, 29), 92(28, 29), 93, *101*
Neugebauer, H. N., 151, *188*
Neuschel, S. K., 81(28), *87*
Newell, R. E., 204(66), 215, *220, 221*
Niblett, E. R., 34, *41*
Nicolet, M., 197(27), *218*
Nikolsky, G. A., 203(62), *220*
Nikonov, A. I., 80(2), 86(2), *86*
Norrish, R. G. W., 197(32), *219*
Northend, C. A., 114(22), *138*
Noxon, J. F., *165*
Nudelman, S., 106(20), *110*

O

Oblanas, J. W., 129(4, 5, 23), *138*
O'Brien, D. P., 42(53), 51(53), *79*
O'Donnell, J., 42(54), *79*
Ohring, G., 202(42), *219*
Oliver, D., 106(29), *111*
Oskam, H. J., 230(7), 233(16), 234(17), *253*
Overhauser, A. W., 24, *40*

P

Pack, J. L., 172(33), 180(33), *189*, 228(4), 232(4, 12), 241(29), *253, 254*
Packard, M., 16, *40*
Pahl, M., 233(15), 238, 241, *253*
Parasnis, D. S., 41, *77*
Parham, A. G., 204(65), *220*
Parker, A. K., 109(31), *111*

Parry, J. R., 42(53), 51(53), *79*
Parsons, L. W., 27, *40*
Paterson, N. R., 32(35), *40*, 42, 77, 78, 98, *101*
Patterson, P. L., 233(16), *253*
Pearce, J. B., 200(41a), *219*
Peirson, D. H., 79(25), 85, *87*
Pemberton, R. H., 41, 42, *78*, 83(26), *87*
Penner, S., 163(23), *188*
Perlman, I., 82(22), *87*
Perov, S. P., 204(67), *220*
Perschy, J. A., 129(18), *138*
Phelps, A. V., 172, 180(33), *189*, 228(4), 231, 232(4, 10, 12), 241(10, 29), *253*, *254*
Phillips, G., 17, *41*
Pickup, J., 79(6), 81(6, 27), *87*
Pirart, M., 106, *111*
Pitkin, J. A., 81(28), *87*
Plass, G. N., 192, 208(5, 6), *217*
Podolsky, G., 42, 72, *78*
Poloskov, S. M., 193(14), 194(14), *218*
Popov, E. I., 88(31, 34), 89(33), 95(31, 32, 34), 96(34), *101*, *102*
Poultney, S. K., 129(18), *138*
Pound, R. V., 100(35), *102*
Powell, J., 152, *188*
Powles, J. G., 17(36), *40*
Pradel, J., 246(33a, 33d), 250(33a, 33d), *255*
Prasad, A. N., 174(36), 176(38), *189*
Pringle, R. W., 82(29), *87*
Puckett, L. J., 243(37), *255*
Pugh, B., 166(28), *189*
Purcell, J. D., 193(11), 194(11), 200(40), *218*, *219*

R

Ragotzkie, R. A., 107(27), *110*
Randhwa, J. S., 195(24), *218*
Rapp, D., 177(44), *189*
Rassat, A., 24(29), *40*
Ratcliffe, J. H., 98(26), *101*
Rattew, A. R., 42, *78*
Rawcliffe, R. D., 193(13), 194, *218*
Rayle, W. D., 144, 145(3), 146(3), *188*
Rayleigh, Lord (Strutt), 152, *188*
Rebka, G. A., 100(35), *102*
Rees, J. A., 232(11), 234(11), 239(11), *253*

Rees, M. H., 200(37), *219*
Reford, M. S., 31(37), 32(38), *40*
Reid, M., 109(30), *111*
Reinhardt, P. W., 79(11), 80(11), *87*, 243(38), *255*
Rempel, R. C., 109(31), *111*
Renoux, A., 246, 250(33f), *255*
Rinker, J., 104(10), *110*
Rivera, R., 42(54), *79*
Roberts, B. C., 41, *77*
Robinson, J. M., 104(10), *110*
Rodewald, M., 147(6), *188*
Rodgers, C. D., 202, 207, 213, *219*
Rogers, E. H., 193(13), 194(13), *218*
Romberg, F., 92, *101*
Rosenzweig, A., 79(30), *87*
Ross, E. M., 31(39), *40*
Roulston, K. I., 82(29), *87*
Rounthwaite, C., 166(29), *189*
Rubinstein, J., 150(10), 152, 168, *188*
Rumbaugh, L. H., 14, *40*
Rydstrom, H. O., 109(32), *111*
Ryzko, H., 176(39), *189*

S

Sakamoto, R. T., 107(11), *110*
Saker, E. W., 31(39), *40*
Salmela, H. A., 203(56), 204(56), *220*
Salvi, A., 24(29), *40*
Samson, J. A. R., 238(21), *254*
Sandford, M. C. W., 122(24), 125(24), 129, *139*
Saporoschenko, M., 230(8), 237(8, 20), 241, 242, *253*, *254*
Savet, P. H., 100(41), *102*
Sawatzky, P., 21(50), *41*
Schaub, Yu. B., 42, *78*
Scheps, B. B., 108, 109(26), *110*, *111*
Schiff, H. I., 242(31), 243(32b, 32c, 32h), *254*, *255*
Schiffer, J. P., *100*
Schlumbohm, H., 237(20), 242, *254*
Schmeltekopf, A. L., 242(31), 243, *254*, *255*
Schmidt, S. J., 109, *111*
Schofield, K., 200, *219*
Schonstedt, E. O., 13, *40*
Schotland, R. M., 113, *139*
Schulz, G. J., 230(9), *253*

AUTHOR INDEX

Schulze, R., 94(13), 95(36), *101*, *102*
Seaborg, G. T., 82(22), *87*
Seeley, J. S., 203(57), *220*
Seigel, H. O., 83(26), *87*
Serson, P. H., 12(43, 44a), 16(42), 21, *40*, *41*
Shafer, C., 232(11), 234(11), 239(11), *253*
Shafranov, V. D., 151, *188*
Shahin, M. M., 238, 250, *254*
Shale, C. C., 232(13), *253*
Sharpless, R. L., 177(42), *189*
Shved, G. M., 192(4), 203(62), 208(4, 72), 209, 211(4), *217*, *220*, *221*
Siksna, R., 240, *254*
Silberg, P. A., 146, 151, *188*
Sinnott, G., 235(18), 238(18), 239(18), 243(18), 249(18), 250(18), *254*
Sipler, D. P., 232(11), 234(11), 239(11), *253*
Sissenwine, N., 203(56), 204, *220*
Slichter, L. B., 42, 66(42), *78*
Smullin, L. D., 114, 129, *138*
Snuggs, R. M., 230(8), 237(8), *253*
Solomon, I., 23(1, 2), *39*
Sorem, A. L., 104(10), *110*
Spilhaus, A. F., 114, *138*
Stead, F. W., 79(31), *88*
Stein, H. P., 179(47), *189*
Stekol'nikov, I. S., 151, *188*
Steljes, J. F., 79(17), 80(17), *87*
Stewart, D. H., 193(16), 194, *218*
Stockdale, J. A., 243(38), *255*
Strome, W. M., 27, *41*
Suits, G. H., 106(20), *110*
Sumner, J. S., 32(38), *40*

T

Tafeev, G. P., 80(2), *86*
Tarakanov, Y. A., 97(37), *102*
Taylor, H. W., 82(29), *87*
Taylor, W. C., 174(37), 176, 177, 183, *189*
Teweles, S., 193(10), *218*
Thomas, J., 24(46), *41*
Thomatsu, T., 195(22), *218*
Thompson, L. G. D., 88(38, 39, 40), 80(39), 92(39), 95, 100, *102*
Thompson, N. A. C., 31(39), *40*
Thorton, W. M., 151(16), *188*
Tickner, A. W., 239, 241, *254*

Tikhonov, A. N., 42, *78*
Tomoda, Y., 97(43, 44), *102*
Tornquist, G., 42, *78*
Tousey, R., 193(11), 194(11), 200(40), *218*, *219*
Tsuboi, C., 97(44), *102*
Tsuruta, H., 195(22), *218*
Twomey, S., 117, 123, *139*
Tyndall, A. M., 228(3), *253*

U

Uman, M. A., 152, *188*
Uthe, E. E., 113(25), *139*

V

Vacquier, V. V., 37, *41*
Vallence Jones, A., *165*
van Allen, J. A., 193(12), 194(12), *218*
Van de Hulst, H. C., 117(29), 119(29), *139*
Van Wijk, V., 129(18), *138*
Varian, R., 16, *40*
Varney, R. N., 228(5), 235, 238, 239, 243(18), 249, 250, *253*, *254*
Varshneya, N. C., 172(32), *189*
Vening Meinesz, F. A., 89(15), 98(15), *101*
Venkatswaran, S. V., 197(28), *218*
Verzar, F., 247, *255*
Vest, W. L., 108(6), *110*
Vigoureux, P., 17, *41*
Von Engel, A., 177, *189*
Voshall, R. E., 232(12), *253*

W

Waite, A. H., 109, *111*
Walker, J. C. G., 200(35, 37), *219*
Wallace, L., 197(29), 215(29), *218*
Wallace, R. E., 106, *111*
Walshaw, C. D., 202, 207, 213, *219*
Ward, S. H., 34(18), *39*, 41, 42, 46, 49, 51, 71(50), 73(51), *78*, *79*
Ware, G. H., 42(54), *79*
Washkurak, S., 21(50), *41*
Watanabe, K., 193(11), 194(11), *218*
Waters, G. S., 17, *41*

Watson, K. M., *165*
Watson, R., 113, *138*
Weissler, G. L., 238(21), *254*
Welch, J. A., *165*
Wentwink, T., *165*
Wertheim, G. K., 88(45), 100(45), *102*
West, G. F., 41, 69(16), *77*
White, P. S., 42, *79*
Whitehead, A. B., *100*
Whitham, K., 7(52), 16(42), 18(52) 31(52), 34, 37(52), *40*, *41*
Whitlock, R. F., *165*
Wiatr, Z. M., 27, *40*
Wickerham, W. E., 33(55), 37, *41*
Wieduwilt, W. G., 42, *79*
Wilkins, E. M., 214, *221*
Williams, D., 79(6), 81(6, 32, 33), 82(34), 85(34), *87*, *88*
Williams, W. J., 203(60), *220*
Williamson, E. J., 204(63), *220*
Wilson, A. W., 200, *219*
Wilson, R. A. R., 33(56), *40*
Wisner, W. W., 242, *254*
Wold, R. J., 19(57), *41*

Wolfe, W. L., 106(20), *110*
Woodman, D. P., 116(15), 126(15), *138*
Workman, E. J., 149(8), 172(8), *188*
Wormell, T. W., 148, *188*
Wright, A. C., *165*
Wright, R. W., 122(14), 125(14), 129(14), *138*

Y

Yakovleva, A. V., 193(15), 194(15), *218*
Yang, J., 228, 232(4), *253*
Yanovskii, B. M., 23(58), *41*
Young, C., 215, *221*
Young, G. A., 149, *188*
Young, P. A., 200(41), *219*
Young, R. A., 165, 177, *188*, *189*

Z

Zissis, G. J., 106(20), *110*
Zucker, M., 152(19), *188*

SUBJECT INDEX

A

Aerial photography, 104
Aero Canso rigid boom AEM system, 48
Aeromagnetic airborne methods, 4–39
   aircraft for, 31–32
   data compilation, 34–37
   geomagnetic field in, 4–6
   gradiometers for, 37–39
   magnetometers for, 6–31
   survey techniques, 31–34
Airborne electromagnetic methods (AEM), 41–79
   basic principles, 42–64
   coil configurations, 57–64
   data interpretation, 68–73
   future developments, 74
   noise in, 52–68
   signal sources, 53–68
   summary of, 50
   survey procedures, 73–74
   system design, 49–68
   types of, 46–49
Airborne geophysical methods, 1–112
   aeromagnetic methods, 4–6
   applications, 3–4
   gravity types, 88–102
   infrared methods, 105–107
   low-frequency electromagnetic methods, 41–79
   radar methods, 108–109
   radiometric types, 79–88
   remote sensing methods, 102–103
   ultraviolet methods, 104–105
Aircraft, for aeromagnetic surveys, 31–32
Air motion, lidar studies of, 129–132
Atmosphere, middle, *see* Middle atmosphere
Atmospheric turbulence, backscattering by, 121–122

B

Ball lightning,
   electrohydrodynamic forces, 183–184
   experimental evidence for, 152–167
   model of, 167–185
   observations of, 144–148
   stability of, 184–185
   structure of, 141–189
   theories, critique of, 149–152
   thunderstorm conditions for, 148–149

C

Clouds, lidar studies of, 125–127
Cultural noise, in AEM systems, 53
Curtis' method for calculation of heating rates, 205–207

D

Disturbance field noise, in AEM systems, 52–53
Dust cloud, lidar observations of, 132

E

Earth inductor magnetometers, 6–13
Electron-beam magnetometers, 29–31

F

"Figure of merit," of aircraft, 32
Fluxgate magnetometers, 7–13
Fog, lidar studies of, 133

G

GAL gravimeter, 95
Gaseous ions in atmosphere,
   accuracy of mobility determinations, 234–236
   forces influencing behavior, 224–225
   ion mobility and ion identification, 233–234
   Langevin and intermediate ions, 244–252
   mobility theory and equations, 225–229
   nature and properties of, 223–255
   normal ions, 236–243
   reactions between ions and molecules, 229–233
   reaction rates of ions, 243–244
Geologic noise, in AEM systems, 53, 67–68

Geomagnetic fields, components of, 5
Gradiometers,
  for aeromagnetic surveys, 37–39
  airborne gravity types, 97–100
Graf-Askania gravimeter, 94–95, 96
Gravimeters, for aerial surveys, 91–97
Gravity methods of aerial survey, 88–102
  gradiometers for, 97–100
  gravimeters for, 91–97
  Mössbauer effect, 100

H
Hunting gravity gradiometer, 99

I
Ice, thickness measurement, 109
Infrared methods in aerial surveys, 105–107
Infrared radiative transfer, in middle atmosphere, 204–214
INPUT towed bird AEM system, 49
Insecticide clouds, lidar observations of, 130–131
Instrument noise, in AEM systems, 52
International Gravity Formula, 90

J
Johnson noise, in AEM systems, 65–66

L
LaCoste-Romberg gravimeter, 91–93
Laser, use in radar system, see Lidar
Lidar, 113–139
  air motion studies by, 129–132
  atmospheric optical parameters, 117–122
    Mie scattering, 118–121
    Rayleigh scattering, 117–118
  atmospheric studies by, 133–136
  basic technique, 114–116
  characteristics, 136–137
  "clear," lidar studies of, 127–129
  cloud studies by, 125–127
  derivation of term, 114
  fog studies by, 133
  future applications, 135–136
  operational applications, 135
  optical parameters measured by, 122–124
  evaluation, 123–124
    meteorological significance, 122–123

  in studies of "clear" air, 127–129
Lightning, ball type, see Ball lightning

M
Magnetic field noise, in AEM systems, 66–67
Magnetometers (airborne), 6–31
  earth inductor types, 6–7
  electron-beam types, 29–31
  fluxgate types, 7–13
  Hall-effect types, 31
  optical absorption types, 24–29
  proton free- and spin-precession types, 16–24
Microwave radiometers, in aerial surveys, 107–108
Middle atmosphere,
  energetics of, 191–221
  heat sources and sinks in, 214–216
  infrared radiative transfer, 204–214
    atmosphere dynamics and, 212–214
  ozone distribution, 193–202
    atmospheric dynamics, 200
    photochemical theories of, 196
  vibrational relaxation in, 208–211
  water vapor in, 202–204
Mie scattering, in lidar, 118–121
Mineral prospecting, aeromagnetic methods in, 4
Mössbauer effect, 100

N
Noise, in airborne electromagnetic systems, 52–68

O
Optical absorption magnetometers, 24–29
Overhauser magnetometer, 23–24
Ozone, distribution in middle atmosphere, 193–202

P
Proton free- and spin-precession magnetometers, 16–24

R
Radar, laser use in, see Lidar
Radar methods, in aerial surveys, 108–109
Radiometers, microwave type, 107–108
Radiometric airborne methods, 79–88
  scintillation counters for, 80–81

scintillation spectrometers for, 81–84
   survey techniques, 84–86
Rayleigh scattering, in lidar, 117–118
Rigid boom AEM system, 46, 47

S

Scintillation counters, for airborne radiometric methods, 80–81
Scintillation spectrometers, for airborne radiometric methods, 81–84
Surveying, aeromagnetic type, 32–34

T

Thermometers, airborne radiation type, 105–106

Thunderstorms, ball lightning occurrence in, 148–149
Towed bird AEM system, 46, 47

U

Ultraviolet methods in airborne surveys, 104–105

V

Vibrating-string gravimeter, 97

W

Water clouds, volume backscatter and extinction coefficients for, 119
Water vapor, in middle atmosphere, 202–204